Developments in Earth & Environmental Sciences, 6

CLIMATE, ENVIRONMENT AND SOCIETY IN THE PACIFIC DURING THE LAST MILLENNIUM

By

PATRICK D. NUNN

Professor of Oceanic Geoscience
The University of the South Pacific
Suva
Fiji Islands

ELSEVIER

Amsterdam – Boston – Heidelberg – London – New York – Oxford
Paris – San Diego – San Francisco – Singapore – Sydney – Tokyo

Elsevier
Radarweg 29, PO Box 211, 1000 AE Amsterdam, The Netherlands
Linacre House, Jordan Hill, Oxford OX2 8DP, UK

First edition 2007

Notice
No responsibility is assumed by the publisher for any injury and/or damage to persons
or property as a matter of products liability, negligence or otherwise, or from any use
or operation of any methods, products, instructions or ideas contained in the material
herein. Because of rapid advances in the medical sciences, in particular, independent
verification of diagnoses and drug dosages should be made

Library of Congress Cataloging-in-Publication Data
A catalog record for this book is available from the Library of Congress

British Library Cataloguing in Publication Data
A catalogue record for this book is available from the British Library

ISBN: 978-0-444-52816-2
ISSN: 1571-9197

For information on all Elsevier publications
visit our website at books.elsevier.com

Printed and bound in The Netherlands

07 08 09 10 11   10 9 8 7 6 5 4 3 2 1

Working together to grow
libraries in developing countries
www.elsevier.com | www.bookaid.org | www.sabre.org

ELSEVIER   BOOK AID
International   Sabre Foundation

DEVELOPMENTS IN
EARTH & ENVIRONMENTAL SCIENCES 6

CLIMATE, ENVIRONMENT AND SOCIETY IN THE PACIFIC DURING THE LAST MILLENNIUM

In equal measure,
for
Rachel, Warwick and Petra
With love

CONTENTS

This book describes the changes that occurred in the climates, the environments and the societies of the Pacific Basin during the past 1,000 years or so. It is largely descriptive, reporting what is known, but also makes deductive links between changes at particular times. The book concludes that for most of the last millennium, climate change was the main reason for profound societal change in the Pacific Basin.

I hope this book will be read by people who aspire to understand the interconnectivities of the sciences and the humanities, be they trained as archaeologists, anthropologists, social and environmental scientists, geographers, geologists or pre-historians, and by anyone who has a curiosity about the past and the open-mindedness to realize that there are benefits to modern societies to be gained from rigorously interrogating it and trying to make sense of the answers. To help make the text more accessible to particular readers, there is a "Key Points" section at the end of each chapter.

Time in this book is expressed using various notations. Years BP is the commonest for times before about A.D. 0, where BP refers to Before Present, Present being the calendar year A.D. 1950. During the last three to four millennia, years are sometimes also expressed in Years B.C. (Before Christ, equivalent to BCE [Before Common Era]) where these were used in the original citation. For the past 2,000 years, all dates are given as calendar years (A.D. [anno Domini], equivalent to CE [Common Era]) with calibrated years BP (cal BP) where these were the units used in the original citation. Sometimes with older dates, there is no way that a third party can calibrate them, so they are quoted as years BP with approximate calendar equivalents denoted as such.

Much of this book was written during short attachments at the Research Center for the Pacific Islands at Kagoshima University in Japan, and the Institute of Island Studies at the University of Prince Edward Island in Canada. I am grateful to the University of the South Pacific for granting me a Strategic Research Fellowship to work on this book. I thank the following research assistants for help: Ledua Traill Kuilanisautabu, Margaret Mizzi, Maria Ronna Pastorizo and Tammy Tabe.

For various reasons, I would like to thank William Aalbersberg, Steve Athens, Godfrey Baldacchino, Mike Carson, Keith Crook, Rod Dixon, Julie Field, Roland Gehrels, Simon Haberle, Rosalind Hunter-Anderson, Ian Hutchinson, Phil Jones, David Lowe, Michael Mann, Bruce Masse, Mike Page, Frederic Pearl, Jean-Francois Royer, Christophe Sand, Daniel Sandweiss, Matthew Spriggs, James Terry, Frank Thomas, Noel Trustrum, Sean Ulm and Colin Woodroffe.

The immediate impetus for writing this book was provided by reading the 2004 book *"Climate Change – Environment and Civilization in the Middle East"* (by

A.S. Issar and M. Zohar), which argues essentially what I argue here for a different part of the world. But I have my heroes among the environmental determinists of the Pacific region, particularly the much-maligned J.T. Holloway and Patrick J. Grant, whose work highlighted climatic influences on New Zealand pre-history, and Michael J. Rowland, who has done much the same, albeit more recently, for Australian pre-history. And I must mention Brian Fagan's marvellously common-sense books that link climate and society, for I suspect my conviction in seeing such links in the Pacific Basin has become stronger as a result of reading them. My distinguished colleague, Ron Crocombe, has also for many years been badgering me to write this work up as a book.

Among my other inspirations in writing this book have been some of my fiercest critics, who have constantly reminded me that my arguments are only as good as the data on which they are based. Among this number I affectionately include William Dickinson, Geoffrey Hope, and my wife Roselyn Kumar, who has nevertheless been unfailingly supportive of this work being written.

CHAPTER 1

Climate, Environment and Society: Global and Regional Perspectives

A major reason for writing this book is to describe the evidence for last-millennium climate change and its effects on environments and societies from the Pacific Basin, a commonly marginalized region in global syntheses. The paucity of palaeoclimate datasets from the southern hemisphere has often been remarked upon, and many palaeoclimatologists have cautioned that their statements about global last-millennium (and earlier) climate change should be regarded as preliminary until a better balance is achieved. Yet there is also a marked west–east hemispheric imbalance between the Pacific third of the Earth and elsewhere, largely because the Pacific is mostly ocean, but also because there have been fewer scientific investigations of its recent palaeoclimatic history.

In seeking to redress these imbalances, this book utilizes data about Pacific Basin palaeoclimate from numerous sources. Many of these data are imprecise, their relationships to presumed climate drivers often uncertain. For such reasons, some of the conclusions reached about last-millennium Pacific Basin climates are less compelling than would be ideal, not just for comparing with other parts of the world but also for testing whether or not particular changes were global in extent, globally or hemispherically synchronous, or otherwise constrained in time or space. Yet there are more than adequate data to produce a first synthesis for this vast and poorly researched region, a synthesis that the author regards as suffi-ciently compelling to command the attention of geoscientists, climatologists, geo-graphers, archaeologists, anthropologists, sociologists and historians interested in the Pacific.

It is important to recognize that in the Pacific Basin – as in other parts of the world – the study of last-millennium climates is constrained by the effective reach of particular techniques back in time. Obviously, the last hundred years or so, for which there are instrumental records of climate in many places, is the least controversial time period. Beyond the reach of such directly monitored data series, it is necessary to use proxy data. There are three types of proxy used: those that directly proxy climate variables (typically temperature and precipitation), those that proxy climate through an environmental filter and finally those that proxy either climate or climate-driven environmental changes through changes in human societies.

High-resolution climate proxies, such as those from tree-rings and corals, commonly reach back only a few hundred years and depend on a correct understanding of how particular variables (tree and coral growth in these examples) respond to climate change. Such techniques are also commonly dependent on radiocarbon dating, the accuracy of which for this time period was once dogged by potential errors arising from non-uniform variations in atmospheric ^{14}C. As a result, many radiocarbon dates acquired more than 15–20 years ago cannot be precisely converted to calendar years.

In addition to climate proxies, environmental indicators of climate change have been used to reconstruct palaeoclimates. In the Pacific, some of the most common environmental proxies are associated with sea-level change, evidence for which naturally abounds in this ocean-dominated region. Others include changes in glacier-front position, lake levels, forest fire frequency, and the character of deposited sediment and rate of sedimentation.

Third, there are societal proxies of climate change, altogether a more controversial area of study. Changes in human societies can obviously take place for a variety of reasons, some of which have nothing to do with climate – or indeed any external – forcing mechanisms. Yet numerous recent studies have shown climate change to have been involved in societal collapse and other profound and enduring societal changes at various times and in various parts of the world before ~200 years ago. A central theme of the present book is that, for parts of the Pacific Basin, climate change within the last millennium led to widespread and fundamental societal change.

This chapter begins with a general discussion of global climate, environmental and societal change within the last 5,000 years, the later part of the Holocene Epoch (Section 1.1), followed by a more detailed look at the centuries leading up to the start of the last millennium (Section 1.2). Some of the many ways of dividing the last millennium are discussed in Section 1.3 in which the scheme adopted in this book is also explained and justified. The four divisions of the last millennium are then discussed in Sections 1.4–1.7.

This chapter then turns to the Pacific Basin, describing its geography and then giving an overview of its climates, environments and societies and their evolution (Section 1.8). Finally, Section 1.9 describes how this book is organized.

1.1. GLOBAL CLIMATE, ENVIRONMENT AND SOCIETY OVER THE PAST 5,000 YEARS

The last 5,000 years provide an adequate context for any discussion of last-millennium changes in climate, environment and society.

In terms of climate, the last 5,000 years take us away from the earlier influence of Last-Glacial (last ice age) temperature rise and into an interglacial climate system dominated by non-orbital external forcing and by internal system adjustments, often regional rather than global in effect. This is discussed further in Section 1.1.1.

Causes of environmental change during the past 5,000 years were dominated by climate forcing and internal system changes. For coastlines, the principal cause of environmental change during the last 5,000 years was sea-level change, many parts of the world (including the Pacific Basin) experiencing the effects of net sea-level fall within this period. This is discussed in Section 1.1.2.

Finally, for most of the past 5,000 years, human societies were far less resilient to external forces like climate and environmental change than they are today. It seems that in most parts of the world for much of this period, profound changes in human lifestyles were brought about by external driving forces. The nature of human–societal change across the world is discussed briefly in Section 1.1.3.

In terms of understanding interactions between climate, environment and society, the last 5,000 years provide numerous examples that are important to understand given what is proposed for the last millennium in the Pacific Basin. This is a theme that is developed in Chapter 3.

1.1.1 Climate change during the late Holocene

During the second half of the Quaternary Period (the last 850,000 years), global climate change has been dominated by the orbital-eccentricity cycle with glaciations (or ice ages) occurring every 100,000 years or so. The Last Glacial followed the Last Interglacial. From its maximum (coldest time) ~18,000 years BP, the Last Glacial came to an end ~12,000 BP when the Holocene Interglacial – in which we are still living – began (Fig. 1.1).

Climates during the first half of the Holocene were dominated by an overall temperature rise attributed to orbital change and the lingering effects of glaciation. Principal among the latter effects were the existence of large masses of ice on the land and the ocean that continued to affect climate longer after the interglacial commenced. By 5,000 years BP, orbitally forced warming had almost come

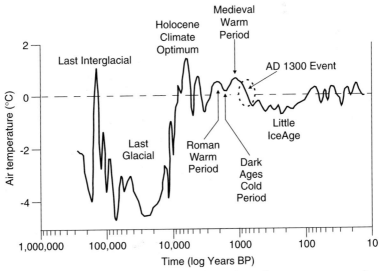

FIGURE 1.1 Climate change during the late Quaternary shown by air-temperature changes in the northern hemisphere expressed as departures from the 1951 to 1980 average. Figure adapted from the compilation of Demezhko and Shchapov (2001). Note the logarithmic time scale.

to an end and climates generally stabilized, system adjustments without orbital forcing accounting for most of the subsequent climate changes.

Some such adjustments may have originated internally: for instance, the development in their current forms of interannual climate phenomena like the North Atlantic Oscillation (NAO), the Pacific Decadal Oscillation and Inter-decadal Pacific Oscillation (PDO/IPO) and even the El Niño-Southern Oscillation (ENSO) phenomenon (Sandweiss et al., 2001; Cronin et al., 2005). Other adjust-ments post-5,000 BP were clearly externally forced. This applies to the tempera-ture fall that has dominated late Holocene climate change in most parts of the world as well as the 1,300/1,500-year oscillation that explains the alternation between warm and cool periods during the late Holocene (Bond et al., 1997; Perry and Hsu, 2000).

Sea-level change during the Holocene, discussed in the following section, also had important effects on climate during the Holocene, especially in flooding low-lying areas and thereby changing ocean circulation locally, and more generally in altering – in association with temperature change – heat transport around the Earth's surface.

While relatively unimportant in the context of the past 5,000 years, it is rele-vant to note that anthropogenic forcing associated with an accelerated rate of fossil-fuel combustion and associated factors is generally held responsible for part of the warming experienced in the past century or so in most parts of the world.

1.1.2 Environmental change during the late Holocene

Large-scale long-term changes of environments worldwide took place during the Holocene in response to climate change. Among the most important were shifts in insolation-dependent climate zones, both latitudinally (north–south) and altitu-dinally (upslope), that set the scene for shifts in biotic zones where these shifts were not precluded by topography. For shorter time periods, Holocene environ-mental change, particularly within the last 5,000 years, has generally been more localized and has come about largely because of local forcing mechanisms. These range from rapid landscape denudation in places where precipitation is unusually high and/or where tectonic (land-level) movements render landscapes more vulnerable to denudation, to places where desert areas are spreading (the process of desertification) because of subtle changes in precipitation regime modulated by the clearance of stabilizing vegetation by humans.

Extreme climatic (and other) events, such as tropical cyclones and severe pro-longed droughts, often drive rapid environmental change which may overwhelm the normal environmental change that occurs between these events. Particularly within the past few centuries, anthropogenic causes of environmental change have become far more important as humans have increasingly modified natural environments and the processes that mould them.

For the purpose of this book with its focus on an ocean-dominated region during a single millennium, the principal process driving region-wide environ-mental change is sea-level change. For this reason, the discussion of environ-mental change in this section and elsewhere in this book focuses primarily on sea-level changes and their role in changing coastal environments and coastal processes.

For most of the past 120,000 years, sea level (the ocean surface) has lain well below its current level (Fig. 1.2). The Last-Glacial oscillation was marked by a long slow sea-level fall followed – after the Last-Glacial Maximum (22,000–17,000 BP) – by comparatively rapid sea-level rise up to a maximum level (at least in the Pacific) of perhaps 1.5–2.0 m above its present level ~5,000 BP.

Sea level has continued to change during the past 5,000 years and these changes have had significant effects on coastal (and lowland) environments worldwide. Glacio-eustatic changes (sea-level changes resulting from ice melt associated with deglaciation) are believed to have ceased ~4,000 cal BP (Peltier, 1998), so any subsequent sea-level changes have been either a result of equatorial ocean syphoning combined with hydroisostasy (eustatic changes) or steric changes.

Equatorial ocean syphoning is the principal manifestation of the movement of the low-latitude geoidal anomaly to higher latitudes in response to the viscoelastic rebound of the continents to deglaciation, and is the process that explains why low-latitude coasts in the Pacific Basin experienced a fall in sea level during the late Holocene (Mitrovica and Peltier, 1991; Nunn, 1995). Hydroisostatic changes of land level are produced locally by the meltwater loading of the ocean floor around larger (>10 km in radius) oceanic islands which leads to a greater degree of emergence (relative sea-level fall) than around smaller islands (Nakada, 1986; Grossman et al., 1998). The late Holocene fall of sea level from its mid-Holocene maximum in the low-latitude Pacific was driven largely by equatorial ocean syphoning with local variations ascribable to hydroisostasy.

Steric sea-level changes are lower order, and are driven largely by temperature (and salinity) changes through the expansion and contraction of the upper part of the ocean; sea-level changes during the last millennium are regarded as

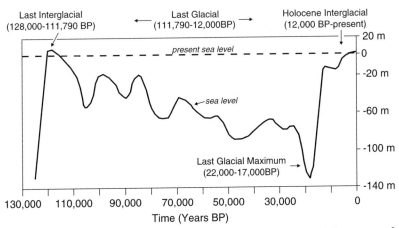

FIGURE 1.2 Sea-level changes over the past 150,000 years reconstructed from ages of emerged coral-reef terraces along the Huon Peninsula, Papua New Guinea (after Nunn, 1999, using data provided by John Chappell).

overwhelmingly steric in causation (van de Plassche et al., 1998; Munk, 2002). An example of this relationship is shown in the record from Bering Island (Fig. 1.3A), one of the only few where there are sufficient late Holocene temperature and sea-level data to show synchroneity. A well-constrained last-millennium sea-level record from the eastern seaboard of the United States is shown in Fig. 1.3B to demonstrate the fact of sea-level variation within the last millennium.

For the Pacific Basin, there is some information – far from sufficient to construct an entirely persuasive picture for this vast region – about sea-level change during the past millennium. The available evidence suggests that sea level rose as did temperatures during the early part (Medieval Warm Period), fell during the middle (A.D. 1300 Event), remained low (but variable) during the penultimate part (Little Ice Age) before rising more recently (Recent Warming). It is emphasized that these sea-level changes appear to have been only approximately synchronous and of variable magnitude (yet apparently the same sign), just as for steric sea-level changes monitored for the region during the past 100 years or so (Casenave and Nerem, 2004). Sea-level data are discussed in each of Chapters 4–6

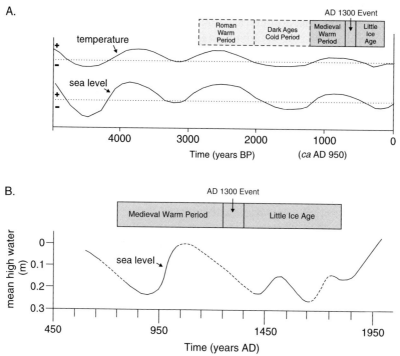

FIGURE 1.3 Temperature and sea-level changes during the latest Holocene. (A) Temperature and sea-level changes on Bering Island, Komandar group, northwest Pacific (after Razjigaeva et al., 2004). The closeness of the two trends shows the dominance of steric (temperature-driven) sea-level change within this period. (B) Sea-level (mean high-water) changes at Hammock River marsh, Clinton, CT, east-coast USA (after van de Plassche et al., 1998). This shows the high sea level during the Medieval Warm Period and the low sea level during the Little Ice Age preceding the period of recent sea-level rise.

for the respective divisions of the last millennium in the Pacific Basin, and play a critical role in the speculative model of the "A.D. 1300 Event" outlined in Chapter 7.

1.1.3 Societal change during the late Holocene

For reasons associated with individual survival, humans have organized themselves into societies – broadly defined – throughout the Holocene. By 5,000 BP, the global range of societal complexity was probably as great as it has ever been. Long-standing hunter-gatherer societies existed in those parts of the world where the environment could sustain this lifestyle although, as we come closer to the present, many such societies became partly sedentary as other lifestyle options (such as agriculture) became part of their subsistence arsenal. Yet in a few parts of the world, complex stratified societies were in existence by 5,000 BP, societies in which food producers produced surpluses that could sustain non-food producers such as tradespeople, military, religious and titular leaders, sometimes in sizeable urban enclaves.

For some social scientists, the appearance of complex societies heralds the start of an era when human resilience to external forcing (such as climate change) started to grow. Yet this view seems untenable, for many such complex societies burgeoned during times of stable climate, suitable – usually through the provision of adequate precipitation – for optimal food production. Yet rather than increasing resilience, these developments increased the vulnerability of complex societies to external change, particularly at times when comparative climate stability was replaced by more variable conditions. When such external changes happened – whether they were short-lived extreme events or more enduring climate (or sea-level) changes – these complex societies were invariably impacted more severely than a simpler society would be (Tainter, 1988).

A good analogy is found in modern times. Having reached an almost unimaginable degree of complexity, large cities the world over are popularly regarded as highly resilient to external forces of potential change. Yet this view – taken almost as implicit, even by many urban planners – fails to acknowledge the impossibility of effectively cocooning any human society from external change. Numerous natural disasters have occurred or threaten some of the world's largest megacities, and have the potential to thoroughly disrupt the way of life of their inhabitants. Yet more than this, large cities create their own vulnerability. This additional vulnerability comes from the challenges involved in effectively managing large cities, ranging from supplying their occupants with adequate amounts of drinking water to ensuring clean air. In this context, both external and internal factors can bring about change.

No scientist who works with human societies and cultures needs to be reminded of their complexity, and the manifest, often unidentifiable interactions that contribute to it. But at the same time, complex societies are vulnerable to external change – witness the effects of the 2004 Indian Ocean Tsunami on communities around the Indian Ocean – and it is pointless in the author's view to assert that complexity is a shield against the effects of external change. If anything, complexity in the form of societal stratification makes a society more vulnerable to externally driven changes; if the levee breaks, then the community sheltering behind it may be more vulnerable than a community that never had a levee (the "risk-compensation hypothesis" of Adams, 1995).

Finally, in support of the view of the enhanced vulnerability of complex societies to external (climatic and environmental) forcing compared to more simply organized societies, it could be noted that what we might regard as normal in the context of modern societies is far from being normal in a long-term sense. Sea level – although the same applies to temperature and other variables – is currently close to its highest level for ~120,000 years (see Fig. 1.2). For most of this time period, the ocean surface has been an average 50 m lower than it is today. We are living in exceptional times, and our societies are therefore as vulnerable in a long-term sense as were those that existed during the Last-Glacial Maximum 22,000–17,000 BP.

It is one of the basic arguments in this book that during the last millennium, it was climate change, commonly filtered through the environment, that brought about the principal and most enduring societal changes in the Pacific Basin. The corollary to this argument is that human influences on Pacific societies independent of ultimate climate drivers were subordinate. Applied worldwide, this argument applies more to earlier (commonly pre-historic) rather than modern human societies.

This is where the argument in this book differs from that in the impressive book "Collapse" (Diamond, 2005). Subtitled "How societies choose to fail or succeed", Diamond argues that the ultimate causes of pre-historic societal collapse are found in the development of the society itself, especially the growth of population above levels that can be sustained without innovations that maximize food output and depend – as many societies during the Medieval Warm Period came to – on a comparatively constant climate, especially adequate seasonal rainfall. Naturally, once climate constancy was replaced by variability – as it was during the transition from the Medieval Warm Period to the Little Ice Age in the Pacific Basin (and elsewhere) – then food production fell and population levels could no longer be sustained. To Diamond, climate is the proximate cause of societal collapse, which would – in his view – never have occurred had societies not made themselves vulnerable, an argument that he and others have transferred to the situation of the world today. For Diamond, societal change is the ultimate cause of societal collapse.

Now this may be applicable to some situations but there are many pre-historic societies in the Pacific Basin that exhibit the effects of climate-driven collapse without exhibiting any evidence of the required societal vulnerability. Some writers have become so trusting of what might be called the Diamond model that they attempt to fit their limited data to it, showing, for example, population-growth curves reaching well above island carrying capacity during the Medieval Warm Period when in fact the only evidence for such a scenario is actually the collapse that followed.

So in this book, climate is argued as having been the ultimate cause of pre-historic societal collapse in much of the Pacific Basin, with a variety of proximate causes, some of which can only be speculated about, but others which are environmental (sea-level fall and water-table fall) for which – unlike reconstructions of population growth – there is ample independent evidence.

1.2. THE APPROACH TO THE LAST MILLENNIUM

Even the most cursory look at the millennia preceding the start of the last millennium suggests that warmer climates were associated mostly with a comparatively rapid expansion of humans and the complexification of human cultures while colder periods marked times when cultures hibernated and sometimes fell apart. From most parts of the world, there is evidence of two distinct climate periods preceding the start of the last millennium. The first is named the Roman Warm Period, the second the Dark Ages Cold Period (see Figs. 1.1 and 1.3A).

The Roman Warm Period has been variably dated, mostly in Europe. In Iberia, for example, it lasted 250 B.C. to A.D. 450 (Desprat et al., 2003), perhaps significantly earlier in the northwest Pacific (Fig. 1.3A).

A wet Roman Warm Period is recorded along the Pacific coast of Canada (Nederbragt and Thurow, 2001). In central America, during the later part of the Roman Warm Period, the Pre-Classic Maya civilization was flourishing, but as the Dark Ages Cold Period began, a number of prolonged droughts saw the beginning of its final collapse around A.D. 900 (see Fig. 3.1). Evidence from China shows that the transition from the Roman Warm Period to the Dark Ages Cold Period here began around A.D. 490 and involved a temperature fall to 1°C below the 1951–1980 mean (Ge et al., 2003). In northwest Europe, deforestation during the Roman Warm Period was accompanied by the expansion of both agriculture and pasture land (Berglund, 2003).

Archaeology-driven research has uncovered the existence of an A.D. 536 Event, when a succession of years without summers led to repeated crop failures and consequent societal stress across Asia, Europe and central America (Gunn, 2000). The A.D. 536 Event marked the start of a period of climate variability, broadly congruent with the Dark Ages Cold Period.

The Dark Ages Cold Period has also been variably dated, also mostly in Europe. In Iberia, it occurred in the period A.D. 450–950 (Desprat et al., 2003). Recent work in Andean Peru has also shown that a cool wet Dark Ages Cold Period occurred here (ending in the period A.D. 700–1000), shown by higher lake levels (Hansen et al., 1994) and glacier expansion (Seltzer and Hastorf, 1990), during which low temperatures precluded agriculture (Chepstow-Lusty and Winfield, 2000). A similar record of cool conditions in the period A.D. 240–800 comes from China (Yang et al., 2002) while in Japan and the northwest Pacific Islands the period A.D. 250–650 (1,700–1,300 cal BP), approximating the Dark Ages Cold Period, is named the Kofun cold stage and was associated with lowered sea level (Sakaguchi, 1983; Razjigaeva et al., 2004).

In Europe, the Dark Ages Cold Period was heralded by a time of rapid cooling and marked by generally increased precipitation and storminess. A recession of agriculture allowed the natural reforestation of large areas of central Europe and Scandinavia (Andersen and Berglund, 1994). A link between the collapse of the Roman Empire about A.D. 480 and the Justinian Plague about A.D. 540 and the adverse climate of this period in Europe was proposed by Berglund (2003).

The purpose of describing these widespread global climate oscillations predating the last millennium is to strengthen the case for recognizing similar oscillations within this period (Section 1.3).

1.3. DIVIDING THE LAST MILLENNIUM

It may appear incredible to a dispassionate observer that so much vitriol could be expended on something as apparently benign as the division of the last millennium for the purposes of understanding its climatic evolution. But there is no escaping the issue, and there are undoubtedly scientists who will bang shut this book when they read how the author has chosen to divide the last millennium. Yet in almost every part of the world, empirical evidence has been reported suggesting a warm period early in the last millennium and a cooler period later on. As a recent authoritative global survey concluded,

> "although substantial divergences exist during certain periods, the timeseries display a reasonably coherent picture of major climatic episodes: 'Medieval Warming Period', 'Little Ice Age' and 'Recent Warming'". (Esper et al., 2005, p. 2164)

Another recent study also argued in favour of a warm period for the northern hemisphere peaking around A.D. 1000–1100 and consistent with the Medieval Warm Period, and a subsequent cooler period, consistent with the Little Ice Age, having minimum temperatures ~0.7 K below the average (for 1961–1990) during the sixteenth and seventeenth centuries (Moberg et al., 2005; Fig. 1.4).

Within the Pacific Basin – the geographical focus of this book – there is ample evidence to suggest a similar division, and there seems no good reason to deny it.

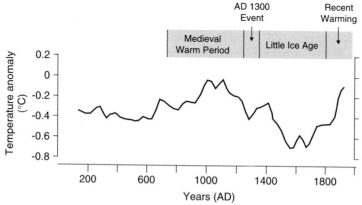

FIGURE 1.4   Estimated northern-hemisphere mean temperature variations in the period A.D. 133–1925 (after Moberg et al., 2005). Eighty-year running means are shown.

Most detractors of this system of division have implied that it is localized – typically occurring only along North Atlantic coasts (which is incorrect) – or that the data used to make these observations are biased in terms of recognizing such divisions *per se*, and that actually last-millennium climate was largely unchanging up until ~200 years ago. Certainly many global or regional compilations of last-millennium temperatures have shown such a pattern but it has also been suggested that the methods used to make such compilations tend to reduce actual variability. This is a subject that is revisited in Chapter 10.

In addition to the Medieval Warm Period and the Little Ice Age, a few scientists – the author included – have drawn attention to the significance of the transition between the Medieval Warm Period and the Little Ice Age (Kreutz et al., 1997; Nunn, 1999; Meyerson et al., 2003; Morrill et al., 2003). It has been named the A.D. 1300 Event by Nunn (2000a), a label intended to elevate its status from mere transition to a period of rapid climate change that – in many parts of the Pacific Basin and elsewhere – probably drove rapid environmental and societal changes.

Most scientists also recognize that the most recent period of climate, in which we are still living, is distinct from the preceding Little Ice Age and can be called the period of Recent Warming, during which anthropogenic forcing of climate became a significant factor in climate change.

In accepting these divisions, it is necessary to be somewhat cavalier about what is meant by the last millennium: in this book, it is taken as having begun about A.D. 750 (the suggested start of the Medieval Warm Period in the Pacific Basin) and still continuing. While not generally agreed upon – something expected, not something controversial – the dates for these various periods in the Pacific Basin are taken as follows:

- Medieval Warm Period – A.D. 750–1250;
- A.D. 1300 Event – A.D. 1250–1350;
- Little Ice Age – A.D. 1350–1800;
- Recent Warming – A.D. 1800 to present.

The recognition of distinct periods of climate like the Medieval Warm Period and Little Ice Age is helpful for understanding the contemporaneous environmental and human–societal changes, but we must be wary of recognizing such periods just for this reason. For example, the Pre-Classic Maya Civilization, which thrived during the Roman Warm Period, collapsed spectacularly during the transition (about A.D. 250) to the Dark Ages Cold Period, and recovered only slowly thereafter (see Fig. 3.1). Similarly, many Pacific Island societies changed profoundly during the transition (A.D. 1300 Event) from the Medieval Warm Period to the Little Ice Age, plausibly because of rapid climate change (Nunn, 2000a; Nunn et al., 2007). In both examples, if one accepts that the climate periods on either side of the societal collapse were discrete periods of largely similar climatic characteristics (in which human societies changed only slightly), then clearly the short-lived transitions are likely to have been responsible for societal collapse. In other words, the nature of the climatic interpretation is key to the explanation of the societal impacts.

A related point is that, just because the Medieval Warm Period and the Little Ice Age are recognizable as distinct periods of climate within the Pacific Basin, this does not necessarily mean that the Medieval Warm Period and the Little Ice Age are exactly the same in character as they were originally conceived to be – sharply bounded periods of unvarying climate (Lamb, 1965, 1977). More recent studies from many parts of the world show that the Little Ice Age was a time of variable climate in contrast to the relative constancy of conditions during the Medieval Warm Period, and the suspicion must be that societies affected by rapid climate change during the transition (A.D. 1300 Event) continued to be impacted by climatic variability, particularly as it affected regular food supply. Good examples come from Morocco and Iceland, and from the Pacific Islands (Till and Guiot, 1990; Nunn and Britton, 2001; Ogilvie and Jónsson, 2001).

In many recent discussions of global climate change during the past millennium, the disparity between long-term data series from the northern and southern hemispheres is highlighted. One could also argue that there is a significant imbalance between data from the Pacific Basin, which covers around one-third of the Earth's surface, and elsewhere. In this regard, this book, while providing no unpublished hard data about last-millennium climates in the Pacific Basin, presents new evidence and interpretations that favour a similar pattern of climate change within this period for this region as for other parts of the world.

There are many different names given to the Medieval Warm Period, fewer to the Little Ice Age and period of Recent Warming. This book keeps to the more conventional system of names, largely for the convenience of readers.

1.4. MEDIEVAL WARM PERIOD

Also known as the Little Climatic Optimum, occasionally the Neo-Atlantic or Medieval Climate Anomaly, the Medieval Warm Period is usually defined as a warmer-than-present period that endured for ~400 years in most parts of the world around the start of the last millennium. In almost every part of the world, evidence has been reported for a period between about A.D. 750 and A.D. 1250 (1,200–700 cal BP) known as the Medieval Warm Period that was warmer than the period before or that after (Le Roy Ladurie, 1971; Lamb, 1977; Villalba, 1990; Dean, 1994; Grove and Switsur, 1994; Keigwin, 1996; Esper et al., 2002a, 2005). In his survey, Broecker (2001) concluded that a global Medieval Warm Period occurred in the period A.D. 750–1150 (1,200–800 cal BP). A comprehensive review of global evidence for the Medieval Warm Period (Soon et al., 2003) found only two places in the world where this was lacking or ambiguous: in Himalayan ice-core data, where altitude may have moderated the signal, and a tree-ring series from Chile that is contradicted as a regional signal by nearby evidence.

Yet some palaeoclimatologists point to inadequacies of available scientific data, claiming that the scientific evidence for the Medieval Warm Period is not global, and portraying last-millennium temperatures as having fallen almost linearly in the period A.D. 1000–1800 (Jones et al., 1998; Mann et al., 1999; Bradley, 2000). Even the Intergovernmental Panel on Climate Change (IPCC), in stark opposition to their 1990 position, has recently espoused this view (IPCC, 2001). Still others appear more circumspect, pointing out that while there is evidence for warmer and cooler episodes of climate in various places, these do not appear to be inter-regionally synchronous, and more data are needed before final judgement is reached (Hughes and Diaz, 1994).

Many palaeoclimatologists today are somewhat wary of using non-scientific data (such as changes in human societies) to help understand last-millennium climate changes. This is despite the fact that it was primarily these proxies of climate data as expounded in books like those of Lamb (1977, 1982) – in the days before high-resolution palaeoclimate data were widely available – that first drew the attention of scientists to climate divisions of the last millennium. In the same spirit as these early works, this book uses proxy data, particularly the profound changes that took place in various Pacific societies during and just after the A.D. 1300 Event, to support the evidence from the comparatively few long-term palaeoclimate data series for the Pacific Basin.

A full account of what happened in the Pacific Basin during the Medieval Warm Period is given in Chapter 4.

1.5. THE A.D. 1300 EVENT (THE TRANSITION BETWEEN THE MEDIEVAL WARM PERIOD AND THE LITTLE ICE AGE)

Many commentators have failed to acknowledge the existence of a transition between the Medieval Warm Period and the Little Ice Age, let alone its potential importance as a driver of environmental and societal changes. In many discussions of last-millennium climate, it appears that the Medieval Warm Period was followed instantly by the Little Ice Age, but this is an unhelpful generalization. The Little Ice Age did not immediately follow the Medieval Warm Period. It makes no sense to define the Medieval Warm Period as including at its very end a period of rapid cooling. No more does it make sense to include in our definition of the Little Ice Age a period of rapid cooling at its start. It makes more sense to acknowledge a discrete transition, named in this book the A.D. 1300 Event, to emphasize its importance to an understanding of climatic, environmental and societal evolution within the last millennium.

The A.D. 1300 Event is defined by rapid cooling over a period of perhaps 100 years which led to the replacement of Medieval Warm Period conditions by Little Ice Age conditions. Two examples from outside the Pacific Basin are shown in Fig. 1.5, the first a reconstruction of ground-surface temperatures from the Ural Mountains of eastern Russia using borehole thermometry, the other the results of two ice cores through the Greenland ice cap, North Atlantic. In addition, there are many examples of how glaciers advanced during the A.D. 1300 Event and remained stationary during both the Medieval Warm Period and the Little Ice Age (for the Pacific Basin, see Fig. 5.1).

The A.D. 1300 Event was the most rapid period of climate change within the past few thousand years, yet only a few palaeoclimatologists have latched onto its importance (Kreutz et al., 1997; Meyerson et al., 2003). Some examples

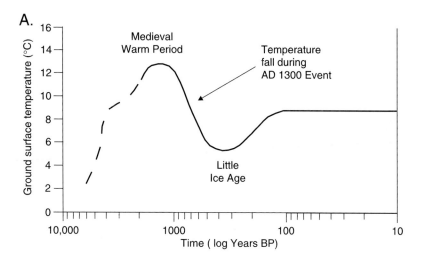

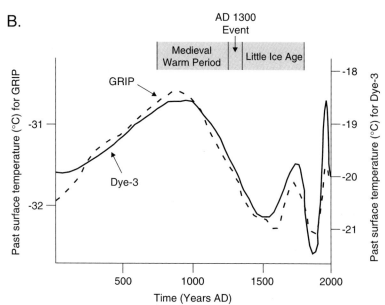

FIGURE 1.5 Examples of the rapid cooling that characterized the A.D. 1300 Event outside the Pacific Basin. (A) Ground surface temperatures within the past 100,000 years reconstructed from the Ural superdeep borehole SG-4 (adapted from Demezhko and Shchapov, 2001). Note the logarithmic scale. (B) Temperature changes for the past 2,000 years reconstructed from two boreholes (GRIP and Dye 3) through the Greenland ice cap (after Dahl-Jensen et al., 1998). Note that the temperature changes recorded at Dye 3 are nearly twice as large as those at GRIP.

illustrate the rapidity and significance of this transition. Chesapeake Bay (eastern United States) has long been known as a sensitive environment to climate change. Mg/Ca analysis of calcitic microfossils showed a rapid cooling of ~2.5°C around A.D. 1350 (Cronin et al., 2003). As shown in Fig. 1.6, a rapid fall of temperature

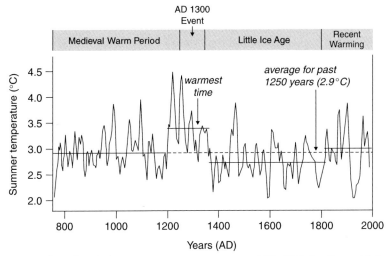

FIGURE 1.6 Northern-hemisphere summer temperature A.D. 750–1998 for Donard Lake, Baffin Island, Canada, calculated from varve thicknesses (after Moore et al., 2001). Ten-year running means.

around A.D. 1300 in the Canadian Arctic marks the transition from the Medieval Warm Period to the Little Ice Age and may have sparked the well-documented southward migration of the people of coastal Labrador at this time (Fitzhugh, 1997). And for the Pacific Basin, the author has argued in a number of papers that climate and associated (non-human) changes during the A.D. 1300 Event were a major cause of environmental and societal changes in the Pacific Basin (Nunn, 2000a, 2000b, 2003a, 2003b, 2007a; Nunn and Britton, 2001; Nunn et al., 2007).

The nature of climate, environmental and societal changes during the A.D. 1300 Event in the Pacific Basin is described in Chapter 5, with a somewhat speculative model of the nature and consequences of the A.D. 1300 Event confined to Chapter 7.

1.6. LITTLE ICE AGE

Far fewer palaeoclimatologists doubt the existence and near-global extent of the Little Ice Age, for which evidence has been obtained from a variety of sources (Grove, 1988; Fagan, 2000). Even Jones et al. (1998), who are sceptical about the Medieval Warm Period, find clear evidence

> "for cooler centuries between 1500 and 1900, particularly the seventeenth and nineteenth centuries which are indicative of a two phase "Little Ice Age"". (p. 469)

The evidence for a near-global Little Ice Age was synthesized by Soon et al. (2003) who reported its absence only in two places: in isotopic measurements from Siple Dome, Antarctica, and from tree-ring records in Tasmania. Both of these locations are likely to be anomalous in a global context, Siple because, like all other

Antarctic sites, it is sensitive to small changes in polar circulation that have nothing to do with global forcing mechanisms (Comiso, 2000), and Tasmania because its climate is significantly moderated by region-specific oceanic influences (Cook et al., 1992).

Compelling high-resolution palaeoclimate studies that show the existence of the Little Ice Age come from the tropical Andes (Thompson et al., 1986), the western USA (Petersen, 1994), the Sargasso Sea (Keigwin, 1996), throughout South Africa (Tyson et al., 2000), Iceland (Ogilvie and Jónsson, 2001), China (Qian and Zhu, 2002) and the eastern USA (Cronin et al., 2003). The influential record from the GISP2 ice core in Greenland shows that the Little Ice Age occurred there in the period A.D. 1350–1800 (Stuiver et al., 1995).

Those who doubt the existence of a near-global Little Ice Age include Mann et al. (1999) who interpret the period from about A.D. 1000 to A.D. 1900 as a time of uniform cooling "possibly related to astronomical forcing" (p. 762). A similar interpretation by Crowley (2000) involving low climate variability within this time period was attributed not to astronomical forcing but to volcanic forcing, and to changes in solar radiation and greenhouse gases.

Compared to the situation during the Medieval Warm Period, the non-climatic proxy evidence for the existence of the Little Ice Age is overwhelming and has deservedly been the subject of several targeted books including Grove (1988) and Fagan (2000). For the Pacific Basin, it is proposed that the nature of environments and societies that characterized the Little Ice Age was a direct result of the climate changes that had taken place during the preceding A.D. 1300 Event. The persistence of retrograde societal developments during the Little Ice Age is attributable to the cool, variable climate conditions that characterized the period, at least its early part. The societal improvements noticeable towards the end of the Little Ice Age result from the gradual recovery of environments from the disruption effected by rapid climate change during the A.D. 1300 Event.

A full account of the changes during the Little Ice Age is given in Chapter 6.

1.7. RECENT WARMING

There is no argument among palaeoclimatologists that Earth-surface temperatures have risen sharply within the past 200 years or so, marking a distinct period of Recent Warming. There is some debate about when it began, some authors preferring around A.D. 1800 (as in this book), others closer to A.D. 1900 (Soon et al., 2003). The issue is not especially important, for we are dealing here with the transition from the Little Ice Age to the period of Recent Warming that appears to have been more nebulous and less abrupt than the A.D. 1300 Event.

There has been much discussion about how this period of Recent Warming, particularly its most recent part, compares to the Medieval Warm Period. Specifically, it is debated whether or not the most recent warming (particularly since the 1970s) is exceptional and therefore demands an explanation without precedent, or whether the latest recent warming was something that was achieved – perhaps even surpassed – during the latest part of the Medieval Warm Period. This debate has to do with the cause(s) of recent warming and the effectiveness of solutions proposed to counter it. Specifically, if the most recent warming is unprecedented, then it is likely to be a result of the human enhancement of the greenhouse effect, and strategies to reduce greenhouse-gas emissions and deforestation stand a good chance of succeeding. Conversely, if the most recent warming is similar to that attained during the Medieval Warm Period, then perhaps the anthropogenic contribution to climate change has been overstressed, and we can safely continue to burn more fossil fuels and deforest our planet's surface.

Not surprisingly in view of its economic implications, this debate has become politicized in many parts of the world, and with this the rhetoric of opinionated politics has entered the relevant scientific literature. The debate has also become surprisingly polarized, with many scientists openly declaring an allegiance to a particular camp, and then remaining there despite new data and analyses being announced.

When this author first became interested in climate change in the Pacific, he had no particular allegiance, and still has none. The data available for the Pacific Basin, as will be seen, show that both the Medieval Warm Period and the Little Ice Age can be clearly distinguished in this region, the implication being that there has been more climate variation in the last millennium than one school of thought maintains. Yet in the Pacific Basin, most palaeoclimate data series also suggest that the most recent (post-1970) warming is unprecedented although there are a few that suggest otherwise. In this context, the author stresses that the interpretations given in this book and the positions taken on various issues derive from the data presented, and might one day change if contradictory data prove persuasive.

1.8. THE PACIFIC BASIN

The Pacific Basin occupies more than one-third of the Earth's surface, and is divided in this book into the fringing continental Pacific Rim and the Pacific Islands and Ocean (Fig. 1.7). The Pacific Basin is a region that exhibits geographical integrity in many ways. It is mostly ocean dotted with islands surrounded by a continental rim marked in most places by mountain ranges trending parallel to the coast that neatly delineate the region. In many parts of the Pacific Rim, these mountain ranges act as formidable barriers to the movement of air masses and various living things (including humans), thus giving the Pacific Basin both an unmatched climatic and biotic integrity. The point is well illustrated by the eastern Pacific Rim where the mountain barrier is effectively continuous and, as a result, areas to the west (along the western seaboard of the Americas) have climates influenced by Pacific air masses (Thompson et al., 1993; Hallett et al., 2003). In contrast, the areas to the east have climates that seem to be in phase with Greenland climate changes and may be driven by changes in Arctic circulation (Yu and Ito, 1999).

In addition to integrity deriving from the configuration of Pacific land areas, the almost continuous Pacific Rim gives this vast region an unmatched oceanic

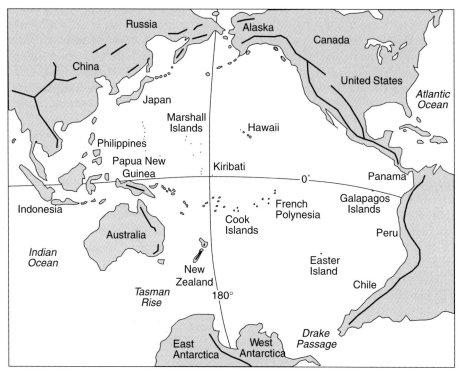

FIGURE 1.7 Map of the Pacific Basin and adjacent regions (equal-area projection). Thick lines represent mountain belts.

integrity. Outside the southernmost Pacific, the gaps in the Rim in the northern Pacific (Bering Strait) and western tropical Pacific (through the islands of Southeast Asia) are generally shallow and insignificant influences on the Pacific Ocean regional circulation.

The integrity of the Pacific Basin lends itself to regional geographical description in the classical sense. Also, inasmuch as particular regions of the Earth were targeted for synthetic geographical study 50 years ago, so the Pacific Basin meets one of the other criteria for such study – it is not well known. For many different facets of modern enquiry, the Pacific Basin is a knowledge void, and is often discussed only briefly, frequently in a marginal context, in supposedly global surveys of particular phenomena. The common perception of many people, accustomed to flying over the Pacific, is of this third of the Earth as empty (Ward, 1989). Of course, this perception is faulty, and one that is likely to change in the next few decades if the twenty-first century truly proves to be the Pacific century.

There follows an overview of Pacific Basin climates and their evolution (Section 1.8.1). Then there is a discussion of Pacific Basin environments and the main controls on their evolution (Section 1.8.2), and finally a brief explanation of the nature of Pacific Basin societies (Section 1.8.3).

1.8.1 Climate of the Pacific Basin and its evolution

The main control of climate within the Pacific Basin is solar radiation (insolation), the distribution of which is regulated by latitude. Within low latitudes, more solar radiation is absorbed in the western equatorial Pacific, and the resulting oceanic warm pool is critical to the formation of tropical cyclones (hurricanes or typhoons). Tropical cyclones will develop only in ocean areas where surface temperatures are greater than 26–27°C, a condition usually met only in the western Pacific warm pool but one that extends across much of the equatorial Pacific during El Niño events.

Northeast and southeast tradewinds meet along the Inter-Tropical Convergence Zone (ITCZ) where air rises and moves polewards, descending to complete the Hadley Cells ~30° away from the ITCZ. This belt of descending air creates the subtropical high-pressure zone, polewards of which is found a zone of mid-latitude westerlies, the subpolar low-pressure belt, and finally the zone of polar easterlies around the poles. The ITCZ moves seasonally north and south, especially in the western Pacific. In the southwest Pacific, the South Pacific Convergence Zone (SPCZ) forms where southeast tradewinds meet southerly air moving north from New Zealand.

When the continents along the western Pacific Rim (Australia and Asia) are seasonally heated and cooled, wind reversals known as monsoons occur. In the southern-hemisphere summer, the ITCZ moves across northern Australia bringing the northeast tradewinds in its wake and causing the summer monsoon here. For much of Asia, including that portion within the Pacific Basin, the summer monsoon is the most critical aspect of climate for its inhabitants. The Asian summer monsoon occurs when Pacific winds are pulled onto the continent by the development of a strong low-pressure cell in its centre.

Two important sources of interannual climate variability in the Pacific Basin are ENSO and the PDO/IPO. ENSO involves alterations to ocean–atmosphere circulation in the Pacific, and probably became significant as a source of inter-annual climate variability only ~5,000 BP when west–east sea-surface temperature gradients in the Pacific increased markedly and tradewinds attained their present strength (Rodbell et al., 1999). During El Niño (ENSO-negative) events, tradewind strength decreases markedly, affecting both sea-surface temperature and sea level. As the major cause of interannual climate variability in the Pacific Basin, the ENSO phenomenon, particularly the recurrence times and intensities of El Niño events, has a great potential for disrupting human societies. This is as true today as it was in the past, and thus this book pays special attention to reconstructed El Niño periodicity at particular times during the last millennium. Less is known about long-term variations in the PDO/IPO, a 20–30-year cycle with warm and cool phases that affects the incidence of both tropical cyclones and ENSO, and which exhibited variations within the last millennium that are broadly consistent with other indicators (MacDonald and Case, 2005).

The main regional control on Pacific Basin climate (and ocean circulation) is the configuration of land and sea, the main elements of which were established several million years ago following the build-up of the Antarctic ice sheets, particularly after the gaps in the Pacific Rim effectively closed (Nunn, 1999).

1.8.2 Environments of the Pacific Basin and their evolution

The most important determinant of the ways that terrestrial environments appear and the ways in which they change is their composition. Within the Pacific Basin, the most fundamental distinction is between oceanic and continental settings, the latter being composed of generally far older and diverse types of rock than the former. Tectonic context – whether an area is stable or unstable (rising or subsiding) – is also important. Yet in the context of the last millennium, while considerations such as rock type and tectonics are important in determining antecedent conditions of a particular environment, they are comparatively unimportant to the understanding of such short-term environmental change.

Environments in the Pacific Basin have changed for multifarious reasons during the last millennium, and it is not always easy to establish the causes of these changes. They may be grouped into four main categories:

- climate change, particularly in the short term, the occurrence of extreme events (such as droughts and storms), and in the longer term, the slower change from one climate regime to another;
- sea-level changes, particularly along coasts and in lowland areas, where sea-level rise or fall produces, respectively, drowning (and water-table rise) or emergence (and water-table fall), which affect not only land area available for habitation but also coastal and lowland ecosystems;
- tectonic changes which can produce (rapid) emergence or subsidence; and
- finally, humans who have brought about profound changes to Pacific environments, especially within the past 150 years or so, through a range of adaptations intended to enhance or preserve environmental sustainability, productivity or habitability.

In Chapters 4–7 that deal with aspects of the pre-modern (pre-industrial) Pacific Basin, the effects of climate change on environments (particularly on vegetation) are considered under the "climate" heading. Tectonic changes are localized and infrequent, and only considered in this book where they impacted environments and societies severely. Human-induced environmental change was either localized and/or minimal prior to about A.D. 1900 in most parts of the Pacific Basin, so it is not considered systematically until Chapter 8 in which the period of Recent Warming is discussed. This leaves sea-level changes as the principal cause of widespread environmental changes in pre-modern times in the Pacific Basin during the last millennium, and this is a topic discussed in each of Chapters 4–7.

1.8.3 Societies of the Pacific Basin and their evolution

Modern societies of the Pacific Basin vary in both their level of complexity and the degree to which they are interdependent with societies elsewhere. The pace of societal change in many parts of the Pacific Basin is rapid, elsewhere comparatively slow. Consequently there is no purpose in attempting to generalize about the modern character of Pacific Basin societies.

Yet a few hundred years ago there was far less variation, the waking hours of most people in the Pacific Basin, as elsewhere, being concerned primarily with food acquisition. There were different methods of this, the most profound disparity being between (nomadic) hunter-gatherers and (sedentary) agriculturalists. But in most places, people a few hundred years ago in the Pacific Basin interacted more closely on a daily basis with the natural environment than do their modern counterparts. This interaction made the activities of these people more vulnerable to the vagaries of the environment, expressed as the impact of either short-lived extreme events (such as a tsunami or a tropical cyclone) or more prolonged climate regime shifts of the kind on which this book focuses.

There is a polarization of views regarding the evolution of human societies, between those (cultural determinists) who regard this as a drama played out against the unvarying and benign backdrop of the environment and those (environmental determinists) who see societies as subject more or less to the whims of environmental change. In recent decades, a number of influential studies (e.g. Binford et al., 1997; Haug et al., 2003) have emphasized the critical importance of environmental change, typically driven by climate change and/or sea-level change, in bringing about profound changes in the trajectories of human–societal evolution, and a few archaeologists are beginning to seriously ponder the effects of environment and environmental change on human activities (e.g. Fagan, 1999, 2004; Jones et al., 1999; Anderson, 2002; Anderson et al., 2006).

There is generally more sympathy for the effects of climate change on pre-historic societal change among palaeogeographers and others who approach the issue from the climate side rather than from the social-science side. For example, "the impact of climate on the ancient history of Japan is undeniable" (Sakaguchi, 1983, p. 1); "many lines of evidence now point to climate forcing as the primary agent in repeated social collapse" (Weiss and Bradley, 2001, p. 610); in the Pacific Islands "there is abundant evidence that environmental changes of extraneous causation forced island people to alter a whole range of lifestyle options" (Nunn, 2003a, p. 226).

While admitting a limited role for external (climate) change, a recent bestselling work (Diamond, 2005) tends to follow the cultural-determinist line, pointing out the dangers of contemporary trajectories of societal development by highlighting the role of poor human–societal choices in the collapse of past societies. Two of Diamond's key case studies – of the Pacific islands of Tikopia and Easter Island (Rapanui) – are also discussed in the present book, where it is argued that his cultural-determinist interpretation of their societal collapse is wrong. This is largely because profound cultural transformation on these islands – at opposite ends of the Pacific Ocean – occurred, like numerous other examples from this region, in association with the A.D. 1300 Event. It therefore seems far more likely that it was climate that drove cultural change; maybe this process was temporarily stayed or accelerated by unwise human–societal choices but its eventual outcome was not affected by these.

Environmental determinism has been treated as a philosophical pariah for decades, largely as a continuing reaction to some untenable applications early in the twentieth century. Among these was the assumption that the environment of a particular area determined not only the culture of its human occupants but also a host of their physiographic attributes from skin colour to stature. The comprehensive rejection of this environmental determinism sent the pendulum in the other direction, towards unrestrained cultural determinism in which societal change was explained solely by internal processes and the environment was regarded as a passive backdrop to the human drama. This book favours environmental determinism, defined as the belief that

> "the environment sets certain constraints on what any population can achieve within a given technology, and that natural environmental variation may play a role in changing the course of cultural development ... changes in climate, resources and habitats are not simply background information overlain by cultural change, but are considered a continually changing set of problems and opportunities altering the context for human survival". (Haberle and Chepstow-Lusty, 2000, p. 350)

The debate between the cultural and environmental determinists is not confined to the Western, English-speaking world. In China, for instance, there has been long and intense debate about the reasons for the successive rise and fall of Changjiang (Yangtze) Delta civilizations over the past 8,000 years (see Fig. 3.3). Some writers aver that this was due to social factors, typically the outbreak of conflict (Zhou and Zheng, 2000), while others – with more empirical support – argue that it was either changes in sea level (Chen and Daniel, 1998) or flood frequency and intensity (Zhang et al., 2005) that brought about these changes.

Finally, it is appropriate to draw some distinctions between continental and island societies in pre-modern times. A continental-based society is less inherently vulnerable than an island-based one. On continents, the extent of the land area means that land resources are essentially unlimited, and that no continental-based society ever needs to depend – at least at low population densities – on marine resources. In addition, unlike societies on (smaller) islands, continental-based societies also often have the resources of large river systems available to them. Continental societies in the past, when impacted by a natural disaster, have then the option of moving relatively easily to another area.

Island societies are distinct from continental societies in many ways, perhaps the two most important being the effects of circumscription (island boundedness) and the challenges of routine (interisland) interactions. Circumscription means that terrestrial (and nearshore marine) resources are visibly finite, especially on smaller islands, and that variability in food supply is likely to result as island populations approach island carrying capacity. Typical responses may be to enhance agricultural productivity, implement resource conservation measures and expand dietary range (Johannes, 1982; Spriggs, 1986; Kaplan and Hill, 1992). The impacts of natural disasters on the societies of islands, especially smaller and more remote ones, will be generally amplified compared to societies on continental areas of similar size. In one illustration, this may be why Christianity "is a continental ideology, not an island one" (McNeill, 1994, p. 324) given that it lacks taboos on resource use yet has strong taboos on abortion and infanticide. From what we know of pre-modern Pacific Island societies, the opposite effects seem to have applied.

1.9. ORGANIZATION OF THIS BOOK

This book is organized with the intention of making its content as accessible as possible to readers of different backgrounds. The present chapter explained the context and divisions of last-millennium climate change, before introducing the Pacific Basin and the nature of its climates, environments and societies. Chapter 2 describes how and when modern humans reached the Pacific Basin and spread throughout it. Chapter 3 explains the influences that climate and sea-level changes have had on humans in the past, emphasizing the role of sea-level change in bringing about profound societal change, an argument that underpins the main thesis of this book.

The nature of the Medieval Warm Period in the Pacific Basin is discussed in detail in Chapter 4, which is subdivided – as its successors are – into sections on climates, environments and societies. Chapter 5 details the nature of the A.D. 1300 Event – the transition between the Medieval Warm Period and the Little Ice Age in the Pacific Basin – while Chapter 6 explains what happened here during the Little Ice Age. Chapter 7 is devoted to an account of an explanatory model of the A.D. 1300 Event, showing how climate change may have induced changes in both environments and societies. A series of case studies from both the Pacific Rim and the Pacific Islands is given to illustrate the arguments. In Chapter 8, there is a discussion of the nature of the period of Recent Warming, dominated by rising temperatures and rising sea level but with climate-driven environmental and societal changes intertwined with – and becoming progressively subordinate to – changes arising from globalization, broadly defined.

Chapter 9 looks at areas of the Earth outside the Pacific Basin and remarks on the similarities between the last-millennium histories of these regions and that of the Pacific Basin. Emphasis is placed on the A.D. 1300 Event. The final chapter looks at the possible causes of last-millennium climate change, both ultimate (global) causes and proximate (regional and local) causes.

KEY POINTS

1. Less is known about climate change and its effects on environments and societies in the Pacific Basin than in many other parts of the world.
2. Within the past 5,000 years, temperatures have been generally cooling, sea level falling and societies becoming more complex, thereby increasing their vulnerability to external change.
3. The Roman Warm Period and Dark Ages Cold Period are times of distinct climate preceding the last millennium that have been recognized globally.
4. Within the Pacific Basin, there is ample evidence supporting a fourfold climatic division of the last millennium (A.D. 750 to present). The Medieval

Warm Period (A.D. 750–1250) was a time of warm conditions, the A.D. 1300 Event (A.D. 1250–1350) a time of rapid cooling, the Little Ice Age (A.D. 1350–1800) a time of cool conditions and the period of Recent Warming (A.D. 1800 to present, and continuing) a time of generally rising temperature.

5. The Pacific Basin occupies more than one-third of the Earth's surface, and functions in many ways as a closed system, the effectively continuous mountainous continental Pacific Rim confining much of the air and ocean water of the Pacific Basin so as to produce locally distinctive climates and ocean circulation. Environments and societies in the Pacific Basin both vary depending on whether they are part of its continental rim or on an oceanic island.

CHAPTER 2

Arrival and Spread of Humans in the Pacific Basin

This chapter outlines the pathways – both literal and developmental – by which early *Homo sapiens* (modern humans) reached the Pacific, and then those by which its rim and later its centre were colonized (Fig. 2.1). Throughout this chapter, the relationship between human societies and climate/environmental change is emphasized, a theme that is explored further in Chapter 3.

The first part of this chapter (Section 2.1) looks at how and why modern humans moved out of Africa into – among other places – the continental hinterland of the western Pacific Rim. The second part explains how and when humans dispersed along the Pacific Rim (Section 2.2) and eventually colonized Pacific Islands (Section 2.3).

2.1. HUMAN BEGINNINGS

The first modern humans and their immediate ancestors, appearing first $\sim$5 million years ago in equatorial Africa, were itinerant. Yet just as when we talk of modern humans spreading out from Africa, this should not be taken to imply any intentional long-term choices as might be assumed were we to apply words like itinerant or spread to ourselves. It seems most likely that the movements of the earliest humans were simply responses to the supply and movement of food sources, perhaps results of conflict with competing predators, and perhaps to the same harsh droughts that bedevil the lives of the inhabitants of modern sub-Saharan Africa.

Itinerant people generally have no home, no fixed base. They are therefore less vulnerable than sedentary people to external changes over which they have no control. In fact, the characteristic of itinerancy or nomadism can be seen as an adaptation by humans to existence in an area with few wild-food resources, where movement is therefore essential to survival. The converse – sedentism – requires that sufficient food be available within a circumscribed area in order for the members of the sedentary group to survive (Khazanov and Wink, 2001; Berland and Rao, 2004).

The earliest bipedal human ancestors in equatorial Africa probably had no notion of home beyond that of short-term residency within a particular resource catchment. But as one catchment's resources were exhausted, so they moved to another, and so on. For some groups, possibly driven by resource scarcity or competition with other groups, their wanderings eventually led them out of Africa and into lands beyond. Many such lands were occupied by fewer people

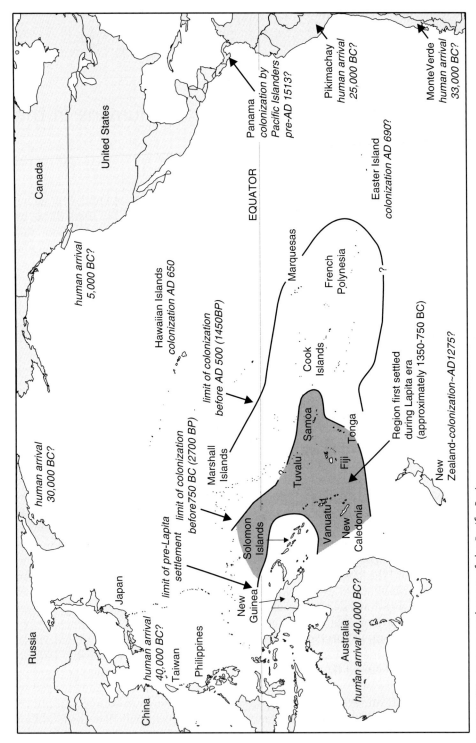

FIGURE 2.1 Human settlement of the Pacific Basin.

and by comparatively naive biota, a combination of circumstances that made food acquisition easier. In this way, *H. sapiens* – following in the footsteps of *Homo erectus* at least 500,000 years earlier – found their way east across the Asian continent and into the upper valleys of the few large rivers – principally the Huanghe (Yellow) and Changjiang (Yangtze) – that drain its eastern part. Although some argue that a strain of *H. sapiens* evolved in East Asia independently of that in equatorial Africa (Pope, 1992), most scientific opinion based on studies of Asian hominid fossils favours the scenario sketched above (Jin and Su, 2000).

Unsurprisingly, while the evidence for early *H. sapiens* in the upper parts of the Huanghe and Changjiang is fragmentary and therefore difficult to understand fully, it seems reasonable to suppose that our species reached first these areas ~200,000 years ago. These wide alluvium-choked valleys – some of the largest in the world – would have acted as magnets for early groups of *H. sapiens*. These valleys would have contained abundant wild foods, particularly cereals, and herds of animals that, like humans, found life on flatter areas like valley floors more agreeable than elsewhere. But the valleys were not without their dangers, for predators with similar dietary preferences to humans also lived there. Among those that not only competed with humans for the same foods, but also posed a danger to them on account of their greater size, were cheetahs and hyenas. In excavations at Zhoukoudien near Beijing, teeth marks on human bones were once thought to be signs of cannibalism, but are now known to have been made by the hyena *Pachycrocuta brevirostris* (Boaz and Ciochon, 2004).

It may seem incredible to us, locked into modern notions of place and space, that *H. sapiens* took apparently 60,000 years or so to move down valleys like those of the Huanghe and Changjiang to reach the Pacific Ocean, but that seems a plausible figure. It did not take this long simply because *H. sapiens* was itinerant and directionless but also because of extraneous changes to human populations, the most important of which was undoubtedly the eruption of Mount Toba on the island of Sumatra in Indonesia. About 74,000 years ago when Toba erupted – a monstrous event that produced more than 200 times as much eruptive material as the 1883 eruption of Krakatau (Krakatoa) – ash and volcanic aerosols were pushed into the lower atmosphere in such huge quantities that sunlight was blocked, plunging the Earth into months of volcanic winter (Rampino and Ambrose, 2000). Most *H. sapiens* could not survive, their numbers dwindling to perhaps just 15,000 across the Earth. This has been described as an evolutionary bottleneck that nearly saw the extinction of our species (Ambrose, 1998). It is unlikely that many survived in East Asia, so close to Toba itself, but there is not enough information available to know this. What we can reasonably infer is that the Toba eruption contributed significantly to the long time it took *H. sapiens* to reach the Pacific Rim and see the Pacific Ocean for the first time 40,000–50,000 years ago.

2.2. HUMANS ON THE PACIFIC RIM

The earliest modern humans to reach the Pacific Ocean probably did so ~40,000–50,000 years ago, and all the indications are that they saw it as a barrier rather than an opportunity. We do not suddenly see the remains of marine foods replacing terrestrial foods in cave deposits like those at Zhoukoudien (Boaz and Ciochon, 2004). We see no signs of tools or the remains of vessels intended to aid the acquisition of marine foods, and we see no sign of permanent settlements in this area at this time. Of course, it may simply be that such evidence has vanished, as you might expect after so long, but at the same time this interpretation is the likeliest given the experience of the humans involved.

Nothing would have prepared them for an ocean. Of course, their long-past ancestors would have encountered oceans but the memory (and the potential) of these would have been long forgotten. For an itinerant people accustomed to terrestrial living, the ocean would have been as much a barrier as would an unbroken range of snow-capped mountains. The imperative of daily food acquisition would have left little or no time to research the potential of the ocean as a food supplier, and probably most groups of H. sapiens who encountered the ocean for the first time put it out of sight as quickly as they could.

But change slowly came. Perhaps ~20,000–40,000 years ago, when vast areas of the continental shelf were exposed by lower (glacial) sea level off East Asia, then people began to realize the potential of these as food sources. Many of the animals that competed with humans for food in the valleys were less common close to the sea, more conservative in their diet, less adaptable. This may indeed have enhanced the attraction of coastal areas for humans. The humans who first encountered the Pacific Ocean in East Asia ~40,000–50,000 years ago are believed to be the ancestors of those who first occupied island Southeast Asia some millennia later (Bowdler, 1997).

Being closer to the equator, the environments first occupied by humans in Southeast Asia were perhaps less attractive than those in temperate and subtropical East Asia. For not only was it hotter and wetter, but also the forest vegetation found in many parts was less conducive to food acquisition, both because mobility was more restricted and because wild foods were less easy to find. Also in these areas there was a whole new range of competing predators, including crocodiles and tigers, that would also have included humans in their diet. It was perhaps a combination of these factors with the proximity of the ocean that led humans dwelling in Southeast Asia ~40,000 years ago to begin to exploit marine foods. And compared to the coast off temperate East Asia, those in archipelagic Southeast Asia, fringed with coral reefs, would have contained a far greater and more conspicuous range of edible foods.

It has been suggested that human skin colour underwent a major change when humans in Southeast Asia first began to exploit marine food resources (Kingdon, 1993). Up until this point, all humans had the same skin colour – light brown. It never evolved a darker colour because humans, particularly those living in hotter parts of the world, tended to keep out of the sun at the hottest time of day, acquiring food mostly in either the coolness of the morning or the later afternoon. But with the collection of nearshore marine foods, it is not possible to select the time of day for food acquisition. You must go when the tide is low, be that in the heat of the day or not. So people, by broadening their diet to include nearshore marine foods, became more frequently exposed to the Sun at the hottest time of day, and thus their skins became gradually darker.

2.2.1 Crossing the Wallace Line: from Southeast Asia to Australasia

It seems reasonable to suppose that increasing dependence on marine foods led to increased interaction with the ocean beyond the reef for humans dwelling in Southeast Asia 30,000–40,000 years ago. Maybe they used watercraft, perhaps bamboo rafts, to carry them out to sea to fish or to reach nearby islands. But however it happened, the evidence suggests that humans crossed the ocean gap from Southeast Asia (Sunda) to Australasia (Sahul) at least 40,000 years ago. Rather than a single crossing of several hundred kilometres, this movement probably involved a number of interisland hops, the longest being ~90 km (Birdsell, 1977; Spriggs, 2000a).

Compared to Southeast Asia, Australia on the other side of the Wallace Line was paradise for the first groups of humans to arrive there. Not only were there no predators to compete with humans for the same foods but the continent was also occupied by a naïve biota, one that had evolved over millions of years in the absence of just such predators. Finding food in such circumstances was easy and quick, leaving *H. sapiens* time with nothing particular to do – what today we might call leisure. It has been suggested that, as a result, this was the place and the time where a great leap forward in human evolution occurred (Flannery, 1994). This great leap forward was represented in physiological terms by an increase in cranial capacity, and in cognitive terms by perhaps the first development of what we now call culture. In a broader sense, climate changes have been regarded as critical in the attainment of higher brain functions by late Quaternary humans (Calvin, 2002).

While Australia may have been largely an attractive destination within Australasia, life was much less easy for the first people in Papua New Guinea. Cloaked in rainforest, with its environment at least superficially resembling those of island Southeast Asia, the absence of large predators was undoubtedly welcomed by the first arrivals but food acquisition was just as difficult, particularly in inland areas. As much as 35,000 years ago, on the island of New Britain, people adapted by deliberately manipulating forest structure to enhance the growth of edible plants (Pavlides and Gosden, 1994). More than this it has been argued that there were never true foragers in the rainforests of island Melanesia (Papua New Guinea–Solomon Islands) and that people were able to settle island interiors only by transplanting useful species from island coasts (Spriggs, 2000b).

Such practices discouraged itinerancy and many human groups in Papua New Guinea may have become sedentary long before those elsewhere. In contrast, the humans who first occupied Australia would have remained itinerant, for although the fauna was naïve, it was scattered and low density. And as these humans came to learn through their long occupation of Australia and intimacy with its natural environment, it is not annual cycles to which most attention should be paid but those interannual cycles, largely El Niño-Southern Oscillation (ENSO), that have the greatest effect on this continent's resource base (Flannery, 1994).

2.2.2 Eastwards and northwards along the western Pacific Rim

Around 40,000 years ago from the coast of East Asia, probably around the mouths of the Huanghe and Changjiang, people began moving eastwards and northwards into other parts of the Pacific Rim (Sato, 2001).

During most Quaternary glacial periods, the three larger islands of Japan – Honshu, Shikoku and Kyushu – formed a single island (Hondo). In contrast to some earlier glacial periods, Hondo was not connected directly to the East Asia continent during the Last Glacial but mammals did cross, probably along an ice bridge, to Hondo from Hokkaido Island, which was itself connected to the Asian mainland *via* Sakhalin Island at this time (Kawamura, 1998; Nunn et al., 2006). Some of the southern islands of the Ryukyu group in Japan were connected during the Last Glacial to Taiwan which was part of the Asian mainland at this time (Kawamura, 1998).

On Hondo there is some evidence – largely proxy evidence from vegetation disturbance – that humans reached the islands of Japan 35,000–50,000 years ago (Sato, 2001). The earliest material evidence is Jomon pottery, dating from ~16,000 years ago (Keally et al., 2003). It is likely that most humans arrived in Japan *via* the Sakhalin–Hokkaido landbridge, although lengthy ocean crossings were undertaken; the obsidian resources of the islands south of Tokyo, which were never connected to Hondo, began to be mined ~30,000 years ago (Hashiguchi, 1994).

After the Last-Glacial Maximum (22,000–17,000 BP) sea level began rising and the human occupants of Japan became increasingly islandized, their largely coastal settlements often being built on bluffs overlooking estuaries rather than along low-lying coastal flats; a classic study is that by Esaka (1943, 1954) on the south Kanto Plain. Sharply increased consumption of marine foods is evident after ~13,000 BP, with shell middens becoming abundant after ~10,000 BP, marking the early Jomon Culture (Ikawa-Smith, 2004).

The first people to occupy the islands of Japan probably reached them in pursuit of herds of animals that were likewise attracted by the relative absence of competition for desirable foods. Similar reasons probably explain why humans gradually moved northwards along the western Pacific Rim, occupying some of the river valleys crossing the steppe of easternmost Russia by perhaps 35,000 years ago; a key site is Diuktai Cave, first occupied ~33,000 BP by people subsisting on wild foods including mammoths and seals (Mochanov, 1980).

Most of the early human sites (Diuktai Culture – 35,000–10,000 BP) excavated in easternmost Russia (western Beringia) also show that people occupying them included marine foods in their diets, suggesting that human spread along the western Pacific Rim and into the America was a process founded on a maritime economy (Ackerman, 1998). Access to the plains of western Beringia, which border the modern Bering Strait, is likely to have been through the Lena River basin (Turner, 1985). The later occupation of this area shows that sea-level rise was responsible, by increasing shoreline length and sinuosity, for making the marine component of diets increasingly important. Sites of the Rudnaya Early Neolithic Culture (7,500–6,000 BP) exemplify the point, with the collection of shellfish, fishing, sea-mammal and shore-bird hunting, and whale scavenging becoming common (Kononenko, 1996) just as they did about the same time in the Jomon Culture of Japan.

2.2.3 Into the Americas

Between 25,000 and 14,000 BP, the Bering Strait, which is today a formidable water barrier between Russia and Alaska (western and eastern Beringia), was mostly dry land (tundra) because of the lower sea levels associated with the coldest time of the Last Glacial (Hopkins, 1982). While some of the earliest humans to have crossed into North America may have done so entirely on foot, most would probably have also undertaken short cross-ocean journeys. For people accustomed to interacting with the ocean, this would not have been especially testing. A similar maritime-linked culture existed across southern Beringia in the earliest period of its human settlement (Laughlin and Harper, 1988).

From these people, the first groups of maritime-adapted hunter-gatherers may have begun to move southwards along the western seaboard of North and South America, initially following caribou, well before inland areas of these continents were settled (Carlson, 1994; Dillehay, 1999). Yet the less contentious view of early human arrival in North America is that people arrived ~12,000 BP and established the Clovis Culture, which seems uncontroversially to have inherited traits, particularly lithic traditions, from Beringia (Haynes, 1980).

For some years it has been uncomfortably clear to many archaeologists working in the Americas that the earliest human settlement sites are all apparently in South America. One of the least controversially dated is Monte Verde, close to the Pacific coast of southern Chile where people may have been living 33,000 years ago, at least 13,000 BP, significantly before the Clovis culture of North America (Dillehay, 1989, 1997). While most archaeologists still regard the crossing of the Bering Strait as the only possible way for humans to have entered the Americas, it has been suggested – on somewhat spare and dubious grounds – that the tropical Pacific Ocean may have been traversed by people during the terminal Pleistocene (Wyatt, 2004).

2.3. COLONIZATION OF THE PACIFIC ISLANDS

The first people to occupy almost every part of the Pacific Islands region came from the western Pacific Rim, most *via* the Bismarck Archipelago of Papua New Guinea. It seems likely that this astonishing colonization process, started by stone-age people, itself began thousands of years earlier when rapidly rising post-glacial sea levels drowned coastal lowlands in what we now call southern China and Taiwan. Some of these displaced people took to the ocean in search of new lands to settle, a good example of the profound effects of sea-level change on human cultural evolution is discussed further in Section 3.4.2.

The daunting size of the Pacific Ocean – roughly one-third of the Earth's surface – makes it difficult to generalize about. For a long time, Pacific geographers have eased their task by talking about three distinct subregions – Melanesia, Micronesia and Polynesia – but these names have led to much misunderstanding because many writers have assumed (incorrectly) their populations to be homogenous and distinct from each other. In fact, the settlement of the Pacific Islands was carried out in successive migrations from similar source regions, the crude racial differences observed today attributable largely to the degree to which later arrivals genetically overprinted earlier ones in particular places.

The first true exploration of the Pacific and the discovery of the islands within it probably came not from random or accidental voyaging, but from the

FIGURE 2.2   Some evidence of Lapita-era settlement of the Fiji Islands. (A) Obsidian talisman found in Lapita-era strata dating to ~900–1000 B.C. (2,850–2,950 cal BP) at the Bourewa settlement in southwest Viti Levu Island, Fiji (Nunn et al., 2004). Obsidian does not occur naturally in Fiji. This piece has been sourced by density analysis (Dr. Wal Ambrose, Australian National University) to the Kutau-Bao area of New Britain Island in Papua New Guinea, 3,250 km in a straight line from Bourewa. This obsidian is proof that people made this journey ~3,000 years Ago. (B) Reconstruction of a substantially complete skull of a woman (named Mana) recovered from a Lapita-era burial dating to 800 B.C. (2750 cal BP) at the Naitabale site on Moturiki Island in central Fiji (Kumar et al., 2004). The head of Mana was reconstructed at Kyoto University under the direction of Professor Kazumichi Katayama (Katayama et al., 2003).

intentional search for new lands (Irwin, 1992; Kirch, 1997a). We can speculate endlessly about the motives of the people involved, but the chances are that they sought new islands to settle for pragmatic reasons, perhaps because population density on their home island was approaching island carrying capacity or because this had been lowered by an environmental stressor like prolonged drought (Hunter-Anderson and Butler, 1995; Anderson et al., 2006). The Lapita people, who began moving out of the Bismarck Archipelago ~1330 B.C. (3,280 cal BP), were the greatest ocean voyagers of their age, able to cross ocean distances of 1,000 km or more. Within a few hundred years, they had colonized most of the islands in the Vanuatu, New Caledonia, Fiji, Tonga and Samoa island groups (Fig. 2.2).

The Lapita diaspora reached as far east into the Pacific as Tonga and Samoa, where the Lapita culture eventually disintegrated ~550 B.C. (2,500 cal BP: Burley and Clark, 2003). The descendants of these islands' Lapita occupants eventually continued the eastwards colonization of the Pacific, reaching central French Polynesia about A.D. 600 and Hawaii[1] perhaps A.D. 650 (Spriggs and Anderson, 1993), Easter Island about A.D. 690 (but perhaps not until A.D. 1200 – Hunt and Lipo, 2006), New Zealand about A.D. 1250–1300 and even the west coast of the Americas some time before Europeans first arrived there in A.D. 1513 (Ward and Brookfield, 1992; Nunn, 1999; Anderson, 2003). The bond between Pacific Island people and the ocean is epitomized by the fact that unbroken ocean crossings of at least 3,000–4,000 km must have been involved in the colonization of this part of the Pacific.

Around the same time as the southwest Pacific islands were being settled, people were also evidently settling those islands in the northwest quadrant, probably from adjacent parts of the western Pacific Rim (Rainbird, 2004). Being generally smaller and more resource poor, these islands lured and sustained fewer people than those to the south, from which the remainder of the Pacific Islands were eventually reached for the first time (Kirch, 2000a).

Most scientists who have pondered the human settlement of the Pacific Islands have ignored or at best under-estimated the role of non-human factors (particularly climate and sea-level change) in this process. Recent thinking suggests that climate change was an important factor in the successful eastward colonization of the Pacific Islands (Anderson et al., 2006). Likewise it seems that sea-level changes during the later Holocene had a profound influence on the chronology of west–east island colonization (Dickinson, 2003).

KEY POINTS

1. Modern human hunter-gatherers reached the western borders of the Pacific Ocean ~40,000–50,000 years ago and, within ~10,000 years, had incorporated marine foods into their diets, and spread along much of the Pacific Rim.

2. The southwest Pacific Islands were colonized mostly by the Lapita people, who moved eastward from the Bismarck Archipelago of Papua New Guinea, ~3,280 cal BP. The descendants of the Lapita people in Tonga and Samoa colonized most of the other Pacific Island groups, evidently even reaching the west coast of Central and/or South America before the first Europeans in A.D. 1513.

---

[1] Note that in this book the island group is named Hawaii and the island (the Big Island) named Hawai'i to distinguish the two.

CHAPTER 3

Influences of Climate and Sea-Level Changes on Humans

We saw in Chapter 2 how modern humans reached the Pacific Basin and then spread through it. These people came to develop sedentary lifestyles that were tethered, more or less, to the ocean and its resources. As a consequence, such people rendered themselves increasingly vulnerable to changes in both climate and the nature of the ocean. In the present chapter, a systematic account is given of climate and sea-level changes when there were few people, particularly in the Pacific Basin, and their effects on humans.

In the past few years, much has been written about the effects of climate changes on cultural change, particularly about the role of climate change in the collapse of complex societies (Tainter, 1988; deMenocal, 2001; Diamond, 2005). The view of climate (and sea level) as a factor in disrupting well-established trajectories of human cultural evolution is an example of environmental determinism, the belief that environment is a major control on human lifestyles. Owing to the odium surrounding some of its more extreme applications, environmental determinism fell into disrepute some decades ago but is implicit in many modern discussions of long-term human cultural evolution. Like many other examples of climate-induced societal collapse (Curtis et al., 1996; Barlow et al., 1997; Gill, 2000), the A.D. 1300 Event is an example of environmental determinism. Critics often prefer the pejorative term "neo-environmental determinism" (Erickson, 1999).

This chapter is organized so as to illustrate the effects of climate and sea-level changes on humans before A.D. 750, when the Medieval Warm Period began. This chapter opens with a general account of climate change and human evolution to emphasize the fundamental role of climate in the emergence of the genus *Homo* (Section 3.1). There follows a section on how late Pleistocene climate and sea-level changes forced humans to adapt their lifestyles (Section 3.2). The effects of Holocene climate and sea-level changes predating the last millennium on humans in the Pacific Basin are discussed in Sections 3.3 and 3.4 respectively.

3.1. CLIMATE CHANGE AND HUMAN EVOLUTION

Very long-period climate and sea-level changes – the kind that persist for perhaps tens of millions of years – can be recognized through most of the Earth's history (Hallam, 1984; Haq et al., 1987; Nunn, 1999). The root cause of many such changes is believed to lie in tectonics, particularly the regular aggregation and breakup of supercontinents (Nance et al., 1988; Duncan and Turcotte, 1994). The connections between tectonics – land-level movements – and climate are many and no purpose would be served here by attempting to enumerate them all. But many scientists agree that it was tectonic change which was the underlying cause of a pathway for hominid evolution distinct from that of apes, unarguably the earliest example of environmental determinism influencing human evolution.

The story begins in tropical Africa ~5 million years ago, a time before the modern Rift Valley had opened (Baker and Wohlenberg, 1971), and where, because of the warm moist climate, woodland covered the entire area. Within that woodland thrived *Ardipithecus ramidus*, the hominid thought to lie near the base of the human family tree close to the point at which hominids and apes diverged (White et al., 1994). *A. ramidus* walked upright through these woodlands eating fruits and leaves. But then the Great Rift Valley, marking the place where two lithospheric plates are moving apart, began opening and the climate of this part of tropical Africa began changing as a consequence. The region to the west of the Rift Valley remained moist and forested and this was where the African apes gradually appeared as distinct groups. But to the east of the Rift Valley, the climate became drier and savanna and grasslands came to replace the woodlands. In this region, the descendants of *A. ramidus* were forced to adapt their lifestyles to these new conditions, and from these adaptations – and indeed from evolving the ability to adapt *per se* – evolved modern humans, *Homo sapiens* (Kingdon, 2003; Mitchell, 2005).

There are numerous other examples from the history of human evolution in which climate has been suggested as the cause of particular evolutionary changes (Calvin, 2002; Fagan, 2004). For example, it is possible that human ancestors began to stand upright to minimize the harsh effects of the Sun's rays on their skin, reduce sweating and thereby retain as much body water as possible in a dry climate (Kingdon, 2003). The possibility that darker skins evolved in Southeast Asia through the adoption of marine-associated diets and consequent prolonged exposure to the midday sun was discussed in Section 2.2.

All these examples are long-term changes, evolution but not adaptation in the sense of responding to short-term climate changes, examples of which abound in more recent times.

3.2. HUMAN ADAPTATION TO LATE PLEISTOCENE CLIMATE AND SEA-LEVEL CHANGES

The Quaternary Period[1] began ~1.8 million years ago and saw a series of uncommonly regular oscillations in Earth-surface temperatures and sea levels that have been attributed to correspondingly regular changes in the orbit of the Earth around the Sun (Imbrie et al., 1984). Temperature oscillations between cool and

---

[1] Comprising the Pleistocene (*ca.* 1.8 million years ago to 12,000 cal BP) and the Holocene (12,000 cal BP to present).

warm gave rise to glacial (ice age) and interglacial conditions respectively. Owing to the direct links between ground-surface and ocean-surface temperature changes and changes in the volume of the oceans, Quaternary temperature changes were followed closely by sea-level changes. Glacial sea levels were lower than interglacial sea levels.

From ~850,000 BP, glacial–interglacial cycles have lasted typically 100,000 years (Frakes et al., 1992). Considerable research has been directed towards understanding the precise temperature changes and sea-level changes that characterized the last cycle or two (Wright et al., 1993; Nunn, 1999; Siegert, 2001; Alverson et al., 2003), particularly in order to reconstruct the environments in which *H. sapiens* (and many other species) evolved (Akazawa et al., 1992; Williams et al., 1993; Bradley, 1999), and to understand the adaptation strategies that humans employed to cope with the effects of environmental changes (Diamond, 1999).

The interplay of changes in climate, environment, sea level and society is important to appreciate. When dealing with the distant past, for which information is comparatively sparse, there is the temptation to simplify cause–effect relationships. But this is sometimes unhelpful, leading to the growth of orthodoxies which must be challenged as new data appear. A good example of the interaction of all these factors comes from the land areas of the northernmost Pacific (Beringia). The most recent inundation of the Bering Strait, separating west and east Beringia, occurred during the later part of the Last Glacial ~14,000 BP, and led to an amelioration of the climate throughout Beringia. This was marked by the replacement of xeric tundra-steppe – an unattractive environment for humans – by scrub birch (*Betula glandulosa*) and willow (*Salix*), a change that was soon followed by the spread of humans throughout the area (Anderson et al., 1994). This is a clear example of how climate change – both directly and through sea-level change – created environments that attracted humans.

To illustrate these points, two examples are given in the following sections. The first looks at the effects of the Younger Dryas climate oscillation – an example of a rapid climate change – that occurred during the terminal Pleistocene (Section 3.2.1). The second looks at the two principal hypotheses surrounding the cause(s) of the latest Pleistocene large-mammal extinctions, how humans may have been involved and how they adapted (Section 3.2.2).

3.2.1 Adaptation to rapid climate change: the Younger Dryas and agriculture

Human adaptation to most late Pleistocene climate and sea-level changes would, on account of their long duration and comparatively slow rates, have probably been largely unthinking. In other words, humans changed their location or their lifestyle gradually in response to changes that were probably not perceived as such on the timescale of a human lifespan. Yet, the end of the Pleistocene and start of the Holocene is marked by a much-studied period of rapid climate change named the Younger Dryas. On account of its comparative rapidity and long duration, the Younger Dryas had major effects on humans in many parts of the world (Peteet, 1995; Nunn, 1999; Andres et al., 2003).

The Younger Dryas was a cold event – the most extreme and enduring of a number of such Heinrich events – that disrupted the warming trend which defines the last deglaciation. The Younger Dryas was also marked by rapid sea-level fall during its very cold beginning, and then rapid sea-level rise towards its end. The Younger Dryas lasted between ~11,000–10,000 years ago (12,900–11,600 cal BP – Alley et al., 2003) although precise dates have been difficult to obtain using radiocarbon dating because a radiocarbon plateau exists around this time resulting in uncommonly (and unhelpfully) large ranges in calibrated dates.

In the Pacific Basin, one of the most detailed palaeoclimatic studies of the Younger Dryas comes from Taiwan where changes in vegetation in the catchment of the Toushe Basin, particularly the amounts of *Salix* (willow) and Gramineae (grasses), have been interpreted as responses to climate change (Liew et al., 2006). The remarkable rise of *Salix* ~13,000–12,500 cal BP is interpreted as marking the onset of cold conditions, the rise in Gramineae (>35%) 11,800–11,600 cal BP as marking cold-dry conditions, which were followed by a return to subtropical warmth ~11,500 cal BP.

In some parts of the world, the apparent synchronous (between 13,000 and 8,000 cal BP) move by humans whose lifestyle had hitherto involved solely hunting and gathering to a lifestyle involving plant cultivation and animal husbandry (agriculture) appears to demand a global explanation (Mannion, 1999; Richerson et al., 2001; Weiss and Bradley, 2001). It certainly seems likely that shortages in wild foods associated with the rapid onset of extreme cold conditions at the start of the Younger Dryas combined with the direct effects of the cold on humans (the need for more permanent shelter, for example) forced many humans to adopt a sedentary lifestyle. Experiments with the domestication of favoured wild foods during this time led, in the aftermath of the Younger Dryas, to agriculture, the domestication of plants and animals (Catto and Catto, 2004). One example is the Natufian communities of Southwest Asia that were accustomed to hunting and gathering in the open woodlands, where wild cereals were plentiful, for several millennia up until 12,000 years ago. The abrupt onset of cooler drier conditions during the Younger Dryas meant that there was now insufficient food to sustain Natufians by hunting and gathering, and they were forced to settle in places where intentional cultivation (horticulture) was possible (Henry, 1995; Sellars, 1998).

The idea that climate change at the end of the Pleistocene forced agriculture in areas of the world where population densities were comparatively high and where agriculture was possible was an idea first advanced by Childe (1951). While it has numerous more recent adherents, such as Bar-Yosef (1998) and Mannion (1999), it is far from being widely accepted. Yet, as the evidence of climate forcing of cultural collapse has become widely known over the past decade or so, more writers are beginning to admit a significant role for climate change in the first appearance of agriculture (Richerson et al., 2001).

3.2.2 Latest Pleistocene large-mammal extinctions: climate or humans?

The effects of Younger Dryas cooling and sea-level fall may also be implicated in the conspicuous large-mammal extinctions that occurred in many parts of the

Pacific Rim around the end of the Pleistocene (Nunn, 1999; Ward, 2001). In North America, more than 73% of megafauna (those weighing more than 44 kg) became extinct within 3,000 years. In South America, 80% of megafauna vanished. In Australia, 86% of megafauna became extinct in the period 25,000–11,000 BP. The megafauna of eastern Japan disappeared after 15,000 BP.

The climate-change hypothesis for these extinctions holds that changes associated with deglaciation during the latest Pleistocene – increased aridity, warming, increased climate variability punctuated by rapid periods of extreme cold like the Younger Dryas – led to a loss of habitats for large mammals. The alternative hypothesis attributes large-mammal extinctions to overkill, the rapid rise in their predation by humans. This hypothesis receives most of its support from North America, where the Clovis Culture (~12,000–10,600 BP) was marked by the production of stone spearheads with fluted points designed to kill megafauna.

A weakness of the climate-change hypothesis is that similar megafaunal extinctions might be expected to have occurred at the end of every Quaternary glaciation, which they did not. A weakness of the overkill hypothesis is that humans and megafauna had co-existed for millennia, so it is unclear why rates of predation increased sharply at this time. A solution to the conundrum is to suppose that it was a combination of climate change and human predation that proved a recipe for large-mammal extinction, although there were some parts of the Pacific Rim – western Beringia, for example – where this did not occur at this time (Sher, 1997).

The polarization of the debate over the causes of large-mammal extinctions during the terminal Pleistocene has drawn attention away from an explanation involving both climate change and overkill. It may well be that climate-driven environmental stress led to an increase in megafaunal vulnerability to predation (for example, by reducing their range) and a simultaneous increase in human demand (driven, for example, by shortages of other foods).

The importance of this debate for the present book is that it parallels in many respects the debate that has surrounded the cultural consequences of last-millennium climate (and sea-level) change, for at first, the nature of a particular phenomenon (extinctions or cultural collapse) was explained by human actions but, as more has been learned about the nature of contemporaneous climate changes, so many scientists have come to regard these as having been key.

3.3. EFFECTS OF HOLOCENE CLIMATE CHANGES ON HUMANS

Since the end of the Last-Glacial Maximum ~17,000 years ago, the world has generally been getting warmer, and sea levels generally rising (see Figs. 1.1 and 1.2). Such changes provided more subsistence opportunities for humans (and most other organisms) and laid the foundations for the success of our species in recent millennia. But there were a number of setbacks to our progress, some of the major ones having been caused by extraneous changes over which humans had no direct control. These changes were of various types and occurred at various scales with correspondingly variable degrees of disruption to human affairs. In addition to environmental changes associated with climate and sea-level changes – the only types considered in detail here – volcanic eruptions and earthquakes were the main sources of extraneous change to have affected humans during the Holocene (Nunn, 1999).

Within the Pacific Basin, a good example comes from the fringes of the Atacama Desert of northern Chile, the late Pleistocene and Holocene human occupation of which was marked by a cultural hiatus – the *Silencio Arqueologico* – between 9,500 and 4,500 BP (Núñez et al., 2002). Climate change is regarded as the main explanation, the earlier time of occupation (11,800–9,000 BP) being wetter and marked by an abundance of vegetation and animals for hunting, the focus towards the end of this period being on seasonal migration between upland lakes (now dry) and lowland wetlands. The area was abandoned as the lakes dried up and wetlands were reduced in number and area. The hiatus ended as wetter conditions resumed, and people resettled shallow lake borders.

It was seen above how agriculture may have developed in East Asia and elsewhere as a result of increasing climate variability during the late Pleistocene and earliest Holocene, particularly the Younger Dryas. Yet for most of the early Holocene (12,000–6,000 BP) rising temperatures generally ensured enhanced opportunities for human existence, exemplified in the Pacific Basin by the development by 7,000 BP in East Asia of towns such as Hemudu that prospered from rice agriculture (Zhao, 1998), and by the spread 7,200–6,000 BP of settlement into areas of this region that had earlier been desert (Feng et al., 1993). In montane environments of Andean South America, climate (particularly precipitation) was the dominant cause of Holocene changes in settlement pattern and, more broadly, human lifestyles (Rigsby et al., 2003). Around Lake Titicaca, for example, in the period 4,500–3,000 BP, increased precipitation and lake level led to an increase in the area's population. Subsequent periods of reduced precipitation saw people move away from areas near the lake.

This section describes the effects of climate change on societal collapse in various parts of the world (Section 3.3.1) before describing the likely influence of climate on long-distance ocean voyaging (Section 3.3.2).

3.3.1 Climate change and societal collapse

The most conspicuous interactions between climate and humans during the Holocene (pre-A.D. 750) were those in which comparatively short-term climate changes disrupted – often spectacularly – trajectories of human cultural development. This especially affected highly organized (complex) societies in which every subgroup is dependent on another, ultimately on agriculture, emphasizing the point that societal vulnerability to extraneous change often increases with societal complexity.

One of the earliest Holocene examples comes from the association of cooling and drying during the 8,200-BP Event (Alley et al., 1997) with the adoption of irrigated agriculture in parts of western Asia (Weiss, 2000). Later in western Asia, urban late Uruk society ~3500 B.C. (5,450 BP) was sustained by irrigated cereal agriculture and colonies were established across the drier areas of the Near East.

Uruk society collapsed spectacularly 3200–3000 B.C. (5,150–4,950 BP), probably because of the effects of a severe drought lasting less than 200 years (Lemcke and Sturm, 1997; Rothman, 2001). Similar effects were experienced by societies along the fringe of the Sahara Desert at this time, resulting in the migration of a pastoral civilization into the Nile Valley ~5,500 BP (Hsu, 2000).

Various cultural transformations are recorded within the period 2450–1850 B.C. (4,400–3,800 BP), the coldest period on Earth since the Younger Dryas (Perry and Hsu, 2000). In central China, collapse of the impressively organized Neolithic cultures during the late third millennium B.C. has been widely noted but its cause(s) is disputed (Wu and Liu, 2004). Recent work suggests that an abrupt reduction in the intensity of the Asian monsoon ~2450 B.C. (4,400 BP) over several decades reduced available water to this region and caused societal collapse (Wang et al., 2005). More widespread was the societal collapse ~2200 B.C. (4,150 cal BP) in western Asia and Egypt that was triggered by abrupt climate change heralding a prolonged period of aridity and warmth (Neumann and Sigrist, 1978; Dalfes et al., 1997).

Similar effects led to the collapse of major civilizations in monsoon Asia some 500 years later (Yasuda and Shinde, 2004). The Roman Empire, which expanded during the favourable climate of the Roman Warm Period, collapsed with the onset of the drier Dark Ages Cold Period at the end of the fifth century A.D. (Fagan, 2004), the period that also saw the greatest historically recorded drought in tropical Africa (Thompson et al., 2002). It has been claimed that intense cold initiated the migrations that led to the Germanic tribes overrunning the Roman Empire and the northern Asiatic tribes overrunning the Chinese Empire around A.D. 400 (Hsu, 2000).

In central America, a "clear link" has been found between "the chronology of regional drought and the demise of Classic Maya culture" (Haug et al., 2003, p. 1732). A proxy record of precipitation was obtained from studies of the titanium concentrations in sediment laminae in the Cariaco Basin in northern-most South America. This record showed that periods when Mayan culture faltered and that when it finally collapsed coincide with prolonged drought (Fig. 3.1). Confirmation of this scenario was obtained from analyses of sediments in Yucatán lakes (Hodell et al., 2005).

In the Pacific Basin, changes in El Niño frequency explain the major cultural changes in Peru during the middle and late Holocene (Sandweiss et al., 2001). In the Andes, the raised-field (agricultural) system of the pre-Inca Tiwanaku civilization (300 B.C. to A.D. 1100) located on the *altiplano* around Lake Titicaca might have sustained 760,000 people. Today the same area sustains only ~40,000 (Binford et al., 1997). The Tiwanaku civilization also collapsed as a result of a prolonged severe drought (Kolata et al., 2000).

If there appears undue emphasis on the negative effects of climate change on human societies – understandable given that these effects are often profound and enduring – there is another side to the issue, in which generally warmer and wetter climates often led to the burgeoning of civilizations. For example, the first Egyptian Empire commenced with a resumption of warm conditions ~3050 B.C. (5,000 BP). At the same time, climate amelioration saw the emergence of

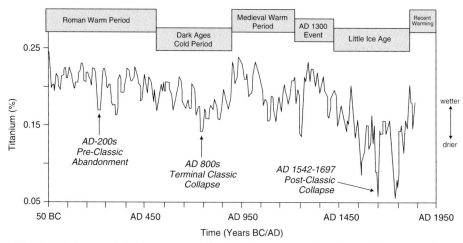

FIGURE 3.1 Palaeoprecipitation measured by titanium content of sediment laminae in the Cariaco Basin as a proxy for climate change and Mayan cultural collapse (adapted from Haug et al., 2003). The Pre-Classic Abandonment (A.D. 150–250) was significant but populations recovered and cities were reoccupied. The Terminal Classic Collapse (A.D. 750–950) saw the permanent abandonment of densely populated cities (Webster, 2002). The prolonged drought around A.D. 750 is likely to have been a critical factor in this. The Post-classic recovery of Mayan civilization came to an end during the period of extreme dryness between about A.D. 1542 and 1697 (Chase and Chase, 2006).

civilizations in northern Mesopotamia and India (Weiss et al., 1993). A global warm period beginning ~3,800 BP saw the start of the Bronze Age, during which many civilizations developed and prospered (Hsu, 2000).

   All these examples – and numerous others given by Lamb (1977, 1982), Fagan (1999, 2004), Issar and Zohar (2004) and Diamond (2005), for example – underline the important (and traditionally much underrated) role of climate change in human cultural evolution. Recent interest in abrupt climate changes has been driven in part by a wish to understand how these have affected human societies (Cronin, 1983; Ambrosiani, 1984; Adams et al., 1999; National Research Council, 2002).

3.3.2 Climatic control of long-distance voyaging

The colonization of the Pacific Islands, depicted in Fig. 2.1, involved unbroken ocean crossings of several thousand kilometres. Incredible as this feat appears to us today, it may have been less so at the times of these voyages for there is considerable evidence of repeated return voyages across long distances. Yet, by any yardstick, the long-distance voyages involved in the colonization of the Pacific Islands are impressive, leading to suggestions that it may have been accomplished at times when the climate was more favourable, especially involving constant winds and a low incidence of storminess (Finney, 1976; Bridgman, 1983; Anderson et al., 2006).

Consider that, for example, the long-distance interisland voyages of as much as 1,000–1,500 km carried out by the earliest (Lapita) people in the Pacific Islands were within the time period 1330–550 B.C. (3,280–2,500 BP), the later part of the Holocene Climatic Optimum when warmer conditions prevailed in the tropical Pacific (Nunn, 1999). This would have lowered Equator–Pole temperature gradients compared to today, meaning that the tradewind belt would have been wider at the time, resulting perhaps in more constant winds and making cyclonic storms easier to evade.

Recent data have refined earlier views of the chronology (and technology) of long-distance voyaging in the Pacific, it now being clear that this was episodic, "consistent with the operation of substantial constraints on maritime migration activity or, at any rate, success" (Anderson et al., 2006, p. 3). Key among these new ideas is the rejection of the traditional idea that the watercraft of the earliest Pacific Islanders had a windward capability and were able to be sailed deliberately against the wind (Irwin, 1992). If this is rejected, then the colonization of the Pacific Islands – largely against the direction of normal wind flow – must have been possible only when these normal conditions were temporarily suspended, as they are during El Niño events.

As late Holocene cooling began in earnest ~450 B.C. (2,400 BP), so long-distance voyaging appears to have become less frequent, perhaps because of the steepening Equator–Pole temperature gradient and the reduced extent of the tradewind belt. It is possible that the colonization of island groups east of Samoa and Tonga occurred accidentally around A.D. 150 (1,800 BP) as a result of unexpected encounters with anomalous westerlies associated with an increased incidence of El Niño events (Finney, 1985).

The later phase of intentional long-distance voyaging in the Pacific Islands region, which began with the colonization of Hawaii and Easter Island about A.D. 650 and ended with the colonization of New Zealand about A.D. 1250–1300, involved far greater distances than during the Lapita era. Voyages from central Polynesia to New Zealand were probably at least 3,000 km, yet there is evidence that there were return voyages along this route within the first few decades of New Zealand's first settlement (Anderson and McFadgen, 1990). For Pacific Islanders to reach Panama from central Polynesia may have involved a journey of nearly 6,000 km (see Section 2.3).

The success of such voyages must have been regarded as likely or they would never have been undertaken with such regularity. Yet around A.D. 1300–1400 in the Pacific Ocean, such routine contacts between islands far apart (and not so far apart) ceased in almost every case, and the following few hundred years were marked by isolation rather than interchange across the Pacific. It is likely that the agreement in timing between the great era of voyaging, about A.D. 700–1280 and the Medieval Warm Period, is not coincidental, but that the climates of the latter period facilitated successful long-distance voyaging in ways that those of the ensuing Little Ice Age did not (Bridgman, 1983; Nunn, 1999, 2000a; Nunn and Britton, 2001). This question is explored further in Section 6.3.6.

It is worth noting that long-distance Norse voyaging in the North Atlantic during the last millennium has an almost identical chronology to that in the

Pacific (see Section 9.3.3). This suggests that the key elements of last-millennium climate that affected the success or otherwise of long-distance voyaging were global, not confined to the low-latitude Pacific.

3.4. EFFECTS OF HOLOCENE SEA-LEVEL CHANGES ON HUMANS

The effects of Holocene sea-level changes on humans have been the subject of numerous case studies but relatively few syntheses (Chappell, 1982; Kraft et al., 1985). This is unfortunate because it has allowed other factors to dominate, somewhat unrealistically perhaps, in the minds of many people interested in questions of cultural evolution.

Three distinct periods of sea-level change can be recognized during the Holocene in the Pacific Basin. The early Holocene (12,000–6,000 BP) was a time of largely sea-level rise, the middle Holocene (6,000–3,000 BP) a time of comparative stability of sea level and the late Holocene (<3,000 BP) a time of largely sea-level fall (Nunn, 1994, 1995, 2001; Grossman et al., 1998; Dickinson, 2001). This pattern was not necessarily the same in other parts of the world (Nunn, 2004).

Owing to the magnitude of early Holocene sea-level rise (120–130 m from ~17,000 to 6,000 BP), the legacy of this dominates coastal landscapes worldwide. Most coasts became submerged (drowned) and still exhibit the signs of this, even though deglacial land-ice melt ceased to contribute to sea-level rise ~4,000 BP (Peltier, 1998). A typical sign of coastal submergence is embayed shorelines, where embayments represent the flooded lower parts of valleys. Post-glacial sea-level rise would also have had displaced humans and other terrestrial organisms from low-lying coastal areas – much as it threatens to do today – but would also have created new opportunities for many biota. Many areas – Japan, for example – became islandized during the early Holocene as land connections to the Asian mainland in the north and south were submerged. Islands have longer shorelines than equivalent areas of continent, so the opportunities for maritime-associated subsistence are proportionately greater. The increase in shoreline length and sinuosity that accompanies islandization also provides opportunities for a range of shallow-water marine biota to diversify and occupy newly created ecological niches. For example, mangroves colonized newly formed bay heads during the early Holocene in many parts of the tropical Pacific, the sheltered shallow-water environment allowing mangrove forest to spread across far wider areas than had been possible along the mainly open (non-embayed) coasts of Last-Glacial times (Woodroffe and Grindrod, 1991).

The sea-level stability that characterized the middle Holocene in much of the Pacific Basin also brought coastal stability. For the first time in perhaps 13,000 years, sea level was stable in relation to the land, so its fringes started to be eroded laterally by marine processes. Broad shore platforms were cut in many places, providing the foundations for the coastal plains that emerged in the late Holocene (see Fig. 3.4A). In tropical areas, mangrove forests spread and those coral reefs that had succeeded in reaching the ocean surface (keeping up with sea-level rise in the sense of Neumann and MacIntyre, 1985) began growing outwards. The accompanying warm temperatures and generally higher levels of precipitation created terrestrial and nearshore environments that provided greater levels of opportunity for expansion in the ranges of many organisms, compared to the early and the late Holocene.

Sea-level fall during the late Holocene was accompanied by cooling and, in many parts of the Pacific Basin, a reduction in precipitation (Nunn, 1999). Sea-level fall transformed many coastal environments at this time as shore platforms cut during the higher sea level of the middle Holocene emerged, becoming covered by sediment and vegetated (see Fig. 3.4A). Such coastal plains attracted human settlers to many Pacific Islands (Dickinson, 2003). In tropical areas, the uppermost parts of those coral reefs that had succeeded in keeping up with early Holocene sea-level rise became exposed by sea-level fall and eroded subaerially. Many coastal embayments shallowed, filling up with sediment and often with mangrove forest; a well-studied example comes from Kosrae in Micronesia (Kawana et al., 1995). Within the past 1,200 years of the late Holocene, sea level rose during the warm Medieval Warm Period, fell rapidly during the A.D. 1300 Event and remained lower than present during the Little Ice Age before beginning (about A.D. 1800) its most recent period of net rise.

Since sea-level fall is believed to have been the main driver of societal change during the A.D. 1300 Event in the Pacific Basin (see Chapter 7), this section deals in some detail with examples of responses by Pacific societies to earlier sea-level changes in order to stress their importance in societal change. Examples are given from the early Holocene (Sections 3.4.1–2), the middle Holocene (Section 3.4.3) and the late Holocene (Section 3.4.4).

3.4.1 Human response to early Holocene sea-level rise

In most parts of the world, the coastline that existed at the start of the Holocene (~12,000 cal BP) is now under several tens of metres of water, so comparatively little can be readily known about the effects that subsequent sea-level rise had on its human occupants. This alters as we approach the time when sea level came within a few metres of present sea level because in many places the contemporary coastlines are accessible, sometimes even exposed by land uplift and/or subsequent sea-level fall. This section gives a few examples from the Pacific Basin of human responses to early Holocene sea-level rise.

In Japan, the process of islandization resulting from sea-level rise brought about many lifestyle changes. For example, when people living in Japan had land access to the Asian mainland they were less dependent on marine foods, less accomplished seafarers and more easily absorbed continental innovations such as pottery making. After Japan became an island nation, ~8,000–9,000 BP (Kawamura, 1998), marine resources began to play an increasingly dominant role in people's lifestyles. For example, shell middens on the eastern Kanto Plain, evidently one of the most desirable locations to live in the archipelago during the early Holocene, date from 9,500 BP (Keally et al., 2003). Later middens are on higher ground suggesting that in response to sea-level rise, people occupied bluffs and smaller islands that allowed them ready access to nearshore foods but removed settlements from the threat of regular inundation (Nunn et al., 2006). In the Ryukyu Islands of southern Japan, it has been suggested that smaller islands were abandoned by people during the early Holocene as sea-level rise reduced their land areas (Takamiya, 1996), a situation similar to that known to have obtained at this time on several islands around the coast of mainland Australia (Jones, 1977).

In places, sea-level rise during the early Holocene would clearly have forced people to occupy less land, probably have concentrated people in circumscribed areas at densities exceeding those to which they were accustomed. Out-migration was clearly one viable adaptation strategy (see following section). Another was to compete for the resources available. One example of the latter comes from Australia where the appearance of fighting scenes in rock art is linked to increased competition for resources arising from "climate change and forced migration and land redistribution" (Tacon and Chippindale, 1994, p. 224) which occurred as a consequence of early Holocene sea-level rise.

Many people in the Pacific Basin during the early Holocene were itinerant or at least not completely sedentary, so that they were able to respond readily to sea-level rise. But in those parts of the region where sedentary coastal communities existed, particularly in parts of western South America and along the East Asian continental margin, the societal effects of sea-level rise during the early Holocene are likely to have been greater (Nunn, 2007a).

The drowning of the continental margin of East Asia during latest Pleistocene and early Holocene times led to the displacement of countless people, some of whom are thought to have set out across the ocean in search of new lands to occupy. This is described in more detail in the following section.

3.4.2 Lowland agriculturalists displaced from East Asia by early Holocene sea-level rise

The precise location of the Asian homeland of Pacific Island people is the subject of some controversy. The original suggestion that Taiwan and southern China were the homeland was based largely on linguistic considerations, principally the observation that of the 10 surviving Austronesian language groups – 1 of which was spoken by all Pacific Islanders prior to European contact – 9 are represented in Taiwan (Bellwood, 1979; Blust, 2004). An opposing view is championed by S. Oppenheimer (1999, 2003) on the basis of a range of evidence, principally that from DNA lineages, an independent study of which also refutes the hypothesis of a Taiwan homeland for Pacific Island peoples (Jin and Su, 2000) in favour of island Southeast Asia.

In a book like this, to which the relationship between humans and their environment is central, it is arguably more important to try and answer the question "why did people leave the relative security of the Pacific Rim for the apparent uncertainty of a vast island-studded ocean" rather than "from which part of the Pacific Rim did the ancestors of the first Pacific Islanders originate". Of course, we will never be sure about the answer to the question "why" although it is plausible to suppose that it was sea-level rise that pushed people out of the low-lying areas of the western Pacific Rim into offshore archipelagoes, thereby giving them the confidence to voyage progressively farther afield (Thiel, 1987; Meacham, 1996; Nunn, 2007a).

At the coldest time of the Last Glacial, sea levels in the Pacific were ~120–130 m below their present levels. Shorelines lay farther out to sea than they do today. Around the mouths of the Huanghe and the Changjiang rivers, which today debouch into the shallow western part of the South China Sea, there was a vast expanse of fertile alluvial-coastal lowland that is today covered by ocean. Yet during the Last-Glacial Maximum (~18,000 BP), much of this area was dry land, and remained so until ~7,000–8,000 BP when it was inundated.

As explained in Section 2.2, throughout most of the time that people have occupied this region, they had a hunter-gatherer lifestyle, one that for many groups became increasingly maritime-adapted as they became more familiar with the ocean's potential. Increasing population densities in this area, probably combined with the impacts of Younger Dryas cooling on the wild-food resource base, led to the appearance of agriculture here – as a means of increasing productivity and making food supply more constant and reliable – by at least 8,400 BP (Higham and Lu, 1998), probably more than 9,000 BP (An, 1991; Harris, 1996; Elston et al., 1997).

A community that comes to depend on agriculture (plant and animal domestication) inevitably has to relinquish an itinerant way of life for one that is largely sedentary, so during the early Holocene many people in coastal-lowland East Asia lived in communities of perhaps 10–30 people. Some were conspicuously larger. The town of Hemudu, the prosperity of which was founded on paddy-rice farming in the lower Changjiang Valley, existed at least 7,000 BP (Zhao, 1998). Equivalent evidence for rice cultivation in parts of Southeast Asia 7,000–8,000 BP is also known (Kealhofer and Piperno, 1994).

All these cultural changes were set against the background of, perhaps even influenced by, rising sea level. The rate of this post-glacial sea-level rise was generally so slow that people had ample time to adapt although clearly a sedentary way of life in particularly vulnerable (lowland) areas necessitates considerable more adaptation than itinerancy. Yet in the Pacific, post-glacial sea-level rise was punctuated by bursts of rapid rise, attributable to breaching of dams holding back large meltwater lakes in North America. The most recent of these catastrophic rise events (CRE-3) occurred ~7,600 BP and involved 6.5 m of sea-level rise in less than 140 years (Blanchon and Shaw, 1995).

As shown in Fig. 3.2, CRE-3 is suggested as having caused a rapid inland movement of the shoreline, which had the effect of displacing many groups of maritime-adapted agriculturalists. The rapidity of CRE-3 is one reason why adaptation may have been particularly difficult in this instance but this may also have been a result of an unprecedentedly high population density in the region, which meant that most neighbouring areas suitable for resettlement were already occupied. A combination of the difficulties of easy adaptation and the familiarity of many coastal dwellers with the ocean may have led to some of them taking to the sea, in search of new lands to settle (Nunn, 2007a).

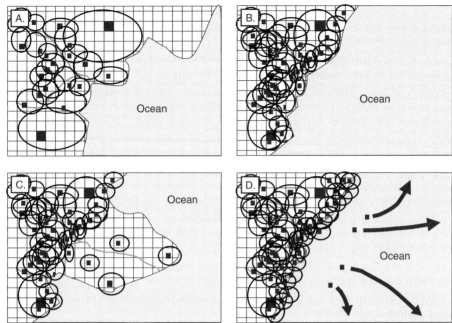

FIGURE 3.2 Palaeogeographic maps of a hypothetical part of the East Asia coastline during the later part of the Last Glaciation and the early part of the Holocene epoch. Filled squares represent groups of people at their (semi-) permanent bases at particular points in time. Circles represent the spheres of influence of the human group in the associated square. (A) 10,500 years BP. People have already colonized the main river valleys in such numbers that, while outlying human groups may still be able to survive from hunting and gathering, the increasing overlap in the spheres of influence of human groups in the most desirable parts of the valley is resulting in increasing sedentism involving the earliest domestication of plants and animals in this region. Note how valley floors are favoured over coasts which presented no special attraction for people at this time. (B) 9,500 years BP. Increasing numbers of people in the area lead to far greater overlap between the spheres of influence of adjacent human groups. This results in the subordination of hunting and gathering as a food-producing strategy and leads to the intensification of food production by agriculture. Note the almost equal favour given to coastal rather than valley sites, which is a result of humans in the former areas beginning to develop strategies for food production from nearshore (but probably not offshore) ocean areas. (C) 8,000 years BP. Slowing of the rate of sea-level rise and increased terrestrial sediment production (associated with increased precipitation) lead to extension of the shoreline around the mouths of large rivers. The new areas are colonized by various pioneer groups. The extension of some spheres of influence across adjoining areas of ocean represents the beginning of development of strategies for offshore fishing including the necessary boat-building and navigational techniques. (D) 7,400 years BP. A rapid burst of sea-level rise 7,600 years BP (CRE-3 of Blanchon and Shaw, 1995) involving 6.5 m in less than 140 years led to rapid shoreline inundation. One consequence of the resulting displacement of people and overcrowding along the new coastline was the out-migration in search of new lands of groups of people familiar with the rudiments of ocean travel.

It is likely that they settled on many of the nearer offshore islands, even within 1,000 years or so to parts of what today we call the Philippines and Indonesia, but some eventually found their way to Papua New Guinea ~4,000–5,000 BP. The new arrivals were Austronesian language speakers. This characteristic and their maritime-dominated economy and seafaring skills distinguished them from the Papuan speakers, who had inhabited the main islands of Papua New Guinea for much longer. The Austronesians became concentrated in the outer islands of the group, particularly in the Bismarck Archipelago where the embryonic Lapita culture appeared ~3,500 BP (Allen and White, 1989; Kirch, 1997a; Green, 2003).

3.4.3 Human response to middle Holocene sea-level stability

Throughout the Pacific Basin, sea level stabilized during the middle Holocene, reaching a maximum level – typically 1.5–2.0 m above today's level – around 4,000–5,000 BP (Nunn, 1995; Grossman et al., 1998; Dickinson, 2001; Nunn and Kumar, 2006). This section gives examples from the Pacific Basin of how its human societies responded to middle Holocene sea-level stabilization, coming as it did after some 13,000 years of mostly sea-level rise. Note that in the Atlantic, unlike the Pacific, sea level continued rising throughout the middle Holocene (Nunn, 2004).

In southernmost South America, particularly along the sides of the Beagle Channel in Tierra del Fuego, mid-Holocene sea-level changes – moderated by tectonic change and ocean-current reorganization – are believed to have been responsible for the pattern of coastal settlement established at that time and persisting subsequently (Isla, 1989). In parts of coastal Peru, such as the Santa Delta, the stabilizing of sea level during the middle Holocene led to shoreline progradation that allowed humans to spread out across coastal lowlands and encouraged adoption of sedentary lifestyles (Wells, 1992; Wells and Noller, 1999). A comparable situation is found on the Kanto Plain in Japan, the mid-Holocene shoreline was as much as 100 km farther inland; as much as 30% of the area exposed today was underwater, the mid-Holocene shoreline being longer and more convoluted (Esaka, 1967). People occupied the coast, commonly around the heads of bays and on smaller offshore islands, exploiting both nearshore and distant marine resources within the shelter of Tokyo Bay.

Similar effects occurred in coastal East Asia, particularly around the mouths of the Huanghe and Changjiang (Saito et al., 2001). Yet during the middle Holocene in the Zhujiang (Pearl River) delta, conditions were also warmer and wetter, and many people appear to have abandoned areas that they farmed in the early Holocene, perhaps because they were too wet (Weng, 1994, 2000). This may seem ideal but, in reality in many parts of the Pacific Basin, the apparent stability of sea level during the middle Holocene was only relative. And in particularly sensitive areas, such as coastal lowlands, short-term variability in mid-Holocene sea level had significant environmental and societal effects.

For example, in the Changjiang Delta of eastern China, the middle Holocene was a time of varying sea level and varying incidence of flooding (frequency and magnitude), both of which have been held responsible for the successive rise and

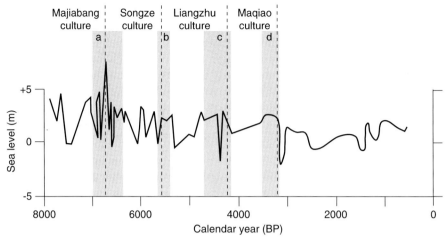

FIGURE 3.3 Relationship between sea-level fall and cultural decline in the Changijiang (Yangtze) Delta (adapted from Zhang et al., 2005). The Majiabang culture collapsed when high sea level caused groundwater flooding (shaded zone a). The Songze culture declined because of increased sea-level variability and flooding (shaded zone b). The Liangzhu culture collapsed as a result of sea-level rise which caused a rise in water tables and expansion of the Taihu Lakes (shaded zone c). Flooding is also implicated in the collapse of the Maqiao culture (shaded zone d).

fall of particular cultures in this area (Fig. 3.3; Chen and Daniel, 1998; Zhang et al., 2005). While the fertile alluvial soils of this vast flat delta plain have long attracted agriculturalists, freshwater appears to have been the critical limiting factor, and one that is implicated in the collapse of the successive civilizations in this area. In addition, the Changjiang Delta plain lies 2–7 m above sea level, making it highly vulnerable to inundation (through storm surges and tsunami) from the sea and flooding from adjacent river channels.

During the Majiabang cultural period (7,800–6,800 BP), the climate of the area was warm and wet, and the sea level was comparatively high, resulting in high water tables conducive to agriculture. Yet when sea level rose sharply ~6,800 BP, there was groundwater flooding and many parts of the delta were permanently inundated, spelling the end for the Majiabang Culture. The Songze culture (6,800–5,700 BP) appeared at the warmest time of the middle Holocene in this area but appears to have collapsed because of both sea-level variability and also wetter conditions that led to the delta being flooded more frequently. Warm yet drier conditions marked the rise of the Liangzhu culture (5,700–4,100 BP), perhaps the most complex of those shown in Fig. 3.3. Its end may have been anticipated by a time of surface-water shortage, shown by low sea level ~4,400 BP in Fig. 3.3, but was eventually brought about by sea-level rise that was associated with widespread flooding and, critically, expansion of the Taihu Lakes around which many people lived. The onset of cooler, wetter conditions provided the context for the rise of the less-developed Maqiao culture (4,100–3,300 BP), the end of which coincided with a period of higher sea level.

A similar pattern of punctuated societal evolution has been suggested for the mid-Holocene societies of coastal Peru, where major El Niño events produced extreme environmental stress that led to profound cultural changes (Moseley, 1997). For example, it has been argued that the transition from the Moche IV to the Moche V cultural periods came about during a major El Niño event about A.D. 550 which itself occurred within a prolonged period of drought (Shimada, 2000). Yet compelling as this may seem, some scientists argue that human adaptation to such catastrophic events in such areas must have been well-established for those societies to evolve as they did, and that it is more likely to be longer term (more enduring) climate changes that brought about societal collapse (Wells and Noller, 1999).

A final, well-documented example of how the stabilization of sea level during the middle Holocene transformed coastal environments, making them more attractive to people who had hitherto eschewed them, comes from New Guinea (Terrell, 2002). During the early Holocene, much of the coast of New Guinea was cliffed and unattractive but the stabilizing of sea level, the upgrowth of coral reefs left behind by rapid sea-level rise and the shoaling of nearshore areas led to the development of coastal lagoons, wetlands and alluvial plains that presented a range of attractive subsistence options for people. The colonization of these areas during the middle Holocene has been portrayed as a response to the environmental transformation resulting from the slowing and then the cessation of sea-level rise.

3.4.4 Human response to late Holocene sea-level fall

In most of the Pacific, sea level fell from its mid-Holocene maximum ~4,000 BP to its present level about A.D. 750 (1,200 cal BP). The sea-level changes that took place subsequently (A.D. 750 to present) are not described in this chapter but in Chapters 4–6.

In terms of creating environments suitable for intensive occupation by coastal-dependent humans in the Pacific Basin, the late Holocene sea-level fall was perhaps the most significant event of its kind. Its effects ranged from exposing fertile deltaic and littoral lowlands available for agriculture (and later, large urban and infrastructure development) to creating the first islands on atoll reefs making these habitable.

Four examples are discussed in the following subsections. In the first, there is a discussion of how sea-level fall enabled human occupation of rapidly outgrowing deltas, illustrated by those in East Asia–Papua New Guinea and South America (Section 3.4.4.1). The second example deals with the shift from upland areas and hillslopes to newly exposed coastal lowlands, by people throughout the Pacific Basin (Section 3.4.4.2). In the next example it is explained how sea-level fall, by creating coastal environments dry enough for permanent human habitation, set the timetable for colonization of many parts of the island Pacific (Section 3.4.4.3). Finally, Section 3.4.4.4 looks at Pacific atolls, and shows how it was sea-level fall that enabled islands to be created on them, thereby rendering them habitable. The purpose of these examples is to stress the point that sea-level change has long been a significant factor in the human history of the Pacific Basin, and that the proposed cultural effects of sea-level fall during the A.D. 1300 Event (discussed in Chapter 7) are not something novel but something for which earlier precedents are legion.

3.4.4.1 Sea-level fall and delta occupation

Examples from across the world are known of delta areas that were first occupied by people only after sea level fell. The inhabited lowlands of the Netherlands (northwest Europe), for example, are mostly subtidal but it seems likely that they were first occupied by humans when they were exposed as a result of a sea-level fall ~800 B.C. (van Geel and Renssen, 1998).

In all but a few places, the Pacific Rim is marked by ranges of fold mountains (shown in Fig. 1.7), a function of the great time that these fringes have been continental-plate margins and the numbers of orogenic events, mostly from plate collisions, that they have experienced (Nunn, 1999). It is in East Asia that the longest river systems – notably the Huanghe, Changjiang and Zhujiang – in the Pacific Basin reach the ocean's edge. Elsewhere, there are occasional large rivers but their catchments do not generally rank among the world's largest, their often high discharges resulting from the fact that they drain mountainous areas with correspondingly high precipitation rates.

The larger rivers of the western Pacific Rim all carry huge quantities of sediment that they deposit around their mouths, building deltas seaward. Yet when sea level was higher during the middle Holocene, the lower parts of these deltas that are now exposed were underwater – the shoreline was farther inland. For example, people occupying the Zhujiang Delta during the middle Holocene (6,500–3,500 BP) lived on high ground – often on islands – but during the late Holocene, particularly the Qin and Han dynasties (221 B.C. to A.D. 220), rapid sea-level fall opened up vast areas for agricultural development. The area became a magnet for people from central China, who brought with them the most advanced agricultural technology of the time, particularly for rice cultivation (Weng, 2000).

The late Holocene in Japan saw the transition ~850 B.C. (2,800 BP) from the Jomon Period, characterized by lifestyles based largely on hunting and gathering, to the Yayoi Period when farming became the dominant method of food production. The coincidence of this conspicuous cultural transition with sea-level fall suggests that the latter may have instigated the former particularly through the rapid emergence of fertile coastal flats suitable for agriculture. For example, studies of the 600-km^2 Kujukuri coastal plain east of Tokyo show that it prograded under the influence of sea-level fall as much as 3.5 m/year during the late Holocene (Matsuda et al., 2001) while the front of the 1350-km^2 Nobi Plain at Ise Bay was extending at ~5 m/year at the time of the Jomon–Yayoi transition (Yamaguchi et al., 2003).

On the other side of the Pacific, studies of the Santa Delta in Peru have revealed a similar series of events (Wells, 1992). During the middle Holocene, a few groups lived along the coast of the deep Santa Bay but significant progradation of the bayhead began by A.D. 200 as a result of sea-level fall. Continued sea-level fall led to delta outgrowth and settlements were established in three places on the delta during the Guadalupito Period (A.D. 400–650). Later the delta had prograded far enough, and sea level had fallen enough, to allow settlement all along the southern front of the delta.

3.4.4.2 Sea-level fall and the shift from uplands to lowlands

Another ubiquitous example of how sea-level fall during the late Holocene influenced human settlement pattern in the Pacific Basin is the way in which upland areas were abandoned in favour of lowland areas. During the middle Holocene because of the higher sea level, many coastal flats and valley floors were too wet for humans to occupy. As sea level fell during the late Holocene, these coastal flats emerged and valley floors became drier as a result of water-table fall.

The resulting change in settlement location from upland (hilltop or hillslope) to lowland is seen in many Pacific deltas, such as the Zhujiang in southern China (Weng, 1994, 2000) and the Santa in Peru (Wells, 1992). But it is also observable on some Pacific islands, such as Aneityum in southern Vanuatu. The earliest settlements on Aneityum were established ~3,000 BP on the slopes above the water-logged valley floors but, as sea level fell, these valley floors became better drained and suitable for horticulture, which resulted in a move downslope marked by the establishment of valley-floor settlements ~2,000 BP (Spriggs, 1986; Nunn, 1990).

Elsewhere in the Pacific Islands region, as discussed in Sections 3.4.4.3 and 3.4.4.4), sea-level fall during the late Holocene evidently created environments for human settlement, some on islands that had previously been spurned by potential colonizers.

3.4.4.3 Sea-level fall and island colonization

The progressive human settlement of the Pacific Islands, largely from west to east (see Fig. 2.1), has long been assumed to have been an outcome of exploration and voyaging reach (Irwin, 1992; Lewis, 1994). But studies of the history of sea-level fall during the late Holocene in the Pacific suggest that island habitability and attractiveness changed profoundly for this reason (Nunn, 1988, 1994; Dickinson, 2003). Late Holocene sea-level fall created attractive coastal environments where none existed before (Fig. 3.4). Well-documented examples of this scenario come from Kunashir Island (Kurile group, northwest Pacific – Korotky et al., 2000) and Viti Levu Island (Fiji – Shepherd, 1990).

The first people who settled in the Pacific Islands followed lifestyles that centred on marine-food consumption. Of course, they supplemented this in many places by terrestrial foods, although many founder island populations are believed to have survived almost solely – perhaps for generations – on marine foods (Burley et al., 2001; Nunn et al., 2004). For such reasons, most Pacific Island people have traditionally preferred to settle on island coasts rather than island interiors. Most of the earliest settlements were on coastal flats, often just behind the contemporary beach ridge, as close as practicable to nearshore (commonly reef) resources. Many early settlements were on smaller, reef-fringed islands, where marine foods were more abundant than on the impoverished reefs that fringed the larger higher islands (Lepofsky, 1988; Nunn, 2005).

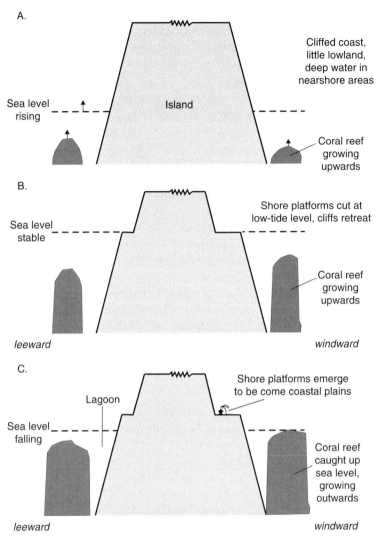

FIGURE 3.4 Transformation of island coastlines as a result of Holocene sea-level changes in the Pacific. The situation applies to most parts of the coasts of larger landmasses, except near where large rivers enter the sea. (a) Early Holocene situation (~7,000 BP) when sea level was rising and coral-reef upgrowth was lagging behind in most places. Island coasts would have been largely cliffed, with little coastal lowland, and deep water close to shore. This type of coastal environment would not have been especially attractive to potential colonizers whose lifestyles centred on marine-coastal foods. (b) Middle Holocene situation (~4,200 BP) when sea level had reached its Holocene maximum ~1–3 m above present. Sea-level stability meant that wave erosion was able to cut shore platforms at low-tide level along many island coasts. There would have been deep water close to shore, as coral reefs strove to catch up with sea level, and little dry flat land along island fringes. This type of environment would have still been comparatively unattractive for potential colonizers. (c) Late Holocene situation (~1,500 BP) when sea level had fallen to close to its present level. The shore platforms cut at

So potential settlers may have chosen not to make permanent settlements on islands that had no emergent (dry, even at high tide) coastal flats. Rather they would have sought islands with emergent coastal flats, maybe an island where uplift had amplified emergence from sea-level fall. Alternatively, they may have noted the existence of an island without emergent coastal flats in a particular place and gone elsewhere, returning to settle that island only when its coastal flats had emerged beyond the reach of high tide as a result of late Holocene sea-level fall (Dickinson, 2003).

The evidence in support of the latter scenario is impressive, principally comprising coincidences between times of initial human settlement of a particular island and times when sea level had fallen sufficiently for coastal flats around that island to become fully emergent (dry, even at high tide) (Fig. 3.5). This time – called the crossover date – reflects the variable character of late Holocene sea-level fall in the tropical Pacific. It has long been known that in the western part of this region, sea level reached a high ~4,500 BP and began falling shortly afterwards, reaching its present level about A.D. 1200 (750 BP) (Nunn, 1995; Nunn and Peltier, 2001), whereas in the eastern part, sea level remained high far longer, typically beginning to fall from its Holocene maximum only ~2,000 BP or even later (Pirazzoli and Montaggioni, 1988; Dickinson, 2001).

For example, as shown in Fig. 3.5A, human settlement began only A.D. 600–800 in the leeward Society Islands of French Polynesia (Spriggs and Anderson, 1993; Green and Weisler, 2002), just after the crossover date of A.D. 500. Likewise in the Cook Islands, the earliest known human settlement was shortly after the crossover date (Fig. 3.5B). The crossover date for the Tuamotu–Gambier Islands of French Polynesia of A.D. 1400 probably represents the time of earliest colonization of the low islands in these groups, although people occupied high Mangareva Island (Gambier Islands) A.D. 700–800 (Green and Weisler, 2002) (Table 3.1).

The corollary to the idea that most island colonization in the central and eastern Pacific took place only after the crossover date was passed is that the people who settled in these islands knew of their existence. This implies that intentional voyages of colonization were preceded by exploratory voyages that established the existence and approximate locations of islands suitable for settlement. This idea is less radical when applied to the islands of the western tropical Pacific colonized during Lapita times because these islands are mostly larger and closer together than those in the central and eastern Pacific.

It is possible that some of the early settlers of the tropical Pacific Islands were (or were descended from) sea nomads, who wandered the ocean, making landfall

---

FIGURE 3.4 *Continued.*

the time of the mid-Holocene high sea level would now have emerged, becoming covered with terrestrial and marine sediments to form coastal plains suitable for horticulture. Coral reefs would have caught up with the sea level, reef surfaces growing outwards, and sediments filling gaps between reefs and island shorelines to form lagoons. The coastal plains and productive shallow-water reef–lagoon ecosystems would have proved attractive to potential colonizers.

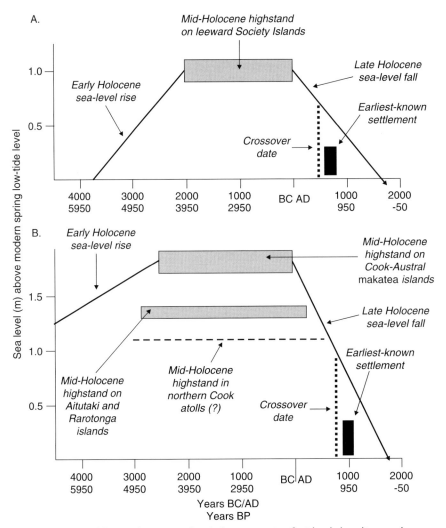

FIGURE 3.5 Effects of late Holocene sea-level changes on Pacific Island shorelines and habitability in time and space (after Dickinson, 2003). In both graphs, the course of Holocene sea-level changes is shown with a shaded box representing the time and elevation of the mid-Holocene sea-level highstand. It was at this time that the shore platforms which subsequently emerged were cut. The time at which late Holocene sea level fell sufficient so that these platforms were no longer awash at high tide is the crossover date. It is no coincidence that the earliest known human settlement in many Pacific Island groups followed shortly after the cross-over date. (a) Data from the leeward Society Islands; sea-level changes based on 25 data points. (b) Data from the Cook Islands and Austral Islands; sea-level changes based on 29 data points.

only when necessary, but otherwise choosing to live on their boats (Nunn, 2007b). The existence of Pacific sea nomads in pre-history – for example, in Taiwan (Chen Chung-Yu, 2002) – and in Southeast Asia today (Sopher, 1977; Ivanoff, 2005) lends weight to this suggestion. One could even go a step further, and propose that it

TABLE 3.1 Sea-level highstand dates, crossover dates and earliest-known human occupation dates for selected island groups in the Pacific Ocean (after Dickinson, 2003).

Island (ocean)	Highstand date	Crossover date	Earliest human occupation
Cook–Austral groups (Pacific)	0 B.C./A.D. 0	A.D. 800	A.D. 1000
Fiji–Tonga groups (Pacific)	1200 B.C.	A.D. 500	950 B.C.
Marianas Islands (Pacific)	1200 B.C.	100 B.C.	1800 B.C.
Society Islands (Pacific)	0 B.C./A.D. 0	A.D. 500	A.D. 600
Tuamotu–Gambier groups (Pacific)[a] (Mangareva)	A.D. 800	A.D. 1400	A.D. 750
Tuvalu–Kiribati–Marshall groups (Pacific)[a]	200 B.C.	A.D. 1100	A.D. 100

[a] Mostly atolls.

was the sea nomads themselves who, realizing that the transformation through sea-level fall of coastal environments on these remote islands would be beneficial to them, gradually exchanged a life on the sea for a life on land. If this is true, then it is another example of how the lifestyles of Pacific people were fundamentally changed by extraneous non-human changes, in this case, sea-level fall.

3.4.4.4 The role of sea-level fall in rendering atolls habitable

The term atoll is commonly used to describe a subcircular coral reef often with low-elevation islands (*motu*) made from largely unconsolidated deposits of reef-derived sand and gravel on it (Nunn, 1994). Some of those atoll *motu* which are large enough for a freshwater lens to exist are inhabited by humans. These communities are regarded as some of the most vulnerable on Earth because of the low-lying nature of their islands and their low resistance to coastal erosion (Roy and Connell, 1991; Barnett and Adger, 2003).

The rate of post-glacial (early Holocene) sea-level rise in many parts of the tropics was too fast for most coral reefs to maintain themselves at the low-tide level (commonly Lowest Astronomical Tide level or Mean Low Water Spring level). Some reefs gave up and today lie dead several tens of metres beneath the present ocean surface. Other reefs were able only to catch up with sea level once its rate of rise decreased (or reversed) after ~6,000 BP. A few reefs – those in the most favourable locations for reef growth – were able to keep up with sea-level rise (Neumann and MacIntyre, 1985) and it is on these that *motu* commonly exist today.

In most parts of the tropical Pacific where atolls are found today, reef upgrowth kept pace with sea level as it rose above its present level ~6,000–4,000 years ago. The fall of sea level since that time has exposed (and killed) the top 2 m or so of these "keep-up" reefs. During the late Holocene, coral growth resumed at a lower level, extending the reef laterally, but where the keep-up reef once was, there exists a core of fossil reef – its surface reduced by subaerial erosion – standing up above the general level of the reef flat (example shown in Fig. 8.5C). It is these cores of emerged reef that became foci for accumulation of sediments scoured up from the submerged forereef slopes and driven onto the reef during storms. The steady accumulation of these sediments led to the development of *motu* on many such reefs during the late Holocene. Yet had there not been a fall in sea level during the late Holocene, no such *motu* could have developed, and the reef – like many of those in the Caribbean where there was no Holocene sea-level highstand – would remain today at the ocean surface.

The development of *motu* on atoll reefs in the central Pacific within the last 1,000–2,000 years made them available for human settlement which soon followed (Riley, 1987; Shun and Athens, 1990; Best, 1998). Atoll archaeology is in its infancy, so we cannot yet answer questions about why particular atolls were occupied, although it seems likely that many were reached by people voyaging intentionally in search of new land who recognized their food-producing potential, particularly for marine foods, horticulture and birds (Irwin, 1992; Di Piazza and Pearthree, 2001a).

KEY POINTS

1. Climate change has long influenced human affairs, having had a seminal role in hominid evolution, and probably having led to both the independent adoption of agriculture by communities in many parts of the world following the Younger Dryas cold event and the mass extinction of large mammals across the Earth during the terminal Pleistocene.

2. During the Holocene, temperature and precipitation variations led to numerous human-lifestyle changes, notably the collapse of complex societies. Changes in climate variability, especially interannual variability, may also have played a role in societal collapse as well as changing the efficacy of long-distance ocean voyaging.

3. Sea-level rise during the early Holocene (12,000–6,000 BP) drowned most coasts, forcing human societal changes ranging from changes in subsistence strategies to cross-ocean migration out of overcrowded areas.

4. The stabilizing of sea level in the Pacific Basin during the middle Holocene (6,000–3,000 BP) led to changes in human settlement pattern and subsistence.

5. Sea-level fall in the Pacific Basin during the late Holocene (3,000–0 BP) led to the increased occupation of coastal lowlands and facilitated island colonization.

CHAPTER 4

The Medieval Warm Period
(A.D. 750–1250) in the Pacific Basin

This chapter explains the nature of climate, environment and society in the Pacific Basin during the Medieval Warm Period, generally here a time of plenty marked by low climate variability, slow rates of environmental change, comparatively few extreme events, and the consolidation and complexification of societies. The period was also notable in the Pacific Basin for lengthy ocean voyages that saw people colonize most major island groups and even traverse the entire Pacific Ocean from west to east.

This chapter is divided into three main parts, the first (Section 4.1) considering the climates during the Medieval Warm Period in the Pacific Basin, the second (Section 4.2) the environments that existed here, and the third (Section 4.3) the societies found at this time in this region. Throughout these sections, connections are made between the effects of various climate drivers on environments and societies where these connections are regarded as uncontroversial and adequately supported by empirical evidence.

4.1. CLIMATES DURING THE MEDIEVAL WARM PERIOD IN THE PACIFIC BASIN

Across the Pacific Basin, there is evidence for warm and dry conditions during the Medieval Warm Period. While there was evidently inter-regional variation and at least four places where the signal of the Medieval Warm Period is weak or absent (part of Chilean Patagonia, monsoonal northeast China, Tasmania and the central equatorial Pacific [Palmyra Atoll]), the generalization appears valid. For example, the general lack of glacier advance during the Medieval Warm Period (shown in Fig. 5.1) is a crude indicator of the warmth of this period throughout the Pacific Basin. Much corroborative evidence of warm dry conditions during the Medieval Warm Period also comes from inferences about how humans in the Pacific Basin exploited these generally favourable conditions.

This part of Chapter 4 begins by describing the nature of climates during the Medieval Warm Period in the Pacific Basin, with geographical subdivisions used for largely organizational purposes. Discussion of climates along the Pacific Rim is divided into east (Section 4.1.1) and west (Section 4.1.2), with those of the Pacific Islands and Ocean discussed separately (Section 4.1.3). The issue of El

Niño-Southern Oscillation (ENSO) variability and its relationship to storminess is discussed separately in Section 4.1.4.

4.1.1 Climates of the eastern Pacific Rim (including Antarctica)

Studies of isotopic thermometry from various ice cores support the general picture of a warm Medieval Warm Period in Antarctica (Benoist et al., 1982; Morgan, 1985). Various analyses of sea-floor cores from the Pacific side of the Antarctic Peninsula showed that environments were quite different here during the Medieval Warm Period compared to the present (Domack et al., 2003). In the Lallemand Fjord/Crystal Sound area, for example, total organic carbon values were notably higher in the period A.D. 800–1250 (1,150–700 cal BP) because decreased sea ice allowed more organic carbon (high $\delta^{13}C$) to reach the sea floor.

An unusual way in which the existence of the Medieval Warm Period in Antarctica has been inferred comes from dating of Adélie penguin (*Pygoscelis adeliae*) rookeries (Baroni and Orombelli, 1994a). Among Antarctic penguins, the Adélies are the most resistant to extremes of cold and therefore the first species to reoccupy areas that have just become ice-free. Abandoned rookeries dated to A.D. 700–1300 suggest that the Medieval Warm Period opened up parts of coastal Antarctica for penguin occupation that had previously been ice covered and uninhabitable. A more conventional study by these authors showed that the Edmonson Point glacier in Terra Nova Bay (part of the Pacific Rim of Antarctica) receded in two distinct phases A.D. 920–1020 and A.D. 1270–1400 (Baroni and Orombelli, 1994b).

Tree-ring data from southern South America show that there was a warm period here A.D. 1080–1250 (Villalba, 1990), although one record (from Lenca, southern Chile) does not register the Medieval Warm Period (Lara and Villalba, 1993). This is regarded as anomalous since the absence of the Medieval Warm Period is contradicted by numerous tree-ring series from the same region (Soon et al., 2003). Off the arid coast of northern Chile, ocean-floor cores from Bahía Mejillones revealed the presence of warmer waters with greater primary productivity during the Medieval Warm Period (Ortlieb et al., 2000; Valdes et al., 2003).

Farther north in the Peruvian Andes, pollen analyses from sediments in the Marcacocha Basin show evidence of a warm dry Medieval Warm Period (Chepstow-Lusty et al., 2003). In addition, all (except one) of the dry intervals registered here by abundances of sedge (Cyperaceae) pollen coincide with "major independently defined cultural boundaries" (p. 500) implying that drought had significant cultural effects. This relationship extends to coastal Peru, where the period between A.D. 1250 and A.D. 1310 was an "intensely dry" one (Shimada, 2000, p. 103).

One of the most influential palaeoclimate studies in the Pacific Basin, well respected for its detail and precision, comes from ice coring of Quelccaya ice cap (and nearby Huascarán and Sajama) in the southern Peruvian Andes (Thompson et al., 1985). All these records show evidence for the Medieval Warm Period (Fig. 4.1). By using ice accumulation as a proxy for palaeoprecipitation, periods of drought in this region have also been recognized. In particular, the end of the

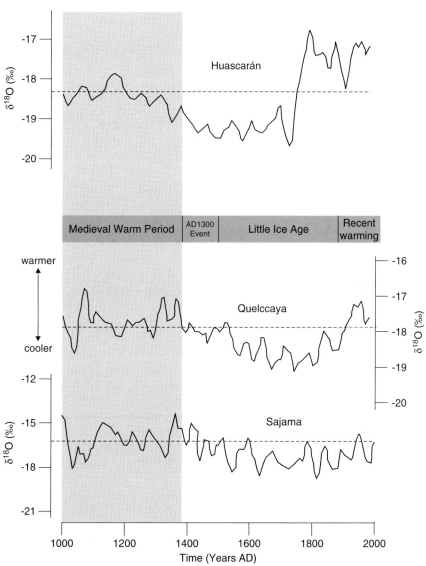

FIGURE 4.1 Changes in $\delta^{18}O_{ice}$ (temperature proxy) for the last millennium from three Andean ice cores. Lines are 30-year running means (after Thompson et al., 2003) with series means (horizontal lines) added graphically. Divisions of last-millennium climate are from Thompson et al. (2003) and differ slightly in timing from those used in other figures in this book.

Medieval Warm Period was unusually dry, a severe drought about A.D. 1150 followed by a period of prolonged dryness A.D. 1245–1310. The ice-core record of this drought has been supplemented by data from coring of sediments on the floor of Lake Titicaca (Binford et al., 1997). Lake level in the period A.D. 1150–1200 lay 12–17 m below its pre-drought level, an outcome of a 10–12% decrease in precipitation from the modern average (Dillehay and Kolata, 2004), something that had major consequences for the people living on the Andean *altiplano* (high-altitude plains) during the later Medieval Warm Period (see Section 4.3.3).

Just beyond the edge of the Pacific Basin in South America, but probably still representative of this part of the region, comes a dendrochronological study of 1,000-year-old alerce (*Fitzroya cupressoides*) trees which shows a warm dry period (the Medieval Warm Period) A.D. 1080–1250 (Villalba, 1990). A similarly peripheral site is the Cariaco Basin in Venezuela from which titanium in fluvial sediments has been used to reconstruct regional palaeoprecipitation over the past two millennia (Haug et al., 2003). This record, illustrated in Fig. 3.1, shows the Medieval Warm Period clearly.

In most of central America, the droughts that are implicated in the collapse of the Classic Maya civilization about A.D. 750 (Haug et al., 2003) were paralleled along its Pacific coast, one study showing the "extended period of stable, favorable climatic conditions" here ended about the same time (Neff et al., 2006, p. 397). In most of central America, the restoration of warm wetter conditions during the Medieval Warm Period was marked by population resurgence, the so-called Late Postclassic recovery of the Maya (Webster, 2002), but along its Pacific border, dry, variable conditions persisted and the area remained lightly populated throughout the Medieval Warm Period (Neff et al., 2006).

Tree-ring data from long-lived trees in the coastal ranges of western North America have confirmed the existence of the Medieval Warm Period in this part of the Pacific Rim. Early studies by Graumlich (1993) and Scuderi (1993) found evidence for the Medieval Warm Period A.D. 1100–1375 and A.D. 800–1200, respectively, the differences attributable largely to site altitude. Graumlich's work in the Sierra Nevada led her to conclude that the two warmest times of the Medieval Warm Period were A.D. 1118–1167 and A.D. 1245–1294 and that the two driest periods were A.D. 1250–1299 and A.D. 1315–1364. More recent work shows that the warmest time was between at least A.D. 1000 (the start of records) and A.D. 1150 (Bunn et al., 2005).

Palaoeoclimatic research in the Canadian Rockies, employing both dendro-chronology and lake-level studies, shows that the warmest time of the Medieval Warm Period was A.D. 950–1100 (Fig. 4.2A), a longer, more regionally represen-tative timeseries showing that the end of the Medieval Warm Period, marked by the sudden onset of cooler conditions, occurred about A.D. 1200 in this region (Fig. 4.2B).

Records of glacier recession during the Medieval Warm Period from the mountains along the Pacific Rim of Canada support the idea that it was also a warm dry time in this region (Luckman et al., 1993). A study of 14 glaciers in the Mt. Waddington area (British Columbia Coast Mountains) bracketed the Medieval Warm Period to A.D. 933–1203 (Larocque and Smith, 2003) while, less precisely, the time of glacier recession that includes the Medieval Warm Period was found to have been within the period A.D. 550–1480 (1,400–470 cal BP) at Lillooet Glacier in the same area (Reyes and Clague, 2004).

Proxy data for growing-season temperature over the past 2,000 years from the northwestern foothills of the Alaska Range showed that the period A.D. 850–1200

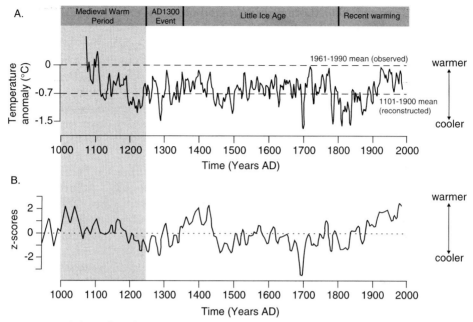

FIGURE 4.2 (A) Northern-hemisphere summer temperature anomalies using dendrochronology from the Columbia Icefield, Alberta, Canada (after Luckman et al., 1997). Ten-year running means are shown. Divisions of last-millennium climate are from this book. Note that the end of the Medieval Warm Period, as defined here, is cooler in this region, suggesting that the dates for the climate divisions used here are not applicable region-wide. (B) Summer temperature reconstruction (RCS2004) for the Canadian Rockies. Series was smoothed with a 20-year spline and standardized to the A.D. 1000–1980 period (after Luckman and Wilson, 2005).

was one of comparative warmth, also marked by drier conditions (Hu et al., 2001). A more recent study of northwest Alaskan tree rings is also consistent with the occurrence of the Medieval Warm Period (D'Arrigo et al., 2005) as is the record of glacier fluctuation, the Medieval Warm Period

> "encompassing at least a few centuries prior to AD 1200 [being] recognized by general retreat of land-terminating glaciers". (Calkin et al., 2001, p. 449)

The onshore record is confirmed by offshore work. For example, sea-surface temperatures in the Santa Barbara Basin A.D. 200–1250 were among the highest within the past 8,000 years (Pisias, 1978). Reconstructed sea-surface temperatures off southern California were found to have risen markedly towards the end of the Medieval Warm Period about A.D. 1150–1300 and were associated with severe droughts. The study of Baumgartner et al. (1992), who examined sardine and anchovy biomass through time in the varved sediments of the Santa Barbara Basin, concluded that temperatures were low at A.D. 800, at their maximum for the Medieval Warm Period about A.D. 1000 and then low again by A.D. 1400. The important implication of this work is that sardines and anchovies in the

California Current ecosystem respond to long-period climate forcing, total sardine–anchovy biomass having been higher during warmer periods (like the Medieval Warm Period) than cooler periods (like the Little Ice Age).

Similar research elsewhere in the Pacific Ocean shows results that can also be explained by changes in ocean circulation and surface temperatures. For instance, studies of salmon-derived nutrients in salmon nursery lake sediments on Kodiak Island in the northeast Pacific suggested that increases in the numbers of Alaskan salmon within the period A.D. 800–1200 were linked to changing ocean–atmosphere circulation in this region (Finney et al., 2002).

We now turn from broad considerations of temperature along the eastern Pacific Rim to precipitation. Against a background of higher temperatures, in a review of last-millennium climate along the Pacific Rim of the Americas, Stine (1998, p. 60) concluded that the climate during the Medieval Warm Period "was, by modern standards, aberrant". He sees no clear evidence regionally for a distinct Medieval Warm Period, preferring to characterize the period as the Medieval Climate Anomaly since most of this region (California, northern Rocky Mountains, southern Patagonia, Yucatan Peninsula and the south-tropical Andes) experienced prolonged droughts while Alaska–Canada underwent periodic increases in wetness.

Summer precipitation in the Sierra Nevada is shown in Fig. 4.3. A definitive study employed tree-ring analyses of bristlecone pines (*Pinus longaeva*) in the White Mountains of California where growth, associated with comparatively high soil moisture, was more rapid A.D. 1080–1129 than at any time subsequently (Leavitt, 1994).

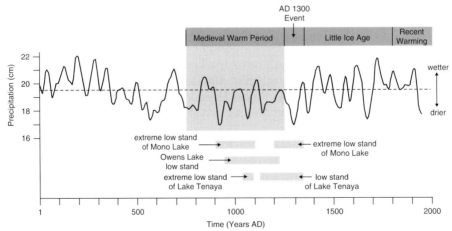

FIGURE 4.3 Relationship between palaeoprecipitation from tree-ring data and lake-level studies in the westernmost United States. Northern-hemisphere summer precipitation for the past two millennia in Nevada (western USA) using dendrochronology from the Methuselah Walk bristlecone pine (after Hughes and Graumlich, 1996). Series smoothed with a bidecadal-scale Gaussian filter. Also shown are the extreme low stands of Mono Lake identified by Stine (1994), the low stands of Lake Tenaya (Stine, 1998) and Owens Lake (Li et al., 2000). Series mean (horizontal line) added graphically. Divisions of last-millennium climate are from this book.

Radiocarbon dating and ring counting of lodgepole pine (*Pinus contorta*) trees rooted 8–19 m below water level in Lake Tenaya (California) show that later part of the Medieval Warm Period was much drier here than in recent decades (Stine, 1998). Between A.D. 1023 and 1093, the lake was at least 13 m lower than today while between A.D. 1192 and 1333, it was at least 11 m lower. Similar conclusions have been reached from studies of other lakes in this region.

Oxygen-isotope analysis of sediments from Mono Lake (California) show that the Medieval Warm Period comprised two periods of lake lowstand (see Fig. 4.3), reflecting dry conditions, A.D. 705–1084 (1,245–866 cal BP) and A.D. 1084–1260 (866–680 cal BP) separated by a short highstand (Stine, 1990). Similar results were obtained from Pyramid Lake (Nevada) (Benson et al., 2002) and show that California was affected by droughts that endured for decades at a time during the later part of the Medieval Warm Period. Studies of Owens Lake in the same area showed that it was at a low level, reflecting dry conditions, A.D. 950–1220 (Li et al., 2000), three peaks of acid-leachable Li and Mg during its later part indicating three periods of prolonged severe drought in the area (see Fig. 5.2).

Aridity also affected the Channel Islands (off southern California) during the later part of the Medieval Warm Period (Davis, 1992) where sea-surface temperatures were also higher at the time (Pisias, 1978), something that undoubtedly affected the people living there (see Section 4.3.4). Studies of palaeosalinity in the San Francisco Bay confirms unusual dryness here A.D. 950–1150 (1,000–800 cal BP – Malamud-Roam et al., 2006).

The aridity marking the later part of the Medieval Warm Period occurred throughout the western United States, severe droughts being implicated in initial Anasazi abandonment of Chaco Canyon A.D. 1130 (820 cal BP) and more widespread and enduring cultural collapse in the region about A.D. 1300 (650 cal BP: Benson et al., 2002). The latter is discussed further in Section 9.3.1. Although few diagnostic data exist in some places, it is reasonable to suppose that the whole of western North America experienced aridity, marked by prolonged droughts, during the later part of the Medieval Warm Period (Laird et al., 1996a). Supporting studies from western Canada come from the Rockies where conditions for tree growth were more favourable in the period A.D. 950–1100 (Luckman, 1994) and southern Alberta where the driest time of the Medieval Warm Period was around A.D. 1100 (Campbell, 2002). The regional synchronicity of late Holocene floods and droughts, including those during the later part of the Medieval Warm Period, was highlighted by Schimmelmann et al. (2003).

Throughout this region, the droughts of the later Medieval Warm Period also evidently caused forest fires. This explanation has been applied to the evidence for more frequent forest fires A.D. 1000–1300 in the sequoia (*Sequoia sempervirens*) forests of the Sierra Nevada (Swetnam, 1993), in southeast British Columbia (Hallett et al., 2003) and Yukon (Yalcin et al., 2006).

One of just a few studies to shed light on the last-millennium climatic history of Beringia (the land areas on either side of the Bering Strait) during the Medieval Warm Period examined the periods of relative activity (indicating dry conditions) and inactivity (indicating wetter conditions) of the Great Kobuk Sand Dunes in northwest Alaska (Mann et al., 2002). Dry conditions during the Medieval Warm Period (A.D. 900–1400) were marked by expansion of the dune field.

4.1.2 Climates of the western Pacific Rim

In the northwest part of the Pacific Rim, there are a few studies of last-millennium climate change that show the existence of the Medieval Warm Period. These include studies of recessed glaciers at the time (Savoskul, 1999) and the spread of forest types indicating warmer temperatures (Razjigaeva et al., 2002, 2004). On Iturup Island (45°N) in the Kurile Group, for example, the Medieval Warm Period saw the spread of cool-temperate forests dominated by broad-leaved genera such as *Quercus* (oak) and *Ulmus* (elm) with birch and maple widely distributed. Recent deep-sea coring in the Sea of Japan led to the conclusion that the warm Tsushima Current invaded the area from south during the Medieval Warm Period, beginning about A.D. 850 (1,100 cal BP: Kuzmin et al., 2004).

Along the western Pacific Rim, most palaeoclimatological information about conditions during the last millennium have come from China and Japan, where written records supplement scientific data. Zhang (1994) used records of the distribution of cultivation of the temperature-sensitive herb *Boehmeria nivea* and various citrus trees to reconstruct temperature maps of China at particular times during the last 1,300 years, concluding that the thirteenth century was the warmest within this period. Annual mean temperatures were 1°C higher than today with winter minimum temperatures as much as 3.5°C higher than today (Zhang, 1994).

A similar synthesis of proxy temperature indicators reached comparable conclusions (Feng et al., 1993). The combination of a $\delta^{18}O$ palaeotemperature record from peat cellulose having a 20-year resolution with various historical records showed that the Medieval Warm Period was at its warmest A.D. 1100–1200 (Hong et al., 2000). A more recent compilation of palaeotemperature data for China showed that the Medieval Warm Period occurred A.D. 800–1100; conditions were warmest in eastern China (Yang et al., 2002). In another compilation for eastern China, the start of the Medieval Warm Period was placed earlier and its warming characteristics described in more detail

"... temperature entered a warm epoch from the AD 570s to 1310s with a warming trend of 0.04°C per century; the peak warming was about 0.3–0.6°C higher than present for 30-year periods, but over 0.9°C warmer on a 10-year basis". (Ge et al., 2003, p. 933)

One of the most compelling scientific studies to support the recognition of the Medieval Warm Period in eastern China is the oxygen-isotope analysis of a stalagmite from Buddha Cave (Fig. 4.4). The Medieval Warm Period (A.D. 965–1475) is seen to have been consistently warm, above or close to the long-term average.

For Japan, the compilations of Fukui (1977) and Tagami (1996) that utilized mostly phenological (flowering and other nature-observing) data concluded that anomalous warmth prevailed throughout this region between the tenth and the

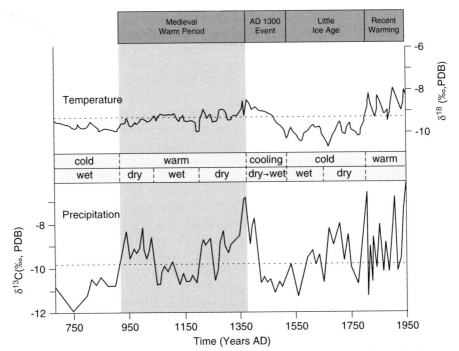

FIGURE 4.4   Isotope analysis of stalagmite SF-1 from Buddha Cave, eastern China, for the past 1,250 years (adapted from Paulsen et al., 2003). The upper curve shows temperature changes (using $\delta^{18}O$) above the average for this period, and the lower curve shows moisture (precipitation) changes (using $\delta^{13}C$) above the average for the period. The climate divisions of the last millennium are based on Paulsen et al. (2003) and are slightly different to those that appear on most other figures in this book. Note that the A.D. 1300 Event is later here than elsewhere, ~A.D. 1375–1475 (575–430 cal BP). Time scale was converted from original years cal BP.

fourteenth centuries. Analysis of tree cellulose in one of the giant cedars (*Cryptomeria japonica*) on Yakushima (island) in southern Japan showed that temperatures averaged 1°C higher than today between A.D. 800 and A.D. 1200 (Kitagawa and Matsumoto, 1995). The synthesis of Sakaguchi (1983) named the Medieval Warm Period the Nara–Heian–Kamakura Warm Period, dated to A.D. 732–1296, its start marked by a sudden temperature rise.

Aridity was also a notable feature of the Medieval Warm Period in parts of the Pacific Rim in East Asia. A good example comes from tropical southern China where the period A.D. 880–1260 was conspicuously dry (Chu et al., 2002). A conspicuous dry period A.D. 880–1260 recognized from geochemical analyses of sediments from Lake Huguangyan in tropical southern China (Chu et al., 2002) supports the written record from this time (Zhang, 1994), suggesting that it was the same warm dry Medieval Warm Period that occurred elsewhere in the world at this time. There is evidence of a progressive increase in aridity during the Medieval Warm Period in eastern China. Lake levels throughout China fell A.D. 950–1250 (Fang, 1993). In addition, a record from near Beijing shows that precipitation decreased more or less constantly between about A.D. 1090 and 1300 (see Fig. 5.3B). The study of the Buddha Cave stalagmite (Fig. 4.4) tells a similar story in this part of eastern China. For Japan, Sakaguchi (1983) showed that drought was common in the eleventh and twelfth centuries. There appears to be a synchrony between the century-long dry spells here and in California that occurred towards the end of the Medieval Warm Period (Paulsen et al., 2003).

An apparent exception to this picture comes from a fossil pollen record from northeast China which suggests that summers were wetter during the Medieval Warm Period (A.D. 950–1270) in this region, likely to be a result of an intensified summer monsoon (Ren, 1998), something also likely to have been an influence on flooding of the Huanghe (see Fig. 5.4).

Turning finally to the southern half of the western Pacific Rim, no high-resolution palaeoclimate data for the last millennium are known to the author from Papua New Guinea, yet some inferential data involving changed agricultural practices in response to changing climate come from the New Guinea highlands (Brookfield, 1989). In particular, the Medieval Warm Period marks the era of dryland cultivation regarded as indicative of warm conditions with less climate variability than during later times.

The Medieval Warm Period in eastern Australia shows up in the spring-fire history of Worimi Swamp, coastal New South Wales, where fires were more frequent during the first part of the last millennium, suggesting warmer conditions in this seasonally dry area (Mooney and Maltby, 2006). Some evidence for the Medieval Warm Period has been reported from tree-ring analysis of the exceptionally long-lived Huon pine (*Lagarostrobos franklinii*) in Tasmania (Cook et al., 1992). Temperatures here were above average in the period A.D. 940–1200 but this climate period was interrupted by an anomalously cold episode in the eleventh century. In general, the signal of the Medieval Warm Period appears to have been muted in Tasmania.

One of the first last-millennium palaeoclimate records from New Zealand (Wilson et al., 1979) was based on oxygen-isotope analyses of a stalagmite from northern North Island and showed that the Medieval Warm Period lasted from approximately A.D. 1200 to 1400 here (see Fig. 5.8). Composite speleothem records suggest the Medieval Warm Period was A.D. 1050–1350 (900–600 cal year BP) in New Zealand (Williams et al., 2004), possibly A.D. 1290–1430 (710–570 cal year BP: Williams et al., 2005).

Higher salinity was also reported in the westernmost equatorial Pacific Ocean early during the Medieval Warm Period (Stott, 2002), suggesting greater runoff from the adjoining part of the western Pacific Rim (principally New Guinea, Indonesia and the Philippines) as a result of high precipitation here A.D. 900–1100.

4.1.3 Climates of the Pacific Islands and Ocean

Most palaeotemperature records in the Pacific Ocean extend from the present only into the Little Ice Age (not the Medieval Warm Period). The few exceptions are described below.

A reconstruction of sea-surface temperatures in the equatorial West Pacific warm pool showed that they were comparatively high (reaching 30°C, about the late twentieth-century maximum) during the period A.D. 900–1100 (Stott, 2002; Stott et al., 2002). For the eastern Pacific, a 6,000-year proxy record of ENSO variations from the Galapagos Islands showed that the Medieval Warm Period, terminating about A.D. 1150 (800 BP), was unusually wet (Steinitz-Kannan et al., 1997). Data from Palau in the western Pacific show that drier conditions prevailed about A.D. 1000–1388 (Masse et al., 2006).

A recent study of central Pacific climate variations during the last millennium was obtained from coral coring at Palmyra Atoll (Cobb et al., 2003). By splicing oxygen-isotope records from corals of different ages, it was possible to obtain an 1,100-year long palaeoclimate record. The tenth century appears to have been unusually dry and/or cool at this location, and there is little evidence for warming in this area during the Medieval Warm Period. This conclusion is important because cool conditions in the central Pacific may be linked to the widespread droughts along the eastern Pacific Rim (particularly in western North America, central America and the Pacific coast of South America – see above) that characterized the later part of the Medieval Warm Period. And while the Palmyra record shows a generally low degree of climate variability for the Medieval Warm Period, the incidence of La Niña events may have increased in the tenth century and during the later part, causing

- periods of prolonged cooling in the central Pacific,
- periods of prolonged drought along the eastern Pacific Rim, except
- in northern South America where increased precipitation was experienced (see Fig. 3.1).

It is important not to attribute undue significance to the Palmyra record, for data are missing and alternative interpretations are possible; indeed, the authors call for more such studies before their tentative conclusions are confirmed (Cobb et al., 2003).

Many data about Medieval Warm Period climates in the Pacific Islands are inferential. The Medieval Warm Period marked a time when most of the successful long-distance voyaging took place, when techniques for water conservation became widespread, and when island societies increased in complexity under the influence of conditions of intensified food production (see Section 4.3.6). Other indicators include the increased concentrations of charcoal in cores through Kawai Nui Marsh (O'ahu Island, Hawaii), a trend that "declines precipitously" about A.D. 1200 (Athens, 1997, p. 267); increased charcoal may signal consistently dry conditions while the subsequent decline may signal generally wetter, perhaps more variable conditions.

4.1.4 ENSO frequency and storminess

Within the past 2,000–3,000 years, the incidence of El Niño (ENSO-negative) events appears to have decreased overall (Moy et al., 2002). This generalization has tended to overshadow the existence, particularly within the last millennium, of significant century-scale fluctuations. It appears likely that while stronger El Niño events occurred periodically during the Medieval Warm Period, there was an overall increase in the frequency of El Niño and La Niña (ENSO-positive) events during the A.D. 1300 Event that endured subsequently. This interpretation is consistent with a low climate variability (more constant climate) during the Medieval Warm Period than in later times.

It is clear that more tropical cyclones – a major cause of storminess – occur in the low-latitude Pacific during El Niño events, so it seems reasonable to posit a crude synchrony between periods of high El Niño frequency and elevated storminess for the past millennium. This relationship appears to hold true for empirical data series such as those for New Zealand (Eden and Page, 1998) and the study of particular concentrations of (mega) El Niño events (Ortlieb, 2000). It follows that during a period of comparatively low El Niño incidence like the Medieval Warm Period, there would be fewer storms, which is exactly what is suggested by various environmental and societal proxies for last-millennium climate in the Pacific (Bridgman, 1983; Finney, 1985; Nunn, 2003a, 2003b).

There is considerable evidence to suggest that during the twentieth century (and the first years of the twenty-first that have passed), there has been an increase in the frequency of tropical cyclones (hurricanes or typhoons) in the low-latitude Pacific Basin (and elsewhere) that is attributable to rising temperatures (Nunn, 1994). The connection seems clear. Higher temperatures result in warming of the ocean surface and the consequent expansion of the areas (oceanic warm pools) in which ocean-surface temperatures exceed 27°C, the lower limit at which tropical cyclones can form. An increased area of tropical-cyclone formation in the Pacific leads to increased development of tropical cyclones, in addition to their greater geographical reach once they leave the area of formation (i.e. >27°C) and their decreased seasonality.

Yet there is no equivalent body of evidence to suggest that rising temperatures during the Medieval Warm Period resulted in an increased frequency of tropical cyclones, a phenomenon that has been explained by both vertical temperature gradients and surface-air heating having been less than today because of the reduced cloudiness (Bridgman, 1983). Even though periods of storminess in New Zealand seem to correlate with periods of warmer temperatures, the only period of storms within the Medieval Warm Period occurred A.D. 865–1015 (1,085–935 cal BP); the remainder of the Medieval Warm Period was warmer yet marked by a comparatively low incidence of storms here (Eden and Page, 1998).

4.2. ENVIRONMENTS DURING THE MEDIEVAL WARM PERIOD IN THE PACIFIC BASIN

It is difficult to discuss environments of the Medieval Warm Period in the Pacific Basin in isolation through time because the available information suggests that they did not vary significantly from those of earlier times. It may simply be that the nature of the environmental changes in various parts of the region, as elsewhere in the world, was too slow or too minor to have registered in palaeoenvironmental archives in which extreme events are most easily recognized.

It is also likely that most of the tools being used to interrogate the past are too crude to be able to reveal environmental changes, although there are manifest differences between the environments of the Medieval Warm Period and those of the Little Ice Age in many parts of the Pacific Basin. Most of these differences occur along coastlines, and are attributable to sea-level fall during the A.D. 1300 Event. The importance of sea-level change as an agent of environmental change during the Medieval Warm Period (and subsequently) is discussed in Section 4.2.1, with two case studies in Sections 4.2.2 and 4.2.3. Section 4.2.4 is devoted to the issue of whether or not sea level reached a maximum during the terminal Medieval Warm Period in the Pacific Basin.

4.2.1 Sea-level changes

In those parts of the Pacific Basin for which there is any information about sea level during the Medieval Warm Period, the general picture seems to be one of sea level rising to a level perhaps above its present level (see Fig. 1.3A), certainly above its level at the start of the Little Ice Age. Evidence for sea-level fall during the A.D. 1300 Event is far more widespread (see Section 5.3) as it is for lower-than-present sea levels during (most of) the Little Ice Age in the Pacific Basin (see Section 6.2). While it is probable that sea-level change was an important driver of societal change across the A.D. 1300 Event, there is no evidence that sea-level change brought about societal change during the Medieval Warm Period in the Pacific Basin.

Some of the beach-ridge sequences found along the lowlands of the Pacific coast of South America indicate a possible higher sea level during the Medieval Warm Period. Two examples from the Bahia de Paita, both now some 500 m inland and significantly above the modern beach ridge, were described by Ortlieb et al. (1995); at Colan, beach ridge 1S formed A.D. 768–1235 (1,182–715 cal year BP) while at Chira, beach ridge Q formed A.D. 1130–1252 (820–698 cal year BP). A reconstruction of Holocene sea-level change from the Santa beach-ridge complex in Peru showed a slowly rising sea level during the Medieval Warm Period reaching a maximum just over 1 m above present sea level at about A.D. 1200 (750 BP: Wells, 1996).

As in many other parts of the Pacific Rim, the record of latest Holocene sea-level changes along the western seaboard of North America is under-researched (Fairbridge, 1992), possibly because it is obscured by tectonic activity. One exception comes from Alaska, where a record of sea-level rise obtained from tidal-marsh sediments in Cook Inlet is considered more likely to have been a precursor of the large earthquake A.D. 1100 (850 cal BP) than a regional eustatic (sea-level) movement (Hamilton and Shennan, 2005). Indeed the human prehistory of the western Alaska Peninsula has been strongly influenced by earthquake-associated (co-seismic) relative sea-level changes that elevated large areas of flat ocean floor creating coastal environments attractive for settlement (Jordan and Maschner, 2000).

On the western side of the Bering Strait, warm conditions during the Medieval Warm Period on Kunashir Island (Kurile group) were accompanied by sea-level rise reaching as much as 1 m above present sea level (Korotky et al., 2000). A similar situation was found on other Kurile Islands and Bering Island in the Komandar group (see Fig. 1.3A), discussed in more detail in Section 4.2.2.

For the western Pacific Rim, on the Korean Peninsula, there is possible evidence for sea-level rise during the Medieval Warm Period (Yi and Saito, 2003). During the preceding Dark Ages Cold Period, lower sea level created a large bay on what is now the Gimhae fluvial plain. The existence of this bay enabled the development of a port that was key to the development of the Golden Crown Gaya State, a powerful substate within the Great Gaya State. Yet around the start of the warmer Medieval Warm Period (before A.D. 560–780) relative sea-level rise drowned this bay, converting it to marsh and a fluvial plain which removed the geographical advantage of the Golden Crown Gaya State and led to its decline. It is possible that this relative sea-level rise was a result of uplift along the Yangsan Fault but there is no independent confirmation of this.

In many parts of Japan, the Heian Transgression spanned the Medieval Warm Period and involved sea-level rise from at or below its present level (during the Kofun cold stage) to a level higher than today around A.D. 1300 (Sakaguchi, 1983). A hint of higher sea level during the Medieval Warm Period was also reported from south Kalimantan, Indonesia (just beyond the Pacific Rim), around A.D. 950 (1,000 BP: Yulianto et al., 2005).

Environmental evidence from the southwest Pacific Rim for sea-level change during the Medieval Warm Period is sparse, although work by Gibb (1986) showed that sea level rose within this period exceeding its present level around A.D. 1300. Gibb targeted estuarine sediments in stable areas of New Zealand finding, for example, that sea level at the mouth of the Weiti River (North Island) averaged 12 cm above modern sea level some time within the period A.D. 1260–1400 (690–550 cal BP).

Over a number of years, investigations of islands in the far northwest Pacific Ocean have resulted in a detailed account of environmental changes associated with sea-level rise during the Medieval Warm Period, as discussed in Section 4.2.2. This is followed by the description of a last-millennium sea-level record from Japan (Section 4.2.3) and finally a review of the evidence for higher sea level during the Medieval Warm Period in the Pacific Islands (Section 4.2.4).

4.2.2 Environmental change driven by sea-level rise in the Kurile and Komandar Island groups, northwest Pacific

Some innovative work on last-millennium sea-level change and environmental change has been carried out in the Kurile and Komandar island groups in the northwest Pacific (Korotky et al., 2000; Razjigaeva et al., 2002, 2004). In the south Kuriles, sea level rose ~2.5 m approximately A.D. 350–1100 while on Bering Island in the Komandar group, there was a 1.6-m sea-level rise perhaps

A.D. 350–950. There is a possibility of tectonic displacement of shorelines in this area but this seems to be a minor contributor to relative sea-level change here given that:

- the Holocene sea-level maximum (~3,900 cal BP here – Razjigaeva et al., 2004) was around +2 m, which is about the same height as it reached elsewhere in the Pacific where the effects of tectonic movement have been subtracted (as in Fiji – Nunn and Peltier, 2001); and
- the timing and magnitude of post-A.D. 350 sea-level change in the southern Kuriles and the Komandar groups, which are likely to have different sets of tectonic influences, are roughly the same.

On Kunashir Island, the period of low sea level that preceded the Medieval Warm Period was marked by the development of sand dunes and widespread grassland and swamp landscapes (Korotky et al., 2000). Warming during the Medieval Warm Period led to the replacement of many grassland areas by broad-leaved tree species (such as oak and elm) and the disappearance of many swampy areas. The shoreline moved inland and upwards by a maximum 2.5 m where it formed a marine terrace as much as 1 m above present sea level.

On nearby Iturup Island, warming led to "a decrease in aeolian processes and soil formation" (Razjigaeva et al., 2002, p. 479). The presence of marine diatoms in a peat bog near the mouth of the Kurilka River shows that sea level rose across such swamps during the Medieval Warm Period accompanied by vegetation changes consistent with warmer conditions.

On Bering Island in the Komandar group, low marine terraces formed during the Medieval Warm Period are found ~2 m above present sea level along the island's Bering Sea coast and 5–6 m at Vhodnoi Mis Cape on its Pacific coast, the differences in elevation being explainable by the higher wave energy at the latter site (Razjigaeva et al., 2004; see also Fig. 1.3A).

Case studies like this one are important because they establish a relationship between sea-level change and environmental change. Many of the examples of environmental change given in later chapters of this book are likely to have been driven by sea-level changes although empirical data to demonstrate the connection are not available.

4.2.3 Lake salinity changes driven by sea-level rise: Nakaumi Lagoon, Honshu Island, Japan

The difficulty of finding palaeosea-level indicators for the last millennium as uncontroversial in their interpretation and as comparable in their precision to those for the earlier Holocene has led to some novel approaches. By studying the relationship between water salinity and the $\delta^{18}O$ and $\delta^{13}C$ values for modern shells in the Nakaumi Lagoon on the Japan Sea coast of Honshu Island, central Japan, it proved possible to derive a palaeosalinity record that spans most of the Holocene (Fig. 4.5).

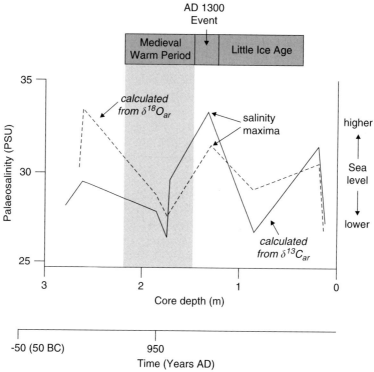

FIGURE 4.5 Palaeosalinity of Nakaumi Lagoon determined from shell, calculated using both $\delta^{18}O_{ar}$ and $\delta^{13}C_{ar}$ (adapted from Sampei et al., 2005). Approximate time periods are shown. High salinity during the Medieval Warm Period is most likely a consequence of high sea level that caused saltwater flooding of the lagoon. The subsequent fall of salinity likely represents sea-level fall and consequent lowering of lagoonal salinity. Imprecision of dating probably explains why the salinity maxima, probably indicative of higher relative sea level, fall within the A.D. 1300 Event rather than the later Medieval Warm Period. Time scale was converted from original years cal BP.

While recognizing that there are other possible influences on long-term salinity changes, it was believed that the salinity maxima represent times of high sea level, when the brackish lagoon was invaded by the sea (Sampei et al., 2005). A good example is provided by a salinity peak ~6,000 cal BP (not shown in Fig. 4.5) which almost certainly is a result of the invasion of the lagoon by the sea during the Holocene Climatic Optimum. The salinity peak shown in Fig. 4.5 about A.D. 1150 (800 BP) may relate to a high stand of sea level near the end of the Medieval Warm Period, for which abundant (yet generally less precise) evidence was reported throughout the Japanese Islands by Sakaguchi (1983). It is possible that other palaeosalinity records spanning the Medieval Warm Period may also reflect higher sea level; a good example comes from the San Francisco Bay–Delta system where higher salinities were reported for the period A.D. 950–1150 (Malamud-Roam et al., 2006).

4.2.4 Evidence for higher sea level from the Pacific Islands during the terminal Medieval Warm Period

In the early 1990s, the author dated an emerged low-level reef near Poloa on Tutuila Island in American Samoa, showing that it formed around A.D. 1305 (645 cal BP) at a relative sea level an average 0.82 m higher than today (Nunn, 1998). The tectonic history of this part of Tutuila Island, as suggested by Dickinson and Green (1998), gives reason to suppose that there may have been recent emergence of this area associated with the structural deformation of the Samoan Island chain, so the Poloa emerged reef may not represent a sea-level highstand during the Medieval Warm Period as first believed. Yet there is considerable evidence of last-millennium sea-level change elsewhere in the Pacific Islands discussed below.

Numerous indicators of higher-than-present sea level during the Medieval Warm Period are found in the central Pacific Island groups of the Tuamotus and the Gambiers. One of the first dates came from Rangiroa Island where sea level was ~60 cm higher some time within the period A.D. 1190–1410 (760–540 cal BP: Pirazzoli and Montaggioni, 1985). Investigations of Tarauru Roa Island in the Gambier group found that sea level here was ~60 cm higher within the period A.D. 960–1240 (990–710 cal BP: Pirazzoli and Montaggioni, 1987). Later work in the Tuamotus found that sea level around Hikueru Island reached some 25 cm above present in the period A.D. 1220–1400 (730–55 cal BP: Pirazzoli et al., 1988).

Located in the Tasman Sea (between Australia and New Zealand) in the southwest Pacific, Lord Howe Island has the distinction of being the southernmost island in the world to have a fringing coral reef (Guilcher, 1973). Studies suggest a history of late Quaternary tectonic stability for Lord Howe Island, making it invaluable as a place for reconstructing Holocene sea levels. Radiocarbon dating of intertidal shells shows that sea level was probably 1 m higher than present perhaps A.D. 1050 (900 BP) although, in the absence of information to the contrary, this was interpreted as a stage in the fall of sea level from its mid-Holocene high level (Woodroffe et al., 1995). Yet, in the light of evidence from elsewhere in the Pacific, it could also be that this represents a transgression during the Medieval Warm Period.

Evidence for a higher sea level during the Medieval Warm Period has also been obtained from Rarotonga Island in the southern Cook Islands group. Rarotonga is a volcanic island, fringed with emerged Pleistocene reef limestone that indicates a history of late Quaternary stability (Spencer et al., 1987). A study of the limestone terraces around Rarotonga led to the identification of emerged microatolls plastered onto the edge of the Pleistocene emerged reef, a situation interpreted as evidence for a short-lived sea-level highstand at the end of the Medieval Warm Period around A.D. 1300 (Moriwaki et al., 2006). Other dates from emerged last-millennium microatolls around Rarotonga show that relative sea level was 1.0–1.3 m higher than present during the Medieval Warm Period perhaps A.D. 850–950 (1,000–1,100 BP, uncalibrated dates from Ian Goodwin, personal communication, 2005).

A comparable study from Niue Island (central South Pacific) suggests that sea level was at least 25–37 cm higher during the Medieval Warm Period than today, but insufficient data are available to confirm this (Nunn, 2003b) (Fig. 4.6).

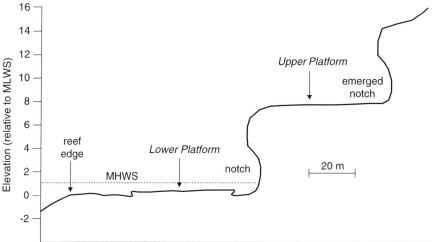

FIGURE 4.6   Photo shows the author and a colleague (Professor Randolph Thaman) sampling the low-level emerged shore platform under adverse conditions at Namukulu on the northwest coast of Niue Island (Central South Pacific). Dating of the platform using cosmogenic Isotopes (courtesy of Dr. John Stone, University of Washington) showed that it was cut within the period A.D. 300–1440 at a time when sea level was at least 25–37 cm higher than today. It may well represent a higher sea level during the later Medieval Warm Period at this location [photo: Patrick Nunn collection]. Cross-section of the site below the photo.

Environmental evidence for a higher sea level towards the end of the Medieval Warm Period also comes from its likely societal effects. One piece of evidence comes from the Palau (Belau) Islands in the northwest Pacific, where by about A.D. 1250 on Babeldaob Island, "the transformation of settlement patterns and subsistence practices ... was made possible by the vastly expanded coastal wetlands" (Masse et al., 2006, p. 111),

probably a consequence of sea-level rise during the terminal Medieval Warm Period.

It cannot be assumed that sea level rose during the Medieval Warm Period by the same rate and to the same relative height in every part of the Pacific Basin. Such an assumption ignores the lesson from studies of recent steric (temperature-driven) sea-level change that rates vary within the Pacific over short time periods. Yet it does seem – from the limited data available – that there was overall sea-level rise during the Medieval Warm Period in most parts of the Pacific Basin. This is what would be expected, given that there is evidence for temperature rise at the same time.

4.3. SOCIETIES DURING THE MEDIEVAL WARM PERIOD IN THE PACIFIC BASIN

As in many parts of the world, the Medieval Warm Period was generally a "time of plenty" for human societies in most parts of the Pacific Basin (Nunn et al., 2007), the conspicuous exception being the eastern Pacific Rim between ~40°N and 30°S where human lifestyles were impacted by prolonged and severe droughts, particularly during the last century or two of this period. But for most of the rest of this vast region, there is evidence that societies broadened their resource base, interacted more than ever before and became more complex during the Medieval Warm Period. Although cultural determinists would not favour the explanation, it seems abundantly clear that the climates of the Medieval Warm Period in the Pacific Basin were generally supportive of all these aspects of human existence. The most parsimonious explanation is that climate change was therefore the main driver of cultural change in the Pacific Basin during the Medieval Warm Period, just as it was a driver of widespread cultural breakdown during the ensuing A.D. 1300 Event.

This section reviews the evolution of human societies in the Pacific Basin during the Medieval Warm Period, looking at the Pacific Rim (Sections 4.3.1–4.3.5) and the Pacific Islands (Section 4.3.6), followed by a section focused on long-distance voyaging across the Pacific Ocean (Section 4.3.7).

4.3.1 Societies of the eastern Pacific Rim

Food-production systems based on wetland agriculture, as practiced throughout the Pacific Basin during the last millennium, are limited to areas with sufficient water supply and are especially vulnerable to drought. Both conditions are well illustrated by Pacific South America. The start of the Medieval Warm Period in the northern Andes was warmer and wetter than earlier times, allowing maize agriculture to spread southwards. But it was the dryness of the Medieval Warm Period that has been regarded as the factor limiting its spread farther south, into the Bolivian Andes, for example (Graf, 1981).

Settlement pattern in Pacific South America was also influenced by climatic amelioration during the Medieval Warm Period. An example comes from the Santa Elena Peninsula in coastal Ecuador (Paulsen, 1976). Having been abandoned by people during the early Medieval Warm Period (approximately A.D. 800–1000), the peninsula was then reoccupied by the Libertad culture in response to warmer conditions. Subsequent abandonment of this area after about A.D. 1400 was due to climate deterioration.

In central America, prolonged periods of intense drought were almost certainly the principal driver of Mayan civilization collapse about A.D. 750 (see Fig. 3.1) but a return to more favourable conditions during the Medieval Warm Period saw its partial (Postclassic) revival. Along the Pacific coast, however, conditions remained dry and the area remained sparsely populated until the later part of the Little Ice Age, the study of Neff et al. (2006) providing

> "a strong empirical case that climate variation dramatically affects human population levels and constitutes an important selective pressure in cultural evolution, even in the relatively favorable, agriculturally rich environment of Pacific coastal Guatemala".
> (p. 398)

It is likely that people occupying those parts of the eastern Pacific Rim adjacent to ocean currents flowing towards the Equator – such as Chile–Peru and California – were dependent, particularly in times of terrestrial food shortages, on the near-shore fisheries. This may have been especially important during the Medieval Warm Period when the productivity of these fisheries was generally much higher than later during the Little Ice Age (Baumgartner et al., 1992).

In many parts of the eastern Pacific Rim, there is evidence that human populations experienced an increasing incidence and duration of drought during the terminal Medieval Warm Period. While there is considerable evidence of impacts on subsistence systems, drought appears to be something to which some societies may have been able to adapt in the short term (see Section 4.3.4).

In the Alaska-Aleutians area of the northeast Pacific Rim, settlements of the Medieval Warm Period were marked by small villages, likely to have been occupied by single families, whose lifestyles were based on marine or riverine subsistence and long-distance exchange: there is clear evidence for increasing societal complexity after about A.D. 900 (Maschner and Reedy-Maschner, 1998). Although poorly dated, there were radical changes in human societies after about A.D. 1200 in this region, some of which such as the dramatic increase in village and house sizes, and the emergence of a class structure, are similar to parts of the Pacific Islands region. In particular, there is a "monumental increase" in the evidence for "violence and war" in this region after about A.D. 1100.

To illustrate some of the key ways in which societies changed along the eastern Pacific Rim during the Medieval Warm Period, three case studies are given in the following sections. The first records the effects of two distinct periods of climate change within the Medieval Warm Period on the inhabitants of the Jequetepeque Valley in lowland Peru (Section 4.3.2). The second explains the effects of drought during the later Medieval Warm Period on one of the most complex societies to have developed at this time in South America: Tiwanaku

(Section 4.3.3). The third details the effects of increasing drought incidence during the Medieval Warm Period on the Chumash people of coastal California (Section 4.3.4).

4.3.2 Climate change and culture, Jequetepeque Valley, coastal Peru

This case study looks at the effects of the Medieval Warm Period on the Moche culture and its successors that existed in the northern part of coastal Peru, particularly in broad fertile valleys fed by meltwater rivers such as the Moche, Lambayeque and Jequetepeque (Dillehay and Kolata, 2004). The Moche culture describes a number of competing states that existed A.D. 100–750. A hierarchical organization defines Moche culture, which supported this through agricultural techniques that produced grain surpluses and cotton for textiles. The proximity of the Moche to the offshore anchoveta fishery also played a major role in the success of Moche culture.

As a result of the earlier periods of drought, the Moche occupants of the Jequetepeque Valley relocated by A.D. 600 from down-valley locations, which had shown themselves vulnerable to sand-dune encroachment at such times, to valley heads. But these locations proved vulnerable to the meltwater floods associated with El Niño events and a succession of these around A.D. 700 led to many valley-head settlements being abandoned by the Moche. These El Niño events also shut down the anchoveta fishery which, combined with the effects of upvalley flooding, led to the disappearance of Moche culture (Fagan, 1999). The next 200 years marked the rise of the Chimú culture, which dominated a large part of this area A.D. 1200–1470.

In the Jequetepeque Valley, a site of both Moche and Chimú settlements, the development of sand dunes has been used a proxy for the timing of droughts (Dillehay and Kolata, 2004). The decline of the Moche culture in this area was associated with droughts A.D. 524–540, A.D. 563–594 and A.D. 636–645 (Shimada et al., 1991) but the rise of the Chimú during the Medieval Warm Period (post-A.D. 800) was marked by an absence of sand-dune encroachment on the agriculturally productive lowlands of the Jequetepeque Valley. It is likely that this signals a period of comparatively unvarying climate (marked by low El Niño frequency) that nurtured Chimú culture and allowed its spread across a large area.

A major cause of stress for the Chimú occupants of the Jequetepeque Valley was the prolonged drought A.D. 1245–1310, registered throughout this region at the Quelccaya ice cap (Thompson et al., 1986). But interestingly, this did not herald the end of the Chimú ascendancy in the Jequetepeque Valley for these people proved more adaptable than their Moche predecessors. Among the adaptations that have been noted for the Chimú of the Jequetepeque Valley was the development of "anticipatory agricultural infrastructure" (Dillehay and Kolata, 2004, p. 4328), comprising a network of canals, overflow weirs and aqueducts intended to regulate and optimize the flow of irrigation water to different parts of the valley. There is evidence for repeated rebuilding of these structures, likely to have been necessary after the impact of floods during El Niño events.

4.3.3 Collapse of the complex Tiwanaku society, Andean South America

Along much of the Pacific Rim in South America, the land rises sharply in the upper parts of lowland valleys like the Jequetepeque until it reaches a high plateau, the *altiplano*, the societies of which experienced climate stresses during the last millennium that were often different to their lowland counterparts. There is no reason to doubt that

"protracted droughts occasioned both significant social disruptions and innovative cultural, demographic, and technical adaptations among indigenous societies in the Andes". (Dillehay and Kolata, 2004, p. 4325)

There is considerable evidence that the Pacific Rim in South America became increasingly arid towards the end of the Medieval Warm Period, a severe drought beginning about A.D. 1150 and a corresponding fall in the level of Lake Titicaca heralding the collapse of the Tiwanaku civilization that occupied its fringes (Binford et al., 1997; deMenocal, 2001). A parallel drop in precipitation registered A.D. 1245–1310 at the Quelccaya ice cap, 200 km away, shows the prolonged drier conditions marking the terminal Medieval Warm Period (Thompson et al., 1986) which saw the final collapse of Tiwanaku society.

The Tiwanaku culture had adapted to high-altitude environments around Lake Titicaca by developing a system of water-dependent raised-field agriculture which at its apogee sustained an urban population of nearly 500,000 people. Drought and associated lake-level fall is likely to have abruptly reduced food production, leading directly to the area's abandonment. The abrupt abandonment of the urban centres and raised-field systems A.D. 1150–1200 was followed by the disappearance of the Tiwanaku civilization within a few hundred years, as formerly urban dwellers became more mobile and adopted a lifestyle based on pastoralism and dry farming (Kolata and Ortloff, 1996).

Societies of comparable complexity in central America collapsed during the Medieval Warm Period. In the central highlands of Mexico, the abandonment of the cities of Cholula and Teotihuacán occurred early in the Medieval Warm Period, probably in part as a result of the onset of drier conditions than those to which their inhabitants had become accustomed (Curtis et al., 1996). Elsewhere in central America, the Medieval Warm Period saw the rise of the Toltec Empire that finally collapsed about A.D. 1200 during a period of severe drought and cold (Gill, 2000).

4.3.4 Drought impacts on the Chumash people of coastal Southern California

High sea-surface temperatures during the later part of the Medieval Warm Period off the California coast may have driven important cultural changes among the Chumash people living there and on the islands offshore. After this idea was initially proposed (Arnold, 1992), little interest was shown in it by archaeologists who habitually attributed cultural change in this region "to intrasocietal forces, particularly social and economic innovations that exploited the rich natural environment in progressively more productive ways". (Raab and Larson, 1997, p. 319)

Yet now the environmental-determinist interpretation is in the ascendant here although, rather than elevated sea-surface temperatures, it now seems more likely that changes in Chumash subsistence manifested more far-reaching cultural changes among Chumash peoples and had their origins in regional climate changes, particularly the increased incidence of drought during the later part of the Medieval Warm Period (Raab and Larson, 1997). A similar study shows that before the later part of the Medieval Warm Period, people on the California mainland had their settlements near both perennial and ephemeral streams, suggesting that water was comparatively abundant, while settlements moved close to perennial streams by the end of this period, implying increasing water shortage (True, 1990).

Various dendrochronological studies from upland areas of southern California have confirmed this general picture independently. The warmest wettest time of the Medieval Warm Period was A.D. 800–1000 but during the later part (A.D. 1100–1250), the climate became very dry; the driest time was A.D. 1120–1150 (Larson et al., 1996). From the study of lake levels in the area, there is evidence for an "epic drought" A.D. 1209–1350 (Stine, 1994). While Chumash society appears to have largely weathered these droughts, it fell apart subsequently during the A.D. 1300 Event (see Section 7.6.2).

4.3.5 Societies of the western Pacific Rim

It is interesting that the evidence for societal change in response to climate (and associated) changes during the last few millennia is so conspicuous along the eastern Pacific Rim but not along the western Pacific Rim (Haberle and David, 2003). There are a few examples from the latter, one being the collapse of the Longshan culture around 2050 B.C. (4,000 BP) linked to the onset of cool wet conditions in the Huanghe (Yellow River) valley (Liu et al., 2000). The answer may lie in lower population densities and more subsistence options for the people along the western rim compared to those along the eastern rim, which was also generally drier than the west during the last millennium.

Evidence of the beneficial effects of climate constancy for human societies during the Medieval Warm Period comes from Japan, where the Heian Period (A.D. 794–1192) was an unequalled time of prosperity and peace. A similar situation obtained for much of the early and middle Medieval Warm Period on adjacent parts of the Asian mainland although here, as in Japan, the terminal Medieval Warm Period was marked by societal breakdown largely driven by the Mongol invasion in the mid-thirteenth century, itself a probable response to climate-driven food stress (Hsu, 2000).

In highland New Guinea, on the equatorial western Pacific Rim, studies of the uncommonly lengthy agricultural history of the Kuk area show adaptations to climate change throughout the Holocene (Golson, 1982). For most of the Medieval Warm Period, when the Kuk area experienced low climate variability, there was no swamp agriculture, presumably because sufficient foods could be obtained from the catchment. But then, as severe and prolonged droughts began to affect the area – synchronously with those along the equatorial eastern Pacific Rim – the Kuk agriculturalists began to use the swamp for agriculture, establishing (or reopening) a grid of field ditches to optimize groundwater use (Haberle and Chepstow-Lusty, 2000).

Little information is available for much of the last millennium from eastern Australia because of both a dearth of relevant palaeoenvironmental studies and because the inhabitants of this landmass were mostly nomadic, and had consequently very little environmental impact detectable beneath the overprint of more recent environmental change here. New Zealand remained unoccupied by humans for most of the Medieval Warm Period, the first Maori arriving there A.D. 1250–1300 (Hogg et al., 2003).

4.3.6 Societies of the Pacific Islands

In the Pacific Islands, the Medieval Warm Period was characterized by stable climate conditions and at least in some places a slightly higher sea level. Combined with the comparatively warm conditions, this stability was associated with the entrenchment and flourishing of human cultures and cultural exchange across this vast region.

The constancy of the climate compared to earlier (and later) times facilitated human efforts to increase food production through agriculture. While it is difficult to find any data that refer directly (and uncontroversially) to agricultural production on Pacific Islands during the Medieval Warm Period, the establishment of often-complex irrigation systems and various landscape modifications evidently intended to boost agricultural productivity were widespread. Examples include the irrigation and water-supply infrastructure on Aneityum Island in southern Vanuatu (Spriggs, 1986), terracing in Easter Island, Palau and the Hawaiian Islands (Heyerdahl and Ferndon, 1961; Earle, 1980; Kirch, 1986; Masse et al., 2006), and the construction of food gardens sunk into atoll water tables (Weisler, 1999).

Throughout the tropical Pacific, the (later) Medieval Warm Period appears to have been a time of increased dependence on artificial terraces for growing root crops, commonly taro (*Colocasia esculenta*), with dated examples known from Easter Island and Palau and elsewhere at this time (Earle, 1980; Lucking, 1984). Dates from O'ahu Island in Hawaii show that the later Medieval Warm Period was a time when agricultural terracing spread; for example, it marks the first use of the upper Maunawili valley A.D. 1200–1400 while at Kane'ohe the Luluku terraced pondfields became more extensive (Cordy, 1996).

The warmer temperatures of the Medieval Warm Period would also have increased the range and availability of wild foods, particularly marine foods, on which most people in the Pacific Islands region depended. In warmer waters, higher temperatures and higher sea level are likely to have enhanced lagoonal water circulation and aeration, thereby boosting the productivity of a range of marine ecosystems exploited by humans in this region. The low incidence of storms would have placed no related hurdles in the way of optimal coral-reef growth (as it does for modern reefs – Dollar and Tribble, 1993). In cooler parts of the Pacific Ocean, the influx of warmer waters would likewise have boosted marine productivity, extending subsistence options for many people living in transitional locations, such as those on the islands of the Japan Sea who benefited from the influence of the warm Tsushima Current during the Medieval Warm Period (Kuzmin et al., 2004).

Estimates of population size during much of the last millennium for the Pacific Islands have been dogged by the problem of circular argument. Some authors, conveniently assuming that climatic (and other non-human) drivers of societal change were negligible, have argued that increased net occupation areas and evidence for apparent increased resource extraction during the Medieval Warm Period equate simply with increasing population (Kirch, 1984a, 1990; Dye and Komori, 1992). Such explanations are fundamentally flawed because there is no independent evidence available in support of them in contrast to explanations involving climate-driven changes, for which there is ample independent evidence (Nunn, 2003b). Approaches to reconstructing former population size and density that involve landscape archaeology (such as on Kosrae – Athens, 1995) or numbers of active population centres (as in the Marquesas – Rolett, 1989) may be persuasive but are nonetheless open to alternative, independently verifiable, explanations.

On some Pacific Islands where the chronology of settlement over the past two millennia or so is comparatively well known, the Medieval Warm Period is marked by a puzzling low visibility of settlement. Good examples are the islands Lakeba and Naigani in Fiji (see Figs. 6.9 and 7.7 respectively) which may have been largely abandoned during the later Medieval Warm Period owing to water shortages brought about by their vulnerability to drought. Other possible examples include islands in Chuuk in Micronesia (Rainbird, 2004), Tonga (Burley, 1998) and Palau (Masse et al., 2006). It is possible that this can be explained – on a region-wide basis – by the existence of mostly smaller nucleated settlements, archaeologically less visible than the larger, more complex settlements that commonly succeeded them (see Fig. 7.3).

Most settlements in the Hawaiian Islands, colonized perhaps around A.D. 650, were coastal. These settlements were "nucleated hamlets" (Kirch, 2000a, p. 293) that gave way to larger settlements later in the Medieval Warm Period when more hierarchical (stratified) societies appeared; "between about A.D. 1000 and A.D. 1300, site numbers increase dramatically" (Kirch, 1990, p. 324). A similar story has emerged in the Kodiak Islands of the northern Pacific where the smaller scattered settlements that characterized Late Kachemak times aggregated into larger settlements averaging 2.5 times larger around A.D. 1200, a change explainable by the need for cooperation in whale hunting (Fitzhugh, 2002).

Cooperation may also explain the replacement of large numbers of small settlements early during the Medieval Warm Period in the tropical Pacific Islands by fewer numbers of larger settlements during its later part (see Fig. 7.3). It has been proposed that this change was driven by increasing aridity during the Medieval Warm Period and the progressive need to conserve available water, particularly for agriculture (Nunn and Britton, 2001). In the landscape, this need was manifested by the construction of artificial terraces intended to optimize the use of available water. Such terraces could not be constructed or maintained by a small group of people, so cooperation was required, which led eventually to the amalgamation of smaller settlements into larger units. The aggregation of large settlements and the development of cooperative enterprise led to a need for greater organization and leadership, which in turn led to social stratification, something that has been widely noted as characterizing the later Medieval Warm Period in the Pacific Islands (Kirch, 1984a, 1997b; Ladefoged, 1995; Burley, 1998).

While most people inhabited coastal settlements on Pacific Islands during the Medieval Warm Period, there were settlements slightly inland on some islands. These were probably to service nearby agricultural production and were dependent, either through trade or because they were the same community, on coastal settlements nearby. A well-studied example comes from the Sigatoka Valley on Viti Levu Island in Fiji (Kumar et al., 2006; see Fig. 6.4). But on a few Pacific islands, coasts were not favoured for settlement during the Medieval Warm Period, typically because coral reefs were absent and because flat areas adjoining the sea were small in area and vulnerable to large-wave impact (unbroken by offshore reefs). Two examples are Easter Island and the islands of the Marquesas (French Polynesia).

The history of Easter Island during the Medieval Warm Period is covered by the Ahu Moai Period (A.D. 1100–1500), the most distinctive expression of which was the carving, transport from inland quarries to the coast and erection of stone statues (*moai*) on temple platforms (*ahu*). Chiefly dwellings adjoined the *ahu*, inland from which on open, windswept plains, were settlements surrounded by fields of sweet potato (Bahn and Flenley, 1992).

On Nuku Hiva Island in the Marquesas group of French Polynesia, the earliest settlements, established a few hundred years before the start of the Medieval Warm Period, were small, nucleated and coastal (Kirch, 1984a), a pattern that probably continued into the earlier part of the Medieval Warm Period. But coastal flats are narrow and discontinuous on most Marquesan islands and during the Medieval Warm Period, favourable conditions led to larger settlements being established on the floors of the "verdant valleys", a time that saw the rise of field cropping (Kirch, 1984a). Examples come from Nuku Hiva (Suggs, 1961) and the Hanamiai Valley on Tahuata Island which was first settled about A.D. 1025 (Rolett, 1998).

4.3.7 Long-distance voyaging across the Pacific Ocean

It was first suggested by Bridgman (1983) that climate change may have facilitated long-distance Pacific Islander ocean voyaging during the Medieval Warm Period, an idea that has been supported by many writers since then (Finney, 1985; Irwin, 1992; Nunn, 2000a). Bridgman (1983) argued that the

"persistent trade winds, clear skies, limited storminess, and consistent Walker Circulation [low El Niño incidence]" (p. 193)

provided the optimal conditions for successful long-distance ocean voyaging in the Pacific at this time. This was in contrast, Bridgman averred, to the Little Ice Age that was characterized by increased variability in tradewinds, an erratic

Walker Circulation, increased storminess and perhaps increased dust from volcanic eruptions which obscured long-distance vision and the star-studded skies essential for navigation. While Bridgman's conclusions about conditions during the Medieval Warm Period may appear somewhat generalized today, it is clear from studies of last-millennium palaeoclimates that he was correct to highlight the contrast between the constancy of conditions during the Medieval Warm Period and the uncertainty and variability of those during the Little Ice Age.

A general impression of the minimum numbers of interisland voyages involved in the colonization of the Pacific Islands and the distances covered can be gleaned from Fig. 2.1. Most of these islands were reached not only once but also, as shown by the record of material culture, again and again, often apparently as a result of return voyaging to and from an island "homeland" (see Fig. 6.7).

Good examples come from the so-called mystery islands of eastern Polynesia (see also Section 7.7.6), particularly the Pitcairn Group. Within this group (comprising Pitcairn, Ducie, Oeno and Henderson), there is ample material evidence of routine interisland voyaging involving one-way distances of 400–600 km during the Medieval Warm Period that ceased shortly after its end (Weisler, 1995, 2002). There is also evidence for return voyaging between Pitcairn and the likely homeland of its first settlers on Mangareva (Gambier Islands) that involved unbroken ocean crossings of ~1,000 km (Weisler, 1996a).

New Zealand was first settled about A.D. 1250–1300 (Hogg et al., 2003). In addition to the oral traditions, there is material evidence for return voyaging between New Zealand and east Polynesia (Anderson and McFadgen, 1990), voyages that would have involved open ocean crossings of ~3,000 km.

Evidence that Pacific Islanders reached the eastern Pacific Rim before European discovery of the Pacific comes from two principal lines of evidence. The first involves the record of coconuts – a genus native to the southwest Pacific – in Panama when Europeans reached the area in A.D. 1513 (Ward and Brookfield, 1992). While coconuts probably colonized most Pacific Island groups independently of people, they could only have reached Panama, which involves an ocean crossing against the persistent Peru (Humboldt) Current, had they been carried there by people. The second line of evidence suggesting that Pacific Island people traversed the entire Pacific Ocean from west to east before its European discovery comes from the presence among people along the Pacific coast of South America of war clubs and blades similar in style and name to those of New Zealand Maori, implying that people from southern Polynesia reached southern South America in pre-European times (Anderson, 2003).

As a final word in this chapter, it is conceivable that the increasing aridity that characterized the later part of the Medieval Warm Period stimulated the long-distance voyages of settlement at this time (see Fig. 6.7). These voyages include:

- the colonization of New Zealand A.D. 1250–1300 (Hogg et al., 2003);
- the numerous back migrations of people from Polynesia into the island groups of the western tropical Pacific (mostly in Melanesia) that resulted in both the colonization of many modern Polynesian outliers (Kirch, 1984b) and the arrival

of migrants from the east into already-occupied islands, particularly in Vanuatu and Solomon Islands (Spriggs, 1997; Kirch, 2000a);

- a possibly large influx of Tahitians to Hawaii about A.D. 1300 who brought with them a new religion that sowed the seeds for profound societal change in Hawaii (Spriggs, 1988); and
- possibly the colonization of Panama from Easter Island or Pitcairn Island and/ or the colonization of the Pacific coast of southern South America from southern Polynesia.

The possibility that drought stimulated the longest ocean voyages in the Pacific before its European discovery is a radical suggestion, not published before to the author's knowledge (although it has been informally suggested by Geoffrey Hope, Australian National University). In its favour, it has not only the solid evidence of the longest distance voyages for settlement during the latest part of the Medieval Warm Period but also the absence of any other cogent explanation for an upsurge in the number of people searching for new lands at this time. It has been suggested that the Pacific Islanders of the Medieval Warm Period were inveterate explorers or that younger sons were commonly thus because they could not inherit land but both explanations still fail to explain why so many deliberate voyages of settlement were undertaken in probably such a short period of time. Some Maori legends explain the settlement of New Zealand by the fact that their island homeland (Hawaiki) in tropical Polynesia was overpopulated (Sorrenson, 1979), something that can plausibly be interpreted in a relative sense to mean that insufficient food under the increasingly arid conditions of the later Medieval Warm Period was available for population levels to be sustained.

It is also relevant to note that Pacific Islanders' responses to climate extremes – typically drought or tropical-cyclone impact – have been similar in more recent times. For example, famines in the Marquesas led to some 800 canoes leaving in search of reputed islands of plenty, only one canoe being heard from again (Handy, 1930). These may be euphemisms for the suicide voyages, undertaken at times of seemingly unending food shortages, by the people of remote islands like Tikopia in Solomon Islands (Firth, 1959).

KEY POINTS

1. The climate of the Medieval Warm Period in the Pacific Basin was marked by warm dry conditions exhibiting a low degree of interannual variability. Storminess appears to have been less in most parts of the Pacific Basin. Prolonged and severe droughts affected much of the eastern Pacific Rim during the latest part of the Medieval Warm Period.
2. Available data suggest that sea level rose in many parts of the Pacific Basin during the Medieval Warm Period, reaching a maximum at its end that exceeded present sea level.
3. Most Pacific Basin societies enjoyed times of plenty during the Medieval Warm Period to which the comparatively constant climate contributed. Many societies also show adaptation to increasing warm and dry conditions. Food crises arising from droughts affected parts of the eastern Pacific Rim.

CHAPTER 5

The A.D. 1300 Event (A.D. 1250–1350) in the Pacific Basin

The A.D. 1300 Event (the transition between the Medieval Warm Period and the Little Ice Age) is worthy of separate and in-depth attention because it appears to have been the most rapid period of climate change within the last millennium, certainly prior to 1960 (see Section 1.5). Data from the GISP2 ice-core record (Greenland) show that the transition from the Medieval Warm Period to the Little Ice Age was the "most dramatic change in atmospheric circulation and surface temperature conditions in the last 4000 years" (Kreutz et al., 1997, p. 1294). More recently, Meyerson et al. (2003) concluded that the A.D. 1300 Event (the Modern Millennial Event) "is the most dramatic climate change event of the last 5000 years" (p. 2) recorded in ice cores from both north and south polar regions.

In most of the Pacific Basin, the A.D. 1300 Event involved – uncontroversially – a drop of temperature and – less so – sea-level fall. In some parts of the Pacific Basin, there is evidence for a temporary increase in storminess at this time. Climate and environmental changes during the A.D. 1300 Event brought about enduring effects on human societies in the Pacific Basin adapted to the relatively unchanging climate of the Medieval Warm Period. This chapter does not claim more than is unequivocally supported by data although speculations based on this foundation (Chapter 7) are needed to make sense of the observations in a region of the world which, being largely ocean, has yielded comparatively few data with which to formulate and test such ideas.

In the Pacific Basin, the key link – that which involves rapid climate and rapid environmental change forcing abrupt societal change – can never be proven, and is likely to remain unpalatable to many social scientists however much inferential evidence is amassed to support it. Yet it is becoming increasing clear that

"... the categorical rejection of environment as a potential cause of cultural change will lead to unsuccessful if not naïve characterizations of prehistoric human behavior". (Jones et al., 1999, p. 137)

Indeed such a link is no more than has been claimed for innumerable other situations, ranging from the effects of drought ending the Uruk and Classic Maya civilizations (Rothman, 2001; Haug et al., 2003) to the effects of sea-level fluctuations on the rise and fall of complex societies in lowland East Asia (Zhang et al., 2005) and the human colonization of the remotest Pacific Islands (Dickinson, 2003), and does not strike the author as especially radical. Compared to alternative ideas for fourteenth-century societal changes in the Pacific Basin, such as the implausibly synchronous attainment of carrying capacity by populations on islands of different sizes in different parts of the Pacific, this idea has at least the merit of being supported by independently derived corroborative data rather than being conjectural and entirely self-supporting.

More controversial are the suggestions made in this chapter (and Chapter 7) that (a) relative sea level throughout the Pacific fell during the A.D. 1300 Event and (b) this sea-level fall precipitated the societal crisis that affected many parts of the region subsequently. There are ample data to show that a fall in relative sea level marked the A.D. 1300 Event in various parts of the Pacific Basin. Whether or not this sea-level fall was a Pacific-wide phenomenon, as the author has proposed (Nunn, 2000a, 2000b, 2007a), cannot be unequivocally demonstrated and may be judged improbable, given our knowledge of the spatially variable pattern of recent sea-level change (Casenave and Nerem, 2004). Yet relative sea-level fall – localized or not – is likely to have been a significant factor in the demonstrable resource depletion along many Pacific coasts which makes it reasonable to argue that sea-level fall was an important cause of societal change at this time in the region.

The view of sea-level change as affecting nearshore food supplies to the detriment or benefit of marine foragers is less widely documented and accepted than the role of climate change in this. Yet there are innumerable case studies showing how sea-level change profoundly affected cultural evolution (see Section 3.4), some of which for the last millennium in the Pacific are discussed later in this chapter.

This chapter is divided into four major parts. In Section 5.1, the nature of the A.D. 1300 Event in the Pacific Basin is described, specifically the magnitude of the climate changes involved, their synchronicity and their abruptness – a key factor in the suggested environmental and societal impacts. Section 5.2 deals with the nature of climate change across the A.D. 1300 Event, particularly the widespread evidence for cooling, changing precipitation (including increased storminess in many parts of the region) and the changes that took place in interannual climate phenomena such as El Niño-Southern Oscillation (ENSO). Section 5.3 deals with the environmental changes that occurred in the Pacific during the A.D. 1300 Event, particularly those most plausibly linked to sea-level fall. Societal changes during the A.D. 1300 Event are discussed in Section 5.4, with emphasis on the changes that took place in the more environmentally sensitive Pacific Island (rather than Pacific Rim) societies.

In some cases, climate forcing during the A.D. 1300 Event often produced no or little discernible environmental or societal response within this period, plausibly because of lags in the various systems involved. These lags mean that responses to climate forcing during the A.D. 1300 Event sometimes occurred during the (early) part of the Little Ice Age which followed; they are therefore described in Chapter 6.

5.1. THE NATURE OF THE A.D. 1300 EVENT

The principal criterion used to distinguish the A.D. 1300 Event from the periods before and after is temperature fall (cooling). In many parts of the world, including the Pacific Basin, the end of the Medieval Warm Period was one of the warmest times of the last millennium – in some places the warmest time – in contrast to the early part of the Little Ice Age that was one of the coolest times.

Data from Greenland ice cores show a temperature fall of ~2.25°C occurred between the warmest time of the Medieval Warm Period and the coolest time of the Little Ice Age (see Fig. 1.5B). The identification of the Medieval Warm Period in Greenland ice cores is paralleled by its recognition in thermal reconstruction from a borehole here, which shows a fall of 2°C during the transition between the Medieval Warm Period and the Little Ice Age (Steig et al., 1998). This record is particularly important, because borehole thermometry is one of the only two methods that can yield past temperatures accurate to within 0.5°C, the other being mountain snowline elevation (Broecker, 2001). Accordingly, considerable weight should also be placed on the conclusion of Huang et al. (2000) who, following an analysis of 6,000 continental thermal records, concluded that between 500 and 1,000 years ago (approximately during the Medieval Warm Period), temperatures were higher than today but that they fell to a minimum 0.2–0.7°C below present levels by ~200 years ago (near the end of the Little Ice Age). Data from the Ural superdeep borehole suggest that temperatures fell as much as 7°C across this transition (see Fig. 1.5A).

For the Pacific Basin, recent work suggests that the A.D. 1300 Event in the tropical Andes was marked by a fall in temperature of 3.2 ± 1.4°C (Polissar et al., 2006). Data from China show that temperatures here fell across the A.D. 1300 Event by as much as 3.2°C (see Fig. 5.3A). Evidence assembled from New Zealand suggests that temperatures fell 1.4°C across the A.D. 1300 Event (between about A.D. 1270 and A.D. 1350 – Grant, 1994), a comparable amount to that estimated from stalagmite palaeotemperatures here (see Fig. 5.8).

Ice-core records apparently show the A.D. 1300 Event began slightly earlier at Siple Dome (Antarctica) than it did in central Greenland, but the disparity (28 years) is within the combined dating error and it is probable that the change was synchronous (Kreutz et al., 1997). Polar synchronicity in the timing of the A.D. 1300 Event is matched by synchronicity from mid-latitudes in the Pacific (Sakaguchi, 1983; Schimmelmann et al., 2003). Implicit in this, but shown independently by Haberle and Ledru (2001), is a synchronicity between the east and west Pacific Rim for the last millennium.

One of the key factors thought to have been responsible for driving enduring societal and – to a lesser extent – environmental change in the Pacific Basin during the fourteenth and fifteenth centuries is the abruptness of the climate changes during the A.D. 1300 Event. Whether dealing with natural or human systems, it is clear that an abrupt event has greater potential to upset evolutionary trajectories, whether these are environmental (geomorphological) or cultural, than one that is less rapid and to which systems can adjust without significant disruption.

There is considerable evidence from the Pacific Basin for the abruptness of the A.D. 1300 Event. "Rapid" cooling marks the A.D. 1300 Event (A.D. 1270–1340) in the southernmost Andes (Villalba, 1990) while the ice-core record from Quelccaya farther north shows that the A.D. 1300 Event – the "onset of the Little Ice Age" – was "abrupt" in tropical South America, lasting only a "few decades" here and marked by an "abrupt" increase in precipitation (Thompson and Mosley-Thompson, 1987). In China, temperatures decreased "rapidly" after the A.D. 1310s at a rate of 0.1°C per century (Ge et al., 2003); similar "rapid" cooling has been identified from $\delta^{18}O$ in speleothems here (Paulsen et al., 2003). In New Zealand, after the end of the Medieval Warm Period, temperatures fell "rapidly" (Williams et al., 2004). The palaeotemperature analysis of a New Zealand stalagmite showed a total fall across the A.D. 1300 Event of ~1.5°C within 60–70 years (see Fig. 5.8). Rapid, often short-lived, glacier advance along the entire Pacific Rim also marks this period (see Fig. 5.1).

In studies from beyond the Pacific Basin, many authors have concluded that the transition between the Medieval Warm Period and the Little Ice Age was rapid or abrupt. For example, diatom assemblages from Moon Lake, North Dakota, were analysed to give a record of palaeoprecipitation which showed that the transition between the Medieval Warm Period and the Little Ice Age (around A.D. 1300) here was abrupt (Laird, 1996). Palaeotemperature data from oxygen-isotope analyses of human tooth enamel in Greenland led Fricke et al. (1995) to conclude that cooling was rapid here A.D. 1400–1700. In the GISP2 ice-core record from Greenland, the transition is "abrupt" (Kreutz et al., 1997), as it appears in other Greenland cores (Mayewski et al., 1993). Polar ice-core data show that the A.D. 1300 Event was "the most dramatic climate change event of the last 5000 years" (Meyerson et al., 2003).

5.2. CLIMATES DURING THE A.D. 1300 EVENT IN THE PACIFIC BASIN

The A.D. 1300 Event was characterized by cooling, as described in the preceding section, together with possibly increased precipitation in parts of the Pacific Basin. It seems likely that the latter increase represented an increase in the numbers of storms affecting particular locations, partly a result of an expansion and possibly a latitudinal shift in storm belts resulting from increasing equator–pole temperature gradients.

In this section there is a description of climate changes during the A.D. 1300 Event for the east Pacific Rim (Section 5.2.1) and the west Pacific Rim (Section 5.2.2). The evidence for abrupt changes in the Asian summer monsoon during the A.D. 1300 Event is described in Section 5.2.3. Climate changes during the A.D. 1300 Event in the Pacific Islands and Ocean are described in Section 5.2.4. In Section 5.2.5, the evidence for increased storminess in the Pacific Basin during the A.D. 1300 Event is described, illustrated by a case study from Tahiti (Section 5.2.6).

5.2.1 Climates of the eastern Pacific Rim (including Antarctica)

It appears that parts of Antarctica experienced climate changes during the past millennium that opposed those elsewhere in the world (Mosley-Thompson, 1992). One record from Taylor Dome was deconvoluted to show that the A.D. 1300 Event at this location involved a temperature rise of 3°C (Clow quoted in Broecker, 2001) but evidence for cooling at this time elsewhere in Antarctica and

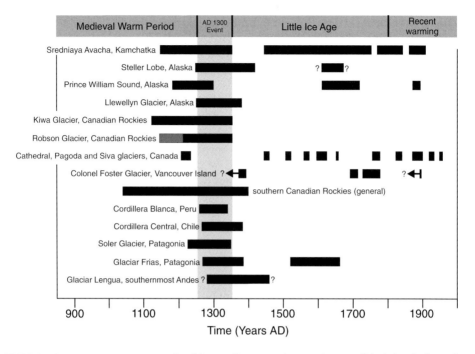

FIGURE 5.1 Representative records of last-millennium glacier advances (black bars) along the Pacific Rim. Data are shown for Glaciar Lengua (southernmost Andes – Koch and Kilian, 2005), Glaciar Frias (central Patagonia – Villalba et al., 1990), Soler Glacier (northern Patagonia – Glasser et al., 2002), representative chronologies from the Cordillera Blanca, Peru, and Cordillera Central, Chile (Grove and Switsur, 1994), Kiwa and Robson glaciers (southern Canadian Rockies – Luckman, 1994), Cathedral, Pagoda and Siva Glaciers (British Columbia Coast Mountains – Larocque and Smith, 2003), Vancouver Island (Colonel Foster Glacier, Strathcona Provincial Park – Lewis and Smith, 2004), Alaska (various glaciers in west Prince William Sound – Wiles et al., 1999a; Calkin et al., 2001), Steller Lobe (Bering Glacier, Alaska – Wiles et al., 1999b), Llewellyn Glacier (Juneau Ice Field – Grove and Switsur, 1994) and Sredniaya Avacha (Kamchatka, eastern Russia – Savoskul, 1999). Note that temperature fall appears to be the main climate driver of glacier advance although short-term fluctuations in precipitation and a range of localized factors are also important to glacier mass balance. The broad synchrony of periods of glacier advance and recession shown here is echoed by other syntheses (Grove and Switsur, 1994; Larocque and Smith, 2003).

adjoining parts of the Pacific Basin is well documented (Mosley-Thompson, 1992; Comiso, 2000). The more representative trend towards cooler conditions along the Pacific Rim of Antarctica during the A.D. 1300 Event is exemplified by palaeoclimate records from the South Pole and Law Dome (Morgan, 1985; Mosley-Thompson, 1992).

Tree-ring data from northern Patagonia show the onset of cool wet conditions A.D. 1270–1340 (Villalba, 1990). In South America, glaciers along the western side (the Pacific Rim) of the Andes appear to have been influenced mostly by precipitation in contrast to the eastern side where temperature is more important

(Koch and Kilian, 2005). With this in mind, a region-wide advance of Patagonian glaciers along the Pacific Rim approximately A.D. 1270–1460 can be linked to falling temperatures and increased precipitation driven by an increased intensity of westerly winds. A selection of records of last-millennium glacier advance from the Pacific Basin is shown in Fig. 5.1.

Andean ice-core records show the A.D. 1300 Event as a period of cooling (see Fig. 4.1). For central America, a record of cooling and increased precipitation A.D. 1380–1522 around Lake Pátzcuaro in central Mexico marks the A.D. 1300 Event in this area and appears from other proxy records to have been a regional phenomenon (Metcalfe, 1987; O'Hara, 1993). The proxy record of precipitation from titanium inputs into the Cariaco Basin shows rapid change from dry to wet to dry conditions across the A.D. 1300 Event (see Fig. 3.1).

From the western seaboard of North America come a variety of proxy indicators of comparatively rapid climate change around the end of the Medieval Warm Period. These include glacier advances across forests in the Canadian Rockies, rising lake levels and decreases in forest fires (Luckman et al., 1997; Hallett et al., 2003). All these indicators are consistent with the start of cooler wetter conditions that led to vegetation changes in the more sensitive environments where human influences were absent or minimal. Reconstructions of temperature (separate from precipitation) are less clear, showing cooling towards the end of the Medieval Warm Period but warming during the A.D. 1300 Event and into the early part of the Little Ice Age (see Fig. 4.2). It is possible that such records incorporate significant time lags in the responses of the proxies used to climate forcing.

Evidence for increasingly wet conditions along the northeast Pacific Rim during the A.D. 1300 Event is widespread (see Fig. 4.3). Yet this trend was punctuated by droughts in places. For example, tree-ring data from the Sierra Nevada show a comparatively dry interval A.D. 1315–1364 (Graumlich, 1993). A more representative regional record may be that from Owens Lake (California) where records of salinity and acid-leachable Li and Mg are proxies for precipitation (Fig. 5.2). The A.D. 1300 Event in this area shows up as a period of consistently higher lake level.

As noted above, wetter conditions developed around the end of the Medieval Warm Period in the Canadian Rockies. An example from the area of Dog Lake, British Columbia, is the change about A.D. 1300 from dry open forests to wet closed forests, a condition that persisted until the end of the Little Ice Age about A.D. 1800 (Hallett and Walker, 2000).

A number of dates from glaciers in the Canadian Rockies and Alaska suggest they advanced at the end of the Medieval Warm Period (see Fig. 5.1), approximately A.D. 1245–1380 (Grove and Switsur, 1994), but here, as elsewhere, evidence for many such advances disappeared during the larger magnitude advances marking the time of greatest cold during the Little Ice Age. An exception is provided by sites in the southern Canadian Rockies where forested sites overrun by glaciers during the A.D. 1300 Event (twelfth to the fourteenth centuries) have been fortuitously exposed by deglaciation associated with recent warming (Luckman, 1994).

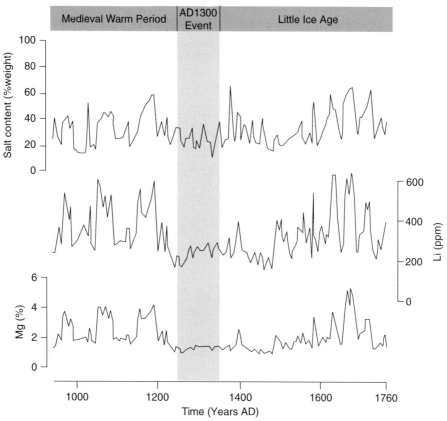

FIGURE 5.2 Depositional history of Owens Lake, east central California, interpreted in terms of palaeoprecipitation (after Li et al., 2000). The record is curtailed by tsunami-associated erosion at A.D. 1760. High salinity of Owens Lake during much of the Medieval Warm Period and Little Ice Age suggests that its level was low, a view supported during the later part of the Medieval Warm Period by the three peaks of acid-leachable Li and Mg. Low concentrations of Li and Mg (approximately A.D. 1220–1480) indicate that Owens Lake was relatively deep during the A.D. 1300 Event as a result of wetter climate conditions. Climate becomes generally more variable during the Little Ice Age, with six wet–dry cycles A.D. 1480–1760.

5.2.2 Climates of the western Pacific Rim

A record of glacier advance from Kamchatka shows the earliest last-millennium advance to have occurred during the A.D. 1300 Event, about A.D. 1150–1350 (800–600 cal BP – see Fig. 5.1). A reconstruction of temperature for the lower parts of the Changjiang and Huanghe valleys in China showed that temperatures began falling after the A.D. 1310s by 0.1°C per century until they reached 0.6–0.9°C lower than the 1951–1980 mean for this area (Ge et al., 2003). The series, illustrated in Fig. 5.3A, shows the rapid cooling that affected this area during the A.D. 1300 Event from the Medieval Warm Period to the Little Ice Age. A similar record from

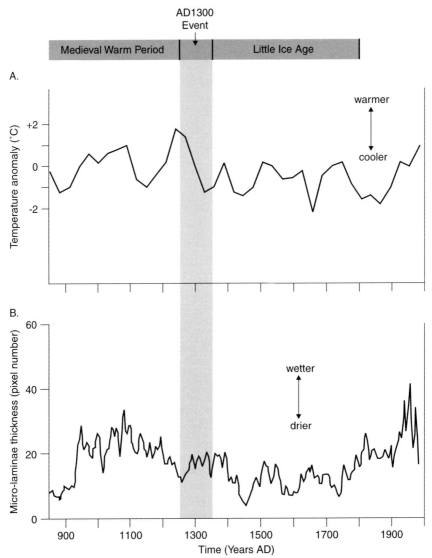

FIGURE 5.3 Temperature, precipitation and water-table changes in eastern China across the
A.D. 1300 Event. (A) Winter temperature anomalies for eastern China with 30-years resolution
(after Ge et al., 2003). (B) Changes in the thickness of laminae in a stalagmite from Shihua
Cave (near Beijing) for the period A.D. 850–1980 (after Qin et al., 1999). Laminae thickness will
increase when conditions are wetter and decrease when they are drier (Qian and Zhu, 2002).

Buddha Cave in eastern China shows that the cooling characterizing the A.D. 1300
Event took place between about A.D. 1375 and A.D. 1520 (see Fig. 4.4).

The temperature-sensitive mollusc *Nerita atramentosa* has been used to track
last-millennium climate changes in northern New Zealand (Szabó, 2001). The
presence of this mollusc in this area up until the mid-fourteenth century can be

linked to warmer conditions (Medieval Warm Period) while its subsequent disappearance until perhaps the end of the Little Ice Age can be explained by cooling.

A proxy record of precipitation across the A.D. 1300 Event in eastern China is shown in Fig. 5.3B. It shows a rise in precipitation at this time. This interpretation is supported by other studies from eastern China, such as the synthesis of lake levels by Fang (1993) which showed these increasing markedly A.D. 1250–1650. For example, the Putian Lakes in the lower Huanghe (Yellow River) formed ~36 small bodies of water in the Northern Song Period (A.D. 960–1127), dried up completely in the J'in Period (A.D. 1127–1279) before increasing in size to form more than 150 lakes during the Wanli Period of the Ming Dynasty (A.D. 1573–1620). A similar conclusion can be reached from the Buddha Cave record illustrated in Fig. 4.4 which shows a rapid change from dry to wet conditions in a few decades around A.D. 1350.

There is a near-complete record of Huanghe flooding for the past 2,500 years (Fang, 1992) yet this is unlikely to be a direct correlate of precipitation. The part of the record illustrated in Fig. 5.4 shows a strikingly low incidence of flooding centred on the period A.D. 1200–1250. While this may be largely the outcome of reduced precipitation, perhaps associated with the weakening of the Asian

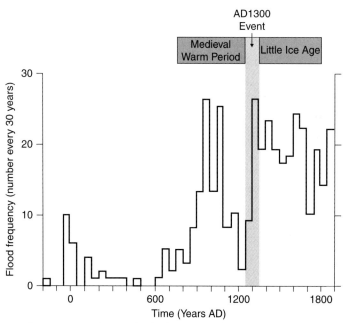

FIGURE 5.4   Secular variation of Huanghe (Yellow River) floods (after Fang, 1992). The incidence of flooding during the Medieval Warm Period and Little Ice Age is very high compared to earlier times and is likely to be primarily the outcome of vegetation clearance and soil erosion in the Huanghe catchment. The conspicuous low incidence of flooding centred on A.D. 1200–1250 may be a result of prolonged drought and/or a short-duration fall in water tables (driven by sea-level fall) that temporarily increased Huanghe channel capacity.

summer monsoon (see below), the general increasing incidence of flooding throughout the entire last millennium is explained by

> "both strong soil erosion after vegetation destruction in historical times and channel aggradation and water-level rise of the rivers as time-lagged responses to the Holocene sea-level rise". (Fang, 1992, p. 509)

If this is so, then the anomalously low incidence of flooding around A.D. 1200–1250 may be the outcome of

- either an abrupt short-lived change in human use of the land in the Huanghe catchment that allowed large parts of it that had been cleared to become re-vegetated, which seems improbable, and/or
- a fall in relative sea level that temporarily reversed water-table rise in the lower Huanghe (see also Section 5.3.1).

5.2.3 Abrupt change in the Asian summer monsoon

From their study of Holocene changes in the (northeast) Asian summer monsoon, based on 36 palaeoclimate records, Morrill et al. (2003) concluded that

> "the most prominent abrupt shift in monsoon strength during the historical period took place at A.D. 1300 ± 50 years". (pp. 468–469)

That this is statistically significant is shown in Fig. 5.5A. The nature of the changes registered in the Asian summer monsoon at this time is depicted on the map in Fig. 5.5B, with most data coming from the synthesis by Morrill et al. (2003) supplemented by data discussed above. The general picture is of wide-spread abrupt change marked everywhere by cooling but with a more variable change in precipitation. For example, in Taiwan, lake records show that the time around A.D. 1300 was characterized by cooling and drying conditions, suggesting that this area may have found itself beyond the reach of the monsoon at this time. In contrast, every place on the East Asian mainland became wetter at this time, consistent with more intense monsoonal conditions here and/or more tropical cyclones making landfall.

Another study of the Asian monsoon during the last millennium shows that it also varied in strength within this period (Wang et al., 2005). Around A.D. 1300, the monsoon was at its weakest for the entire Holocene but has since gathered strength.

It should be noted that, although it does not affect the Pacific Basin directly, palaeoclimate studies of the Asian Southwest Monsoon have shown that it also varied in strength during the Holocene as a result of temperature changes (Gupta et al., 2003). Even during the last millennium, its strength reflected the effects of the warmer Medieval Warm Period (when it was stronger) and the cooler Little Ice Age (when it was weaker).

5.2.4 Climates of the Pacific Islands and Ocean

Little information is known directly about how climates changed in the Pacific Islands and Ocean during the A.D. 1300 Event. Indeed the likelihood that climate

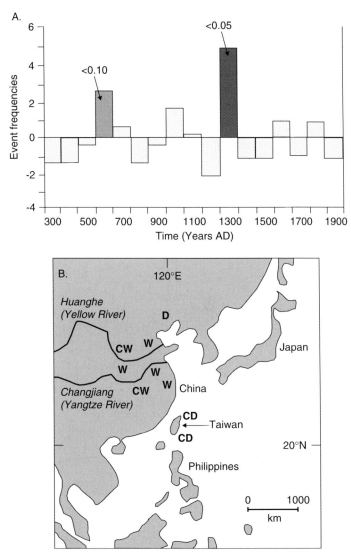

FIGURE 5.5 Abrupt change in the Asian summer monsoon about A.D. 1300. (A) Synthesis of Information from 36 palaeoclimate records covering the last millennium in the region covered by the Asian summer monsoon allowed abrupt events to be identified (after Morrill et al., 2003). The strongest such event within the past 2,000 years was that at A.D. 1300±50. The vertical scale identifies anomalies from a random distribution of these events in time, darker shading indicating the positive anomalies that are statistically different from 0 at the 95% confidence level. (B) Geography of climate change around A.D. 1300 in East Asia, updated from Morrill et al. (2003) with the addition of data from Ge et al. (2003) and Paulsen et al. (2003).

here did change at this time comes mostly from studies along the Pacific Rim (see above) and from studies of environments and societies within the Pacific Islands and Ocean that suggest changes in climate consistent with a rapid transition from the Medieval Warm Period to the Little Ice Age.

Most high-resolution palaeoclimate studies in the Pacific Islands and Ocean do not extend back to the A.D. 1300 Event (e.g. Urban et al., 2000; Moore et al., 2002). Lower resolution studies cannot always detect the evidence for relatively small climate shifts such as the A.D. 1300 Event (Morrill et al., 2003), particularly when there is a large overprint of recent human impact (e.g. Dodson and Intoh, 1999) or where there are significant local modifications of regional climate change (e.g. Finney et al., 2002).

5.2.5 ENSO frequency and storminess

There are few long-term proxy records of ENSO variability that are sufficiently complete and sufficiently high-resolution to capture decadal-century variations around A.D. 1300. Four are shown in Fig. 5.6 and at first glance appear contradictory. In contrast to the other records illustrated, the Laguna Pallcacocha record (Fig. 5.6A) shows an overall fall in El Niño incidence across the A.D. 1300 Event, an anomaly which could be explained by this record's incompleteness, particularly in its less-than-faithful recording of lower magnitude El Niño events. In contrast, the flood record from Laguna Aculeo shows a steady increase in El Niño incidence across the A.D. 1300 Event (Fig. 5.6B). Similar increases are evident in the Sacramento River flow record (Fig. 5.6C) and the event compilation of Anderson (1992) (Fig. 5.6D).

A link between El Niño frequency and storminess was used to explain the comparative climate constancy (low climate variability) during the Medieval Warm Period in the Pacific Basin (Section 4.1.4). In a similar way, El Niño frequency can be used to explain observed and inferred changes in storminess during the A.D. 1300 Event. In many parts of the Pacific Basin, the A.D. 1300 Event is marked by comparatively high storminess.

This has also been observed globally. In her authoritative compilation of evidence for the Little Ice Age, Grove (1988) concluded that the transition from the Medieval Warm Period to the Little Ice Age (the A.D. 1300 Event) was marked by an increase in storminess around A.D. 1200, as inferred from the evidence of the impact of storm surges, coastal sand transport and abnormally high tides on coastal communities worldwide. In the southwest Pacific, increased storminess associated with periods of "extreme" mid-latitude cyclone frequency occurred A.D. 1340–1350 and A.D. 1370–1380 (Goodwin et al., 2004, p. 792).

Using tree rings as a proxy for winter precipitation, Villalba (1990) showed that the period A.D. 1220–1280 was uncommonly wet in central Chile while droughts occurred during the period A.D. 1280–1450. A study of clastic input into Laguna Aculeo on the Chile coast in Pacific South America showed that the period A.D. 1300–1700 was marked by a greatly increased flood frequency in contrast to the Medieval Warm Period (Jenny et al., 2002). This is interpreted as a result of "more winter frontal system activity and possibly ENSO-related variability" (p. 3). This conclusion is echoed by a study of terrigenous (mica) inputs

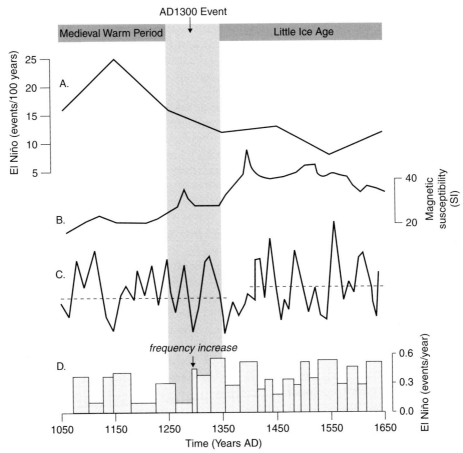

FIGURE 5.6 Reconstructions of El Niño frequency A.D. 1050–1650. Note the apparent contradiction between series A which shows a falling incidence of El Niño across the A.D. 1300 Event, and series B to D which show a rising incidence. (A) El Niño frequency from Laguna Pallcacocha (Ecuador) sediment colour intensity data accessed through IGBP PAGES/World Data Center for Paleoclimatology, Data Contribution Series #2002-76 (Moy et al., 2002). This record is interpreted as one of moderate to strong El Niño incidence (Graham, 2004) showing a fall in such events across the A.D. 1300 Event. (B) Magnetic susceptibility of sediments from Laguna Aculeo (Chile) is a proxy for El Niño-associated floods (Jenny et al., 2002). Note the steady rise in El Niño incidence between the Medieval Warm Period and the Little Ice Age. (C) Sacramento River (Western USA) flow (Meko et al., 2001) is a close proxy for El Niño incidence during the last millennium (Graham, 2004). Two periods of El Niño frequency can be seen (trend lines by author) with increased incidence and temporal variability after the A.D. 1300 Event. The data series is from Graham (2004) and the absence of a vertical scale is deliberate. (D) Reconstruction of El Niño frequency using historical records (Anderson, 1992). Note the sustained frequency increase after A.D. 1300.

into ocean floor sediments off the northern Chile coast around A.D. 1400, which are interpreted as the result of "strong rainy episodes" (Ortlieb et al., 2000). Flooding disrupted lowland societies in Chile and Peru around A.D. 1300 and A.D. 1330 (Satterlee et al., 2000; Magilligan and Goldstein, 2001). By analogy with the recent (200-year) record of precipitation in this area, it is possible that these represent a "mega El Niño" or – more plausibly – an aggregation of strong El Niño events. There is evidence for a similar period of wetness in the northern half of the eastern Pacific Rim. For example, a highstand attributable to increased precipitation occurred at Mono Lake in the Sierra Nevada A.D. 1270–1345 (Stine, 1990).

In a study from Yongshu Reef in the South China Sea, a higher proportion of storm-tossed boulders lying on the living reef front date from the period between approximately A.D. 1200 and A.D. 1450, which is consistent with increased storminess during the A.D. 1300 Event, but sample size was low and this result cannot be regarded as conclusive (Yu et al., 2004).

Increased storminess may also be inferred from increased loss of forest and its replacement by grassland, implying that storms not only remove trees but also significantly thin the underlying soil cover. Examples from the Pacific Basin include Japan where the time around A.D. 1300 was marked by "catastrophic forest destruction" (Yasuda, 1976, p. 56), Guam where increases in upland erosion occurred by A.D. 1375 (Dye and Cleghorn, 1990), and the Hawaiian Islands where the effects of lowland forest clearance are especially noticeable "after about AD 1200" (Kirch, 1986, p. 335). Storminess may also have contributed to infilling of coastal wetlands at places in Hawaii such as Kawai Nui Marsh (see Fig. 6.1B). Societal "crises" on the isolated islands of Mangareva and Pitcairn in the central eastern Pacific were precipitated by "massive episodes of land degradation" around A.D. 1300 (Weisler, 1995, p. 402).

On many Pacific islands, the loss of soil from island interiors during the A.D. 1300 Event was compensated by sediment build-up in lowland and coastal areas. Examples come from Tahiti (see following section), Huahine (Sinoto, 1983) and Mo'orea islands in French Polynesia (Parkes, 1997), Lakeba Island in eastern Fiji (Hughes et al., 1979), Aneityum Island in southern Vanuatu (Spriggs, 1986) and Tikopia in the eastern outer Solomon Islands (Kirch and Yen, 1982). It has been suggested that the development of new methods of food preservation on Tikopia may have been a response to increased storminess around A.D. 1300 (650 BP – M.S. Allen, 1997).

A recent study of Palau provided hints of wetter conditions during the A.D. 1300 Event, perhaps even "a period of especially heavy rains" (Masse et al., 2006, p. 125). The diagnostic layer of redeposited saprolite in the Lake Ngerdok core is constrained by dates for the underlying Layer IV of A.D. 1001–1284 and the overlying Layer II of A.D. 1280–1388.

For most locations, no precise dates for the start of alluviation are available, so it is not possible to demonstrate whether or not this was a synchronous region-wide event or whether it fell within the A.D. 1300 Event or during the early part of the Little Ice Age. For now, it is sufficient to note that evidence consistent with increased storminess during the A.D. 1300 Event is found on many low-latitude Pacific Islands.

For New Zealand, a number of attempts to identify periods of storminess suffer from imprecise dating, but possibilities include the Ruahine Range about A.D. 1180 (770 ± 60 BP: Hubbard and Neall, 1980), areas close to Wellington about A.D. 1230 (720 ± 80 BP: Brodie, 1957) and Auckland about A.D. 1336 (614 ± 67 BP: Grant-Taylor and Rafter, 1971). A lake-sediment record from the east coast of North Island suggests that there was a short-lived burst of storminess, possibly ENSO-linked, A.D. 1445–1450 (Eden and Page, 1998). Increased storminess may also have been a factor in the ultimate abandonment of the Palliser Bay settlements about A.D. 1400 (Leach and Leach, 1979), particularly through lowland alluviation from flooding and the disappearance of filter-feeding shellfish (Grant, 1981). In his synthesis of last-millennium climate for New Zealand, Grant (1994) recognized the warm Waihirere Period (A.D. 1270–1350) as a "very stormy period" during which "gales damaged forests and coastlines" (p. 166).

5.2.6 Increased storminess on Tahiti, French Polynesia

Tahiti is a high volcanic island in the central Pacific tropics that experiences tropical cyclones – the principal cause of storminess in this region – only during El Niño events, when a second "warm pool" forms in the adjacent part of the equatorial Pacific Ocean. The records of increased storminess from Tahiti may therefore refer to times when El Niño events were more frequent. If the interpretation in the previous section is correct, then increased storminess would be expected on Tahiti during the A.D. 1300 Event compared to times before and after. This is exactly what the limited data show.

The pattern of last-millennium settlement on Tahiti (Lepofsky et al., 1996) is similar to that elsewhere in the low-latitude Pacific Islands, with coastal settlement dominant during the Medieval Warm Period. Excavations of lowland sites at the mouth of the Papeno'o Valley showed that many became covered by vast sheets of downwashed material during the fourteenth century, something that has been ascribed to human forest clearance upvalley and upslope. This explanation does not satisfy Orliac (1997) who notes that the events responsible for the downwash of this material

> "can only be the consequence of a soil erosion out of all proportion to any clearing done by Polynesians. Vast fire clearances are quite impossible to carry out in the upper part of the Papeno'o basin, where rainfall totals 10 m per year!" (p. 228)

Rather, Orliac argues that the downwashed material is more likely the product of storms that increased in frequency and intensity at this time. The strongest of these events may even be preserved in a deluge myth (Henry, 1951).

A comparable dialogue surrounds the interpretation of a 5-m core from Lake Vaihiria in inland Tahiti (Fig. 5.7) that shows evidence for "massive erosion" from the catchment from at least A.D. 1500 to A.D. 1700 (Parkes and Flenley, 1990). The explanation favoured by these authors was that this erosion was caused by forest clearance by people, although it was later conceded that it may also indicate an increased frequency of tropical cyclones (Flenley et al., 1991). Orliac (1997) is unimpressed by any explanation involving human impact at this site,

FIGURE 5.7   Lake Vaihiria in inland Tahiti formed when a landslide blocked a narrow part of the steep-sided Tahiria Valley. Coring of the sediments in the floor of the lake suggests a prolonged period of increased catchment erosion during the A.D. 1300 Event and Little Ice Age [photo courtesy of John Flenley].

writing that

> "it would be astonishing if people cleared the abrupt slopes surrounding the lake, as these do not present any possibilities for gardening or settlement". (p. 228)

In the Lake Vaihiria core record, Orliac finds more evidence for natural fluctuations in storminess that could well coincide with the A.D. 1300 Event (Nunn, 2000a).

In all, the available evidence from Tahiti supports a correlation between increased storminess and elevated El Niño frequency during the A.D. 1300 Event.

5.3. ENVIRONMENTS DURING THE A.D. 1300 EVENT IN THE PACIFIC BASIN

In the last decade, there has been an increased realization of the importance of environmental change to the evolution of Pacific societies around A.D. 1300, and especially the understanding that a region-wide forcing mechanism (not a local explanation) is required. For example, recent work in American Samoa concluded that there is "a strong correlation of climate, sea level, and geomorphic change beginning about AD 1300 to 1400", the nature of which "seems to preclude the possibility that the observed landscape change is a localized phenomenon" (Pearl, 2006, p. 64).

As for the Medieval Warm Period, the most widespread climate-driven environmental change to have affected the entire Pacific Basin during the A.D. 1300

Event appears to have been sea-level change. There is abundant evidence – much inferred from societal changes – to suggest that sea level fell in many parts of this vast region during the A.D. 1300 Event.

This section first explains the disparities in the evidence for sea-level fall within the Pacific Basin principally the distinction between the record from the continental Pacific Rim and that from islands offshore (Section 5.3.1). Described next is the evidence for sea-level change across the A.D. 1300 Event from Pacific Islands. This is of two kinds. There is the directly measured evidence, usually employing appropriate palaeosea-level indicators, which is discussed in Section 5.3.2, and then there is the evidence for sea-level fall that can be inferred from coastal-environmental changes, which is discussed in Section 5.3.3.

Coastal-environmental changes during the A.D. 1300 Event in many parts of the Pacific Basin, especially the islands, led to coastal settlements being abandoned in favour of inland/upland settlements. The A.D. 1300 Event thus drove the earliest large-scale sustained settlement of inland areas in many parts of the Pacific Basin, initiating environmental changes that were aggravated in places by climate change. Evidence for environmental changes associated with the occupation of inland areas, especially on Pacific Islands, is reviewed in Section 5.4.

5.3.1 Sea-level change: continental versus island records

Along much of the contiguous Pacific Rim, evidence for comparatively minor changes in sea level during the first half of the last millennium has probably been largely obscured by later coastal change (Nunn and Kumar, 2006). This would be expected because of

- sediment deposition by the comparatively large rivers that drain the Pacific Rim, particularly the effects of commonly increased sediment loads during the past 200 years as human impact has grown;
- the increased human use of this area, particularly the establishment and growth of urban settlements and the consequent artificialization of the coast; and
- the effects of marine deposition, particularly storm surges and tsunami, which more frequently impact the broad target provided by long continental coasts rather than the smaller targets provided by island coasts.

As an example of the kind of slender evidence for sea-level fall along the Pacific Rim during the A.D. 1300 Event, consider the case of stream-channel downcutting in coastal California (Waters et al., 1999). A prominent river terrace was formed as channels downcut ~500 years ago, followed by deposition and then another phase of downcutting. Such episodic downcutting may indeed be "the result of a complex response of the fluvial system to major flooding" (p. 289) but may also register the effects of a two-stage sea-level fall at this time. A comparable situation comes from Garua Island in Papua New Guinea, where evidence for "relative sea-level fall at c.650 BP" (A.D. 1300) may represent "the effects of rainfall on a post-eruption devegetated landscape" following the deposition of the Dakataua tephra (Boyd and Torrence, 1996, p. 274) but may also have been a response to sea-level fall.

A similarly inferential line of evidence for sea-level fall along the Pacific Rim comes from one possible interpretation of the conspicuously low flood incidence of the Huanghe (river) in eastern China centred on A.D. 1200–1250 (shown in Fig. 5.4). It is possible that a short-duration fall of sea level at this location lowered water tables increasing Huanghe channel capacity and thereby reducing flooding during the A.D. 1300 Event. Similar interactions between sea level and river flooding are implicit in the rises and falls of East Asian delta societies shown in Fig. 3.3.

In contrast, the evidence for such a recent and minor sea-level fall would be expected to be more visible on (smaller) islands because

- the comparatively small rivers, if they exist, bring comparatively small amounts of sediment to the coast, and this may not be widely distributed, particularly if other parts of the coast are effectively shielded by coral reefs;
- recent human impact has not been generally so intense on many smaller Pacific Islands as it has in many parts of the Pacific Rim, so many island coastal landscapes are closer to their condition 700 years ago; and
- the imprint of large waves (storm surges and tsunami) is generally localized within islands and archipelagoes, leaving many parts of their coasts comparatively unaltered compared to their condition 700 years ago.

For such reasons, the most compelling evidence for sea-level fall during the A.D. 1300 Event generally comes from island rather than continental coasts in the Pacific Basin.

5.3.2 Direct measurements of sea-level fall

A trend of sea-level change in the Pacific Basin based on various radiometric-dated indicators is illustrated in Fig. 5.8. Original data sources are in Nunn (2000b, 2003b). This compilation shows a two-stage sea-level fall of ~1.8 m (maximum) across the A.D. 1300 Event and into the early Little Ice Age. A palaeotemperature record and an El Niño frequency record are also shown. Wiggle matching suggests that these sea-level changes may be a result of solar forcing with a 90-year time lag (see Fig. 10.4C).

Comparable evidence of sea-level fall at this time comes from parts of the Pacific Rim. On Kunashir Island in the Kurile group (northwest Pacific), sea level fell a maximum of 1.6 m from its highest level during the Medieval Warm Period around A.D. 1100 (850 BP – Korotky et al., 2000; see also Fig. 1.3A). Palaeosalinity data from the Nakaumi Lagoon in Japan are likely proxies for sea-level change, and there is evidence for a fall of sea level across the A.D. 1300 Event here (see Fig. 4.5).

In many parts of the Pacific Basin, there is no directly measurable evidence for a fall in sea level across the A.D. 1300 Event, only the contrast between the evidence for (or the inference of) higher sea level during the Medieval Warm Period and lower sea level during the Little Ice Age. This association is adequate for inferring sea-level fall across the A.D. 1300 Event although it is often silent as to the rate and precise timing of this.

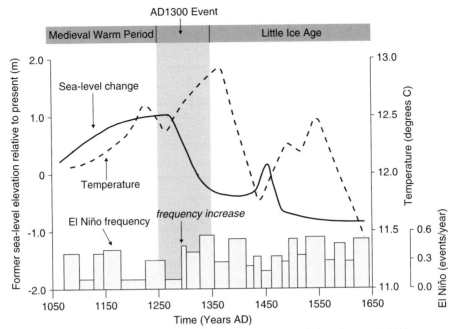

FIGURE 5.8 Sea-level change A.D. 1050–1650 from Pacific Island data (Nunn, 2000b), temperature changes from oxygen-isotope analysis of a New Zealand stalagmite (Wilson et al., 1979) and a reconstruction of El Niño frequency (after Anderson, 1992).

5.3.3 Coastal-environmental evidence for sea-level fall

Throughout the Pacific Basin, there is evidence of changes in coastal environments during the A.D. 1300 Event that are consistent with a fall in sea level. There are legitimate questions about whether this could indeed have been a Pacific-wide sea-level fall, but what is clear is that coastal environments throughout the Pacific Basin register the effects of sea-level fall approximately synchronously during the A.D. 1300 Event. For reasons noted above, the evidence is clearest in the Pacific Islands as the selection in Fig. 5.9 shows. When evaluating such diagrams, the imprecision of many of the dates used should be borne in mind. Many coastal-environmental changes are dated only by a single radiocarbon determination, while some ages given may be estimates.

Much evidence for sea-level fall during the A.D. 1300 Event comes from observed changes in coastal geomorphology. Some of this evidence is imprecise. For example, on Rarotonga Island in the southern Cook Islands, where there is evidence for a sea-level highstand during the (late?) Medieval Warm Period, many of the lowland valleys are conspicuously incised by their river channels, the likely consequence of sea-level fall across the A.D. 1300 Event (Fig. 5.10). An example of more precise evidence comes from the islands of (Western) Samoa, where a fall of sea level some time within the period A.D. 800–1420 (1,150–530 cal BP) was inferred from the restricted vertical reef accretion during this period (Goodwin and Grossman, 2003).

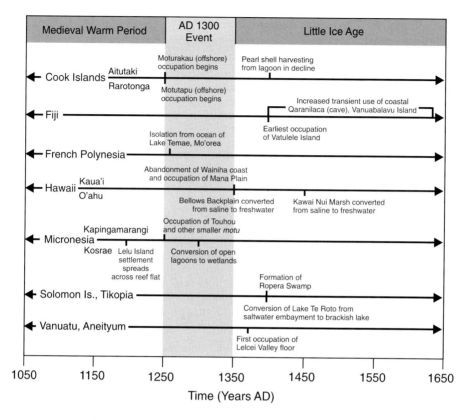

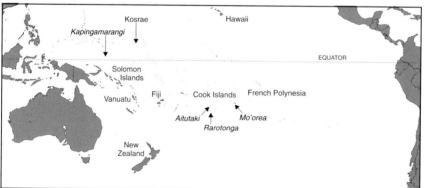

FIGURE 5.9 Evidence for coastal changes on Pacific Islands consistent with a relative fall of sea level during the A.D. 1300 Event (upper) with a map to show locations of places mentioned (lower). Data from the Cook Islands; from Aitutaki (Allen, 1992, 2002) and Rarotonga (Okajima, 1999); from Qaranilaca, Fiji (Nunn et al., 2000; Thomas et al., 2004); for Hawaii, from Kaua'i (Carson, 2003, 2004; Nunn et al., 2007) and O'ahu (J. Allen, 1997; Athens, 1997); Federated States of Micronesia, data for Kapingamarangi (Leach and Ward, 1981) and Kosrae (Athens, 1995); Solomon Islands (Kirch and Yen, 1982); Vanuatu (Spriggs, 1997).

FIGURE 5.10 The Takuvaine Stream on Rarotonga Island, southern Cook Islands, where it cuts through the coastal plain on which the capital Avarua is built. The incised channel may be a result of sea-level fall during the A.D. 1300 Event (Photo by Patrick Nunn).

It is plausible to suppose that, by exposing previously submerged shoreflats, sea-level fall during the A.D. 1300 Event transformed smaller islands located near larger ones into prospects for human settlement. Sea-level fall may even have created habitable islands where none existed before by exposing reef surfaces on which detritus could accumulate without being washed away at high tide. If habitable islands were created in these ways, then humans would not necessarily have occupied these islands immediately, yet the age of this occupation often provides the only age for the appearance of these islands. For this reason, the abundant evidence for the occupation of smaller offshore islands early in the Little Ice Age (see Section 6.3.4) can be taken as supporting evidence for the creation and/or expansion of these islands during the A.D. 1300 Event.

Notwithstanding this, there are some examples of smaller offshore islands probably coming into existence during the A.D. 1300 Event. For instance, archaeological investigations of the island named Motutapu off the east coast of Rarotonga (Cook Islands) show that it came into existence – at least as a habitable prospect – about A.D. 1250 (700 cal year BP: Okajima, 1999). Other smaller islands are inferred to have formed as a result of sea-level fall at this time, including those on atoll reefs such as Kapingamarangi (Federated States of Micronesia) and Bikini (Marshall Islands) where new islands were colonized for the first time during the A.D. 1300 Event and early Little Ice Age.

The situation on Bikini Atoll was discussed in some detail in a percipient paper by Streck (1990). Until the A.D. 1300s, most settlement was concentrated on

Eneu Island but thereafter this was largely abandoned, suggesting that settlement was dispersed A.D. 1400–1700. This may partly have been a consequence of sea-level fall, exposing reef surfaces on which new atoll islands (*motu*) appeared and began growing. In fact, Streck infers a role for sea-level fall in "later prehistoric times" (p. 256) from his study of changes in the form of Eneu.

The size increase of islands such as Lelu off the main island of Kosrae (Federated States of Micronesia) may also have occurred during the A.D. 1300 Event, stimulating their subsequent human occupation. The expansion of Lelu began with the emergence of shallow reef surfaces on which an elite group began to construct walled compounds, imparting an important cultural role to the island within the last 600 years of Kosrae's history. The possibility of sea-level fall being responsible for this emergence was explicitly discussed by Athens (1995), who found evidence for

> "a wetland marsh or swamp formed on top of marine deposits at about AD 500, suggesting a regression at about this time". (p. 360)

Yet, while much attention has been given to Lelu, studies of the main island of Kosrae also suggest that the expansion of Lelu is but one example of a widespread change in coastal environments A.D. 1200–1350 that can be plausibly attributed to sea-level fall. For example, Athens (1995) concluded that

> "significant expansion of alluvial flats as a result of anthropogenic erosion and infilling has not been a significant factor for landscape change on Kosrae". (p. 359)

Rather, he argues, it is likely that the expansion of coastal and valley-mouth flats was in response to sea-level fall around Kosrae, dated in at least some places to the A.D. 1300 Event. Athens also noted that many shallow open-water lagoons around Kosrae were converted to wetland swamps by A.D. 1300, a change also most plausibly explained by sea-level fall.

On the islands of Palau, near-coastal coring on Babeldaob Island revealed changes from dominantly marine-sediment deposition to terrigenous sediment deposition A.D. 1300–1400 that "may reflect lowered sea level" (Masse et al., 2006, p. 127). A similar situation probably affected tidal Lake Hagoi on Tinian Island (Commonwealth of the Northern Marianas Islands) where declining aquatic palynomorph signals across the A.D. 1300 Event (upper Zone C2) suggest "a reduction in areal extent of the wetland" (Athens and Ward, 1998, p. 72) consistent with sea-level fall.

Another example comes from Mo'orea Island in French Polynesia (Parkes, 1997). Brackish Lake Temae in northeast Mo'orea is isolated from the ocean today by a 2-m high beach dune but like other now-dry lakes along this coast "was formerly part of the lagoon system, and that the coral ridge [dune] was once an island on the barrier reef" (p. 185). It seems likely that sea-level fall could explain the conversion of these lagoons to brackish lakes. In cores through Lake Temae sediments, isolation from the ocean is marked by increased amounts of red-yellow nekron mud associated with increased nutrient status, a transition that appears (by interpolation) to have been around A.D. 1250 (700 BP) here.

Similar investigations of coastal wetlands in the Hawaii Islands have reached comparable conclusions that are consistent with the effects of sea-level fall. For example, the Bellows backplain on O'ahu Island was a saltwater wetland until A.D. 1400 when it "quickly" filled with terrigenous sediments to become a brackish or freshwater wetland by A.D. 1500 (J. Allen, 1997, p. 234). Also on O'ahu, the marsh of Kawai Nui (*big water* in Hawaiian) was formerly "a bay and subsequently a lagoon" that became a lake isolated from the ocean by A.D. 1300 (see Fig. 6.1B). By A.D. 1400–1500, "terrigenous sediments filled large areas ... supporting a grass- and sedge-dominated marsh" (J. Allen, 1997, p. 235). A similar history is applicable to the Kahana Valley on O'ahu.

The idea that changes in the availability of particular shellfish species during the last millennium, as implied by changes in their representation in shell middens, could indicate changes in nearshore environments associated with sea-level changes rather than procurement choices was first mooted by Paulay (1996). He concluded that the *Anadara*-to-*Strombus* ratio was a sensitive indicator of relative sea-level change, and interpreted the evidence for changes in this ratio within the last millennium in Micronesia as a result of sea-level fall.

A final line of coastal-environmental evidence for sea-level fall across the A.D. 1300 Event is the appearance of fishponds as significant contributors to human subsistence in the Hawaiian Islands and probably elsewhere. The sudden appearance of fishponds throughout the high islands of the group seems to militate against a ponderously trialled cultural innovation, pointing rather to a simultaneous environmental change (such as sea-level fall) that created the environments suitable for subsequent use – with or without significant human modification – as fishponds. Initial fishpond construction in Hawaii took place "sometime prior to the 14th century AD" (Kikuchi, 1976, p. 295), as the available dates (see Fig. 6.6) confirm.

Some writers cannot credit the role of relative sea-level fall in the manifest infilling of bays along Pacific coasts during the A.D. 1300 Event and cite improbable (yet evidently more comfortable) explanations involving humans. For example, on O'ahu Island in Hawaii, it has been suggested that "the early Hawaiians were responsible for these catastrophic changes in local environments", the alternative explanation of "a several-foot drop in sea level" not considered because "sea level ... has not changed for several thousand years" (Culliney, 1988, p. 325), an assertion contradicted by more recent data (Jones, 1997).

5.3.4 Coastal changes on Tikopia, Solomon Islands

One of the best examples of coastal-environmental evidence for sea-level fall across the A.D. 1300 Event comes from remote Tikopia Island (5 km^2 in area) in the eastern Solomon Islands (Kirch and Yen, 1982). Tikopia has been occupied since the earliest period of island settlement in this region, ~700 B.C. (~2,680 BP) but perhaps the greatest environmental changes in pre-history came around A.D. 1400, almost certainly during the A.D. 1300 Event.

Much of the attraction of Tikopia for its earliest human settlers came from the saltwater embayment that existed during the Kiki and Sinapupu phases of

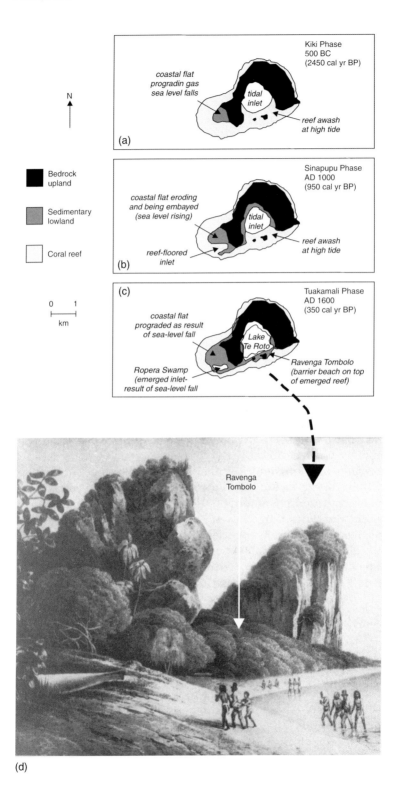

(a) Kiki Phase 500 BC (2450 cal yr BP)
coastal flat progradin gas sea level falls
tidal inlet
reef awash at high tide

(b) Sinapupu Phase AD 1000 (950 cal yr BP)
coastal flat eroding and being embayed (sea level rising)
tidal inlet
reef-floored inlet
reef awash at high tide

(c) Tuakamali Phase AD 1600 (350 cal yr BP)
coastal flat prograded as result of sea-level fall
Lake Te Roto
Ropera Swamp (emerged inlet-result of sea-level fall
Ravenga Tombolo (barrier beach on top of emerged reef)

N

Bedrock upland

Sedimentary lowland

Coral reef

0   1
km

(d)

Ravenga Tombolo

settlement (Fig. 5.11). During the Tuakamali Phase by about A.D. 1400, this embayment was transformed into a brackish lake (Lake Te Roto) separated from the ocean by a tombolo (named Ravenga). In an interpretation that has informed a significant amount of subsequent speculation about the ability of pre-modern humans to manipulate Pacific Island environments, Kirch and Yen (1982) inclined to the view that the Ravenga Tombolo was constructed by humans as the lake edges silted up (as a result of slope wash) to create a new environment suitable for giant taro (*Cyrtosperma chamissonis*) cultivation. This explanation is pure conjecture and unlikely to be correct, given the evidence elsewhere on Tikopia and throughout the tropical Pacific for similar coastal-environmental changes, an observation that demands a regional not a local explanation.

More likely it seems that the conversion of a saltwater tidal embayment to the brackish-water Lake Te Roto was the outcome of relative sea-level fall that caused the reef across the mouth of the embayment to emerge, preventing the regular flushing of the inlet by ocean water. This was fortuitous for the islanders who began planting taro around the lake fringe after this time and occupied its newly emerged coasts.

Comparable changes occurred elsewhere on Tikopia. The focus of much early settlement was the coastal flat in the west of the island, which increased in area as a result of late Holocene sea-level fall. At the start of the Medieval Warm Period (Sinapupu Phase in Fig. 5.11), this coastal flat was extensive but was eroded and embayed subsequently as a result of sea-level rise during this period. When sea level fell subsequently during the A.D. 1300 Event, a prominent embayment in the southwest of Tikopia became the Ropera Swamp. Authors such as Kirch and Yen (1982) have acknowledged that the appearance of Ropera Swamp during the Tuakamali Phase required coastal-environmental change but – locked into the idea of cultural determinism – regard this as having been accomplished by humans rather than by natural processes.

There is no need to cite any human intervention in landscape evolution to explain the observed changes in the landscape of Tikopia in post-settlement times. The danger in doing so is that you then ascribe far more environmental management ability to pre-modern people in such locations than they likely had,

---

FIGURE 5.11   Changing landscape of Tikopia Island, eastern Solomon Islands. (A–C) Palaeogeographic maps reinterpreted from Kirch and Yen (1982); note that reef extent in the Kiki Phase is unlikely to have been as great as in later phases but is shown as such in the absence of information to the contrary. Sea level was high during the Kiki Phase but falling, allowing coral reef to catch up with sea level and grow laterally, and for sediments (both terrestrial and marine) to build up around the emerged edges of the bedrock island. Sea level had stabilized close to its present level by the Sinapupu Phase, allowing coastal flats to extend but sea-level rise (during the Medieval Warm Period) was causing erosion and embayment of these flats. One of these embayments subsequently became Ropera Swamp, isolated from the ocean by sea-level fall during the A.D. 1300 Event. This sea-level fall also caused emergence of the reef barrier (which became Ravenga Tombolo) at the mouth of the main tidal inlet (which became brackish Lake Te Roto). (D) The 1833 lithograph by Louis Auguste de Sainson of the Ravenga Tombolo with the two bedrock islands that it encompassed (courtesy of Alexander Turnbull Library, National Library of New Zealand).

even to the extent of assuming that human societies such as those of Tikopia early in the last millennium made informed choices that led to their downfall (Diamond, 2005). In overlooking the evidence of sea-level fall at the key time in places such as Tikopia, such interpretations are naive.

5.4. SOCIETIES DURING THE A.D. 1300 EVENT IN THE PACIFIC BASIN

In many parts of the Pacific Basin during the A.D. 1300 Event, societies – especially coastal societies – experienced a food crisis associated with the depletion of various food sources resulting from sea-level fall. The responses of particular Pacific societies to food crisis would not be expected to be uniform although there are remarkable similarities within the low-latitude Pacific Islands region where reef-food resources were evidently significantly depleted (Nunn, 2000a).

In the short term, the response of a particular society to a food crisis of the kind that evidently affected many in the Pacific Basin at this time might involve adaptation, in the hope that the situation would resolve itself within a reasonable period of time. In some Pacific Island societies, it has been speculated that the rise in megalithic monument building around the time of the A.D. 1300 Event represents just such short-term adaptation, in this case an appeal to a higher power for intervention. Examples come from Easter Island where *moai* construction continued during the A.D. 1300 Event (Bahn and Flenley, 1992), Pohnpei in Micronesia where the megalithic site of Nan Madol began to be constructed about A.D. 1200 and on some islands in Tonga where monument construction may have coincided with the A.D. 1300 Event (Burley, 1998).

Arguably more pragmatic short-term adaptation for smaller, more isolated subsistence societies of Pacific coasts at this time might have involved diversifying food supply (including the consumption of foods habitually spurned – Currey, 1980), leaving barren marine areas to rejuvenate, and searching for unexploited resources. The latter may have involved encroaching on another group's territory, thereby sparking conflict over resources that became widespread in much of the Pacific Basin during the early Little Ice Age (see Section 6.3.7).

Perhaps only when the enduring nature of the food crisis became unmistakable might longer term strategies for adaptation have been entertained. These might have involved relocation, particularly to distant territories, although by this stage, societal breakdown may have become so acute that proactive response by the original group was no longer possible. In such cases, responses would have been largely reactive, with subgroups perhaps resorting to violence to control the remaining supplies of traditional foods, and subordinate subgroups moving to safer (often marginal) locations elsewhere.

The lag between the impact of the food crisis and the societal adaptation – proactive or reactive – would depend largely on the antecedent conditions, that is how close a particular population was to the carrying capacity of the area that exclusively sustained them before the crisis hit. At one end of the spectrum, it is possible to envisage a small group of people on a remote yet resource-abundant island (or a similarly circumscribed area of the Pacific Rim) weathering such a crisis, or at least there being a comparatively long lag between impact and societal response. At the other end of the spectrum, it is conceivable that many Pacific Island populations were living close to their area's carrying capacity towards the end of the Medieval Warm Period. In this scenario, response to the subsequent food crisis would have been comparatively rapid although the precise timing and the nature of this response would also depend on location, the remoteness and the circumspectness of the area involved, and the effectiveness of short-term adaptation strategies. Case studies are given in Sections 7.6 and 7.7 to illustrate these points.

Many Pacific societies, most of which were coastal at the end of the Medieval Warm Period, are not known to have responded to the food crisis during the A.D. 1300 Event itself, that response becoming evident only during the early Little Ice Age. So, although these responses are plausibly linked to what happened during the A.D. 1300 Event, most are discussed only in Chapter 6. In this regard, the point is also worth making that dating societal change precisely is extremely difficult, because it is commonly a prolonged and multifaceted process that produces manifestations of change at different times for different reasons. It is therefore almost impossible to demonstrate beyond doubt the likely effects of a precisely dated (climate) driver in producing a particular crudely dated societal response. Of course, the difficulties involved are no reason for shying away from efforts to marry climate and society, but their union will always be regarded as less than legitimate for such reasons by those unconvinced that it was ever possible.

Although the evidence for the environmental transformation of Pacific coasts during the A.D. 1300 Event is obscured along most continental coasts and preferentially preserved along (smaller) island coasts, the evidence of societal change at this time is certainly not. For while the proximate causes of these societal changes are undoubtedly complex, their coincidence with rapid climate (and sea-level) changes and coastal-environmental changes during the A.D. 1300 Event suggests a causal link.

The following account first discusses societies of the Pacific Rim and the changes that they underwent during the A.D. 1300 Event (Sections 5.4.1 and 5.4.2) with a case study from eastern Australia (Section 5.4.3). Changes in Pacific Island societies are considered in Section 5.4.4.

5.4.1 Societies of the eastern Pacific Rim

Cultural upheavals attributable to rapid climate change during the A.D. 1300 Event occurred all along the eastern Pacific Rim.

Along much of the northern coast of Peru, the Chimú culture became dominant about A.D. 1200–1470 (Dillehay and Kolata, 2004). In the lower Jequetepeque Valley of northern Peru, the most extensive suites of dunes (marking times of drought – see Section 4.3.2) date from middle Chimú times (A.D. 1245–1310) and represent a severe and prolonged period of drought identified in both the Quelccaya ice core (Thompson et al., 1986) and Lake Titicaca sediments (Binford et al., 1997). The important urban centre named Cañoncillo was abandoned late in the fourteenth century, probably in response to

"massive dune encroachment ... that choked irrigation canals, buried ... cultivation surfaces, and covered residential structures". (Dillehay and Kolata, 2004, p. 4327)

Yet the Chimú also proved resilient to the increasing climate variability associated with the A.D. 1300 Event, optimizing water use and distribution, and rebuilding the attendant infrastructure when it was damaged during El Niño-associated floods.

The later Inca (Inka) civilization in this region not only weathered the dryness of the terminal Medieval Warm Period but extended their (influence) throughout Andean Peru and Bolivia during the A.D. 1300 Event (Morris and von Hagen, 1993). The reason for their comparative success was principally their location, in the mountain valleys where the climate extremes were less than on the *altiplano* around Lake Titicaca, where seasonal rainfall was consequently more regular and was supplemented by meltwater from the glaciers above. Like the Tiwanaku, the Inca developed a raised-field system of agricultural production that allowed the maintenance of large urban populations but also had techniques for soil stabilization and water management that allowed production to be optimized in a more variable climatic context (Chepstow-Lusty and Winfield, 2000).

Yet many societies along the Pacific Rim on South America did not survive the A.D. 1300 Event, perhaps the most compelling example being the Tiwanaku (see Section 4.3.3). Similar stories are those of the decline of the Chiribaya culture and others throughout coastal Peru following the Miraflores flood about A.D. 1330 (Satterlee et al., 2000). A general picture emerges of human societies along the Pacific Rim in South America flourishing during the Medieval Warm Period when there was a generally low incidence of El Niño events, and then having become increasingly stressed, sometimes collapsing, during the transition to the Little Ice Age when a significantly higher incidence of El Niño events began to affect the area.

Along the arid Pacific coast of South America, there were attempts to counter the effects of the unusually variable climate during the A.D. 1300 Event by moving into marginal areas (also a trend on Pacific Islands during the early Little Ice Age – see Section 6.3.4). One example comes from the "arid coastal promontory" at Wawakiki in southern Peru, which was transformed during this time into a productive landscape through the construction of infrastructure such as canals and terraces (Zaro and Alvarez, 2005). The infrastructure made optimal use of the available water – important for surviving droughts – and the high coastal location protected the site from the river floods during the regular El Niño events that characterized this time.

In central America, many complex societies were already in decline as a result of droughts during the terminal Medieval Warm Period. During the A.D. 1300 Event, cold and drought also affected the people of the valleys of Mexico, the periods A.D. 1332–1335 and A.D. 1447–1454 being particularly stressful (Gill, 2000). Along the Pacific coast of Mexico, there is evidence of areas being abandoned by around A.D. 1300 because of sedimentation attributed to the effects of increased flood magnitude and frequency on nearshore shellfish populations (McGoodwin, 1992), a sign of increased precipitation, perhaps during storms.

Societal change among the inhabitants of southern California is notable during the A.D. 1300 Event, partly in response to the severe and prolonged droughts that gripped this region during the later part of the Medieval Warm Period (see case study in Section 4.3.4). Societal change was expressed by mass abandonment of settlements, particularly on offshore islands, and a number of indicators of "acute nutritional stress" found in the skeletons of people who lived here in this period (Arnold, 1992). In particular, the people occupying the smaller islands appeared more prone to *cribra orbitalia* (disruption of the roof of the skull) than those on the larger islands who were in turn more prone to this condition than those on the mainland (Walker, 1986). This has been explained by high nutrient losses resulting from diarrhoea arising from drinking water contaminated with enteric bacteria. As Walker (1986) noted, mainland people would have had more water sources to choose from than those on the islands; people on larger islands would have had more sources to choose from than those on the smaller islands where "aggregation of people around ... heavily used springs would greatly increase the danger of water contamination" (p. 351).

In the Peninsular Alaska-Aleutians region, there were "radical" changes in human societies after about A.D. 1200–1500 including "major" changes in tool technology and the subsistence economy (Maschner and Reedy-Maschner, 1998).

5.4.2 Societies of the western Pacific Rim

Late in the Medieval Warm Period, the Mongols had conquered much of northern China, perhaps in response to a climate-linked food crisis that affected their homeland to the northwest (Hsu, 2000). During the A.D. 1300 Event, the Mongol invasion of China was completed. The Mongols also attempted to conquer Japan, Kublai Khan sending his fleet there in A.D. 1274. This was destroyed by a tropical cyclone. A second invasion attempt in A.D. 1281 was also ended by a tropical cyclone, during which the Chinese fleet was sunk.

There are various indicators from the eastern coast of Australia that Aboriginal societies there experienced comparatively profound changes in at least strategies of food acquisition and settlement pattern during the A.D. 1300 Event. For example, during the Medieval Warm Period, an important component in the subsistence economy of people living around Queen Charlotte Bay was the mud-dwelling shellfish *Anadara granosa*, which disappeared abruptly around A.D. 1350 (600 BP: Beaton, 1985). In their survey of pre-history in tropical Australia, Hiscock and Kershaw (1992) ascribe the profound lifestyle changes that this induced to "extreme climate events" (p. 66).

Another example comes from farther south, along the Curtis Coast of southeast Queensland where the A.D. 1300 Event is marked by

> "a dramatic reordering of land-use settlement patterns, with no precedent in the history of occupation of this region" (Nunn et al., 2007).

Simply stated, the people occupying this area, who had previously been largely nomadic began to settle along the coast and exploit marine resources around this time (Fig. 5.12). This may be because sea level fell, exposing large areas of sea floor suitable for colonization by edible shellfish. The intensive coastal occupation that began at this time led to the development of more complex societies marked by the creation of new identity-conscious groups (McNiven, 1999).

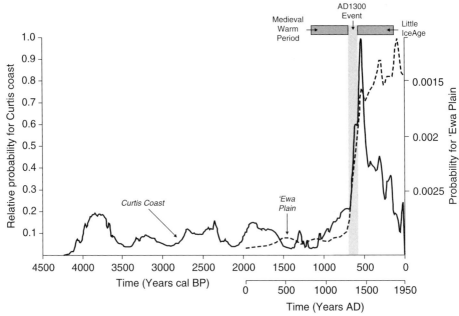

FIGURE 5.12 Rapid increase in settlement during and after the A.D. 1300 Event illustrated by increased coastal settlement along the southern Curtis Coast, Queensland, eastern Australia (after Nunn et al., 2007; original data in Ulm, 2004), and increased settlement of the marginal 'Ewa Plain, O'ahu Island, Hawaii group (after Athens et al., 2002). The summed probability distribution represents the probability that independent event A or B occurred at a particular time and can therefore be used as a proxy for the probability of occupation within a particular period. Periods of low site use are common prior until about A.D. 1200. Within the last 1,000 years marked increases are evident ~700 years ago.

Another example, even farther south along the east coast of Australia, comes from Bass Point and is discussed in Section 5.4.3.

In New Zealand, the first people arrived during the A.D. 1300 Event and soon abandoned their coastal settlements in favour of commonly inland upland settlements named *pa* (see Fig. 6.3). This fundamental change in settlement pattern, discussed in Section 6.3.2, has been commented upon by many writers (e.g. Davidson, 1984; McGlone et al., 1994). A representative study of sustained inland settlement comes from the Waikato region about A.D. 1300 (Lowe et al., 2000).

5.4.3 Nearshore subsistence at Bass Point, New South Wales, Australia

At Bass Point, Bowdler (1976) documented two distinct layers of shell midden deposit, underlain by sands containing occasional cultural materials dating from 17,000 BP. Upwards through the shell midden sequence, a steady decline in numbers of larger gastropods (such as the turban shell, *Turbo torquata*, the triton, *Cabestana spengleri*, and the cartrut, *Dicathais orbita*) was compensated by a rise in the numbers of mussel (*Mytilus edulis*), obtained closer to shore. *Mytilus* represented 50–80% of the shell assemblage in the upper layer compared with 0–10%

in the lower layer. The date between the two layers is around A.D. 1210 (740 BP: Hughes and Djohadze, 1980), similar to that obtained for the widespread transition from gastropods to mussels in middens elsewhere along the New South Wales coast (Sullivan, 1987). The upper midden contained shell fishhooks and blanks, the lower did not.

Bowdler (1976) related the changes in shellfish composition to the introduction of fishhooks and Sullivan (1987) to a regional shift towards a more intensified use of nearshore coastal resources. Mackay and White (1987) and Rowland (1999) found problems with both explanations, in particular because they do not adequately acknowledge environmental factors. It is suggested that the most likely cause of the observed changes in these middens relates to environmental changes, particularly sea-level fall, that created opportunities for expansion of *Mytilus* populations and thereby contributed to the shift from gastropod to mussel gathering.

Studies of shellfish remains from other southwest Pacific Rim coastal middens show changes that are similar in both nature and timing to that from Bass Point and which may also be related to the A.D. 1300 Event. Rowland (1999) and Ulm (2004) have noted shifts in shellfish representation in middens in southeast Queensland that they relate to changing local environmental conditions in the last 1,000 years, in particular an expansion of intertidal habitats. On the North Island of New Zealand, the disappearance at around A.D. 1300–1400 (650–550 BP) of the limpet *Cellana denticulata* from sites where it was previously dominant is also a likely consequence of environmental change (Rowland, 1976). Similarly, the Palliser Bay settlements in the Wairarapa of the southern North Island of New Zealand show significant changes in coastal resource use around A.D. 1350 (600 cal year BP: Leach and Leach, 1979). All these examples could be explained by sea-level fall affecting nearshore ecosystems during the A.D. 1300 Event.

5.4.4 Societies of the Pacific Islands

One of the most widespread and visible societal responses to climate and environmental change during the A.D. 1300 Event is a change in settlement pattern. On many islands at this time and subsequently (during the early Little Ice Age – see Section 6.3.3), coastal and lowland settlements were abandoned typically in favour of inland/upland settlements in readily defendable locations such as hilltops and caves.

One example comes from Nuku Hiva Island in the Marquesas Islands, an island group that has little flat land adjoining the ocean because of these volcanic islands' steeply plunging flanks. Suggs (1961) explained that at the start of the Expansion Period (dated to about A.D. 1200 by Kirch, 1984a) the population "which had been increasing in the large verdant valleys [during the Medieval Warm Period] ... suddenly broke out" (p. 182). On nearby Tahuata Island, this event has been claimed to reflect "rapid population growth" (Rolett, 1989, p. 108) but there is no independent verification of this. Rather one might highlight the "food stress" (Kirch, 2000a, p. 261) together with the warfare indicated by the start of the construction of fortified ridge-top (upland) sites that occurred around the same time (see case study in Section 7.7.1).

In American Samoa, recent work "indicates that mountain residential sites began to be occupied between about AD 1270 and 1310" (Pearl, 2004, p. 340).

In the islands of Chuuk (Micronesia), there is little information about the nature of the islands' undoubted human occupation during the Medieval Warm Period. In contrast, the Tonnachaw phase that began in A.D. 1300 was characterized by the construction and occupation of ridge-top settlements attesting to the onset of warfare (Rainbird, 2004). On the islands of Yap in the same part of the Pacific, "occupation of the inland for residential purposes [was] around AD 1300–1400" (Dodson and Intoh, 1999, p. 18).

A data-rich study of the Wainiha Valley on Kaua'i Island in the Hawaiian Islands showed that occupation began with coastal settlement A.D. 1030–1400 during the Medieval Warm Period (Carson, 2003). After this, residential sites shifted inland and coastal sites were occupied only sporadically. A similar story has emerged on O'ahu Island where "by the late AD 1200s or early 1300s, island-wide political changes were occurring" (Cordy, 1996, p. 597). The earliest sustained land use of the interior lowlands of O'ahu occurred after A.D. 1200 (Athens and Ward, 1997). Inland valley segments, such as Anahulu and Kawainui on O'ahu, were first occupied by humans later, during the fifteenth and sixteenth centuries (Dega and Kirch, 2002). Such inland valleys might be regarded – on account of their distance from the sea – as marginal, less desirable locations for settlement, similar to those in drier parts of the Hawaiian Islands that were settled about the same time. Another study on O'ahu found that occupation of the hot dry and generally inhospitable 'Ewa Plain began in earnest only during the A.D. 1300 Event (Fig. 5.12).

Easter Island was home to a largely peaceful, agricultural society that produced some of the most extraordinary monumental architecture anywhere in the Pacific Islands during the Medieval Warm Period. Yet around A.D. 1300, there are signs of change, notably the start of the exploitation of the island's obsidian for the manufacture of razor-sharp spearheads (*mata'a*) and the start of human occupation of caves, rockshelters and smaller offshore islands. The dating of the eventual collapse of Easter Island society is placed at about A.D. 1500 (Bahn and Flenley, 1992). A case study of Easter Island is given in Section 7.7.5.

Many of the mystery islands of the central eastern Pacific were occupied and perhaps abandoned during the A.D. 1300 Event. They are called mystery islands because, although showing no signs of recent human occupation when first encountered by Europeans, typically in the nineteenth century, these islands nonetheless showed signs of prolonged earlier occupation including house sites, tombs, *marae* (altars) and imported stone tools (Weisler, 1996a; Di Piazza and Pearthree, 2001a, 2004). The available radiocarbon dates for (latest) occupation of these islands suggest that they were occupied during the Medieval Warm Period but abandoned during the Little Ice Age, as long-distance voyaging effectively ceased. Examples of such islands are Kiritimati (Christmas) and Tabuaeran (Fanning) that were occupied, respectively, sometime within the period A.D. 1275–1425 and A.D. 1295–1405. A case study of the mystery islands is given in Section 7.7.6.

Settlement on Lelu, a smaller island off the main island of Kosrae (Federated States of Micronesia), increased markedly A.D. 1200–1350 (750–600 cal year

BP: Athens, 1995), probably as a result of the emergence of its fringing reef surface and the opportunity that this presented for extending lowland settlement in a secure (offshore) location.

There is some evidence of changes in human subsistence during the A.D. 1300 Event in the Pacific Islands. The inhabitants of many low-latitude Pacific Island groups adopted irrigated agriculture during the Medieval Warm Period, perhaps in response to the increasingly dry conditions that appear to have affected this region at this time (see Section 4.3.6). During the A.D. 1300 Event, the addition of large-scale dryland agriculture to the food-production portfolio on many islands likely reflects the increasing wetness of the climate and the realization that irrigation was no longer needed. A good example comes from leeward Hawai'i Island where the establishment of three large dryland field systems "dates to about AD 1300" (Kirch, 1986, p. 333).

There is enough information from some parts of the Pacific Islands to infer ecosystem changes that affected both the nature of marine foods consumed by humans and the strategies used to acquire them. One striking example comes from Kapingamarangi Atoll (Federated States of Micronesia) where significant changes in fish catch, as recorded by the archaeological excavation of fish bones, occurred about A.D. 1250 (700 BP: Leach and Ward, 1981). These changes are illustrated, together with comparable examples from Aitutaki Island (Cook Islands) and Guam Island (northwest Pacific), in Fig. 6.5.

The period around A.D. 1300–1400 (650–550 cal year BP) on Tikopia, Solomon Islands, was when techniques such as pit ensilage for the long-term storage of root crops were used first – a likely adaptation to increased climate variability (Kirch and Yen, 1982) or even extended periods of storminess (M.S. Allen, 1997).

Divisions of pre-history based on archaeological criteria are often chronologically indistinct and therefore crudely or controversially dated. Yet the possible coincidence of many fundamental changes throughout the Pacific Islands region with the A.D. 1300 Event is another line of evidence pointing towards a primary role for climate change in last-millennium societal evolution. For example, in the islands of the Marianas group, there is a fundamental distinction between the Pre-Latte Period and the Latte Period, dated by Craib (1990) to around the A.D. 1300 Event. The three main observed cultural changes in the Latte Period were the presence of latte stones (megaliths), symbols of tribal claims to areas of land at a time of increased resource competition; the dominance of undecorated utilitarian ceramics replacing the more time-consuming Marianas Red Ware; and increasing divergence in ceramic production (Graves et al., 1990; Hunter-Anderson and Butler, 1995).

There are other examples of abrupt changes in ceramic traditions elsewhere in the Pacific Islands around the time of the A.D. 1300 Event. For Rechtman (1992), the increase in the number of ceramic styles that existed in Fiji during the Vuda (Vunda) Phase, which perhaps began during the A.D. 1300 Event (start dates range from A.D. 1100 to A.D. 1450), marks increasing isolation of pottery-making communities. In the islands of central Vanuatu, ceramics ceased being manufactured about A.D. 1250 (700 BP: Spriggs, 1997), a similar date to the end of ceramic use on Tikopia and Vanikoro Islands in the eastern Solomon Islands (Spriggs, 1998).

KEY POINTS

1. Regarded as the most dramatic climate change in the past 5,000 years, the A.D. 1300 Event involved in the Pacific Basin a rapid temperature fall of perhaps 1.4–3.2°C within some 40–100 years.
2. Cooling throughout the Pacific Basin during the A.D. 1300 Event was accompanied by wetter conditions in most parts, likely to have been a result of increased storminess associated with an increase in the incidence of El Niño events.
3. Evidence for sea-level fall of perhaps 0.8–1.8 m during the A.D. 1300 Event exists in most parts of the Pacific Basin. Sea-level fall transformed coastal environments forcing, in turn, changes in the ways in which humans interacted with them in many places.
4. Anticipating the widespread societal crisis throughout the Pacific Basin during the early part of the Little Ice Age, there is evidence that during the A.D. 1300 Event many human groups experienced disruptions to their accustomed ways of life that are explainable by climatic and environmental changes.

CHAPTER 6

The Little Ice Age (A.D. 1350–1800) in the Pacific Basin

The 450-year long Little Ice Age in the Pacific Basin – and indeed in most parts of the world – was generally cooler than the Medieval Warm Period that preceded it (Chapter 4) and the period of Recent Warming that followed it (Chapter 8). The A.D. 1300 Event between the Medieval Warm Period and the Little Ice Age was a time of rapid climate change marked in many parts of the Pacific Basin by cooling, sea-level fall and an increase in storminess linked to an increase in the frequency of El Niño events (Chapter 5).

Yet it is misleading to portray the Little Ice Age as simply a period of cool conditions contrasting with the warmer conditions before and after. For the Little Ice Age was a time of highly variable climate compared to the Medieval Warm Period. It is likely that this variability caused as much of a problem for humans during the Little Ice Age in the Pacific Basin as did the general coolness. For people living off the land and accustomed to the seasonal reappearance or re-growth of key food items, climate variability will inevitably lead to insecurity and may, in places (especially smaller, more remote islands), eventually induce a food crisis, marked perhaps by aggressive competition for a diminished food-resource base.

This chapter is organized into major sections on climate, environment and society. Climate (Section 6.1) is divided into accounts of primarily Little Ice Age temperatures and precipitation in various subregions of the Pacific Basin. The section dealing with environment (Section 6.2) discusses mainly coastal environments and their responses to cool, often stormier, climate conditions, and a lower sea level compared to the Medieval Warm Period. The section about society (Section 6.3) focuses on population dispersal, changes in resource utilization, reduced interaction, and conflict, with subsections on other evidence for societal change such as the cessation of long-distance voyaging. A final section looks at the end of the Little Ice Age (Section 6.4).

6.1. CLIMATES OF THE LITTLE ICE AGE IN THE PACIFIC BASIN

There is abundant evidence that the Little Ice Age was experienced in almost every part of the Pacific Basin. Recent research has shown that there is reason to question whether Little Ice Age cooling affected the equatorial Pacific (Cobb et al., 2003) although some studies suggest it did (Evans et al., 2002).

One of the most persuasive lines of evidence for changed climate in the Pacific Basin during the Little Ice Age is provided by glacier advance (see Fig. 5.1). Glaciers generally advance downslope when both temperatures fall and precipitation does not decrease; cooling by itself is not enough. So Fig. 5.1 shows that cooling and probably increased precipitation across the A.D. 1300 Event were sufficient to cause many Pacific glaciers to advance. Yet these advances were not maintained regionally during the Little Ice Age, plausibly because precipitation fell and there was insufficient moisture available to drive continued glacier advance, irrespective of the cooler temperatures.

The other issue that is key to understanding the causes of last-millennium climate changes such as the Little Ice Age is synchronicity, specifically whether climate change was simultaneous in both north–south and west–east directions across the Pacific Ocean. Synchronous changes imply a dominance of global-scale climate forcing while asynchronous changes point to the influence of more localized factors. There is evidence that decadal-scale climate changes during the Little Ice Age were synchronous between the northern and the southern hemispheres (Kreutz et al., 1997) although that says nothing about their comparative magnitude or their impacts. Synchronicity is also evident latitudinally along the Pacific Rim (Sakaguchi, 1983; Schimmelmann et al., 2003) and in a west–east (longitudinal) sense (Holdsworth et al., 1992; Haberle and Ledru, 2001). Interannual climate variations associated with El Niño-Southern Oscillation (ENSO) and the Pacific Decadal Oscillation and Interdecadal Pacific Oscillation (PDO/IPO) may not be synchronous but they are predictable across the Pacific Basin.

In this section, climates of the eastern Pacific Rim during the Little Ice Age are discussed (Section 6.1.1) before those of the western Rim (Section 6.1.2). There is a case study concerning storm erosion in New Zealand (Section 6.1.3) and then an account of Little Ice Age climates in the Pacific Islands and Ocean (Section 6.1.4). Finally there is a review of ENSO variability during the Little Ice Age that also examines the question of whether or not a two-stage Little Ice Age can be recognized within the Pacific Basin as it is elsewhere (Section 6.1.5).

6.1.1 Climates of the eastern Pacific Rim (including Antarctica)

Reconstruction of palaeotemperatures from isotopic analyses of ice cores supports the widespread recognition of the Little Ice Age along the Pacific Rim of Antarctica (Benoist et al., 1982; Morgan, 1985). A decrease in bioproductivity in the ocean around Antarctica is explainable by increased sea-ice extent during the Little Ice Age, dated off the Pacific side of the Antarctic Peninsula to approximately A.D. 1250–1850 (700–100 cal BP: Domack et al., 2001, 2003).

In the Patagonian Andes, the earliest part of the Little Ice Age is represented by glacier advance, a likely response to the wetness of the A.D. 1300 Event here (see Section 5.2.1). But thereafter until around A.D. 1600, glaciers here did not advance, perhaps a response to drier conditions (Villalba, 1990; Koch and Kilian, 2005). The later part of the Little Ice Age (after A.D. 1600) saw periods of glacial advance and stability, probably reflecting a more variable precipitation regime.

Farther north in central Chile, the period A.D. 1450–1550 was uncommonly wet with a major El Niño event occurring A.D. 1468–1469 (Villalba, 1990). Droughts were especially common A.D. 1570–1650 and A.D. 1770–1820. Ocean-core analyses from offshore northern Chile show that nearshore upper ocean water temperatures were cooler during the Little Ice Age than during either the Medieval Warm Period or at present, perhaps an indication of the prevalence of La Niña conditions at this time (Ortlieb et al., 2000; Valdes et al., 2003).

Across the *altiplano*, the Little Ice Age was cold (see Fig. 4.1) and has been consistently identified in high-altitude palaeoclimate records from this area (Thompson et al., 1986). The Little Ice Age signal at dry Sajama was moderated by evaporation from nearby lakes (Thompson et al., 2003). Recent work has shown that there was a two-phase Little Ice Age – later than in most other parts of the Pacific Basin – that affected the entire *altiplano* (from Quelccaya to Sajama) simultaneously (Liu et al., 2005). The earlier phase (A.D. 1500–1720) was cold and wet, the later phase (A.D. 1720–1880) was cold and dry. The differences are attributable to changes in ENSO periodicity.

Compilations of last-millennium temperatures for the whole of North America suggest that conditions were generally cooler during the Little Ice Age. For example, Bradley and Jones (1992) showed that summer temperatures were ~1°C lower A.D. 1400–1700 than the average for the period A.D. 1860–1959. Tree-ring data from the Sierra Nevada constrain the time of coolness to A.D. 1450–1850 (Graumlich, 1993) although a shorter time of warmth appears in a composite regional record for the Canadian Rockies approximately A.D. 1340–1430 (see Fig. 4.2).

The available data for the western United States suggest that the Little Ice Age was generally wetter than the Medieval Warm Period yet marked by some severe droughts (see Fig. 4.3). Typical of the high-resolution palaeoclimate records from western North America is that from Owens Lake (Li et al., 2000). This was relatively deep during the A.D. 1300 Event (A.D. 1220–1480) but for most of the Little Ice Age (A.D. 1480–1760) it experienced dry/wet cycles with a 30–50-year periodicity (see Fig. 5.2). There are empirical grounds for recognising a two-phase Little Ice Age at Owens Lake, an earlier (cool) wet phase succeeded about A.D. 1550 by a (cool) generally drier phase with more variation.

The variability of precipitation during the Little Ice Age along the eastern Pacific Rim is well shown by the proxy analysis of lake sediments and dendrochronology from British Columbia (Watson and Luckman, 2001; Hallett et al., 2003). Although the Little Ice Age began wet in this region, there were dry periods – marked particularly by low lake levels – approximately A.D. 1470–1510, A.D. 1560–1570 and A.D. 1630–1650, and a conspicuously wet interval A.D. 1520–1550. A major drought in the A.D. 1790s affected much of North America (Watson and Luckman, 2001) but the first half of the nineteenth century was also marked by droughts in at least part of the area; a good example comes from Dog Lake, British Columbia (Hallett et al., 2003). Farther north in Canada, the period from A.D. 1450 was wetter than previous times along the Pacific Rim (Nederbragt and Thurow, 2001).

Palaeotemperatures in northwest Alaska reconstructed from tree rings show that the Little Ice Age here (A.D. 1450–1850) was mostly cooler than at present, but with warming approximately A.D. 1650–1725 (D'Arrigo et al., 2005).

Oxygen-isotope analyses of sediments from Lake Farewell in southern Alaska show that surface-water temperature A.D. 1200–1700 was ~1.25°C lower than during the Medieval Warm Period (Hu et al., 2001). Various high-resolution records extending back into the later Little Ice Age confirm cooler conditions in this part of the Pacific Rim (Moore et al., 2002).

The record of alternate dunefield expansion and stability in Alaska has been used to identify periods of comparative aridity, associated with fewer storms, and wetness, explainable by more frequent storms (Mann et al., 2002). The record from the Little Ice Age shows that the Great Kobuk Sand Dunes field was inactive A.D. 1400–1800 as a result of moister conditions at this high-latitude site. At Farewell Lake, conditions during the Little Ice Age were also generally wetter than during the Medieval Warm Period (Hu et al., 2001).

6.1.2 Climates of the western Pacific Rim

Following the sharp drop in temperature during the A.D. 1300 Event, the onset of the Little Ice Age about A.D. 1400 in lowland China is well marked and culminated in a series of cold spells 0.6–0.9°C lower than the 1951–1980 mean for the area, with a minimum of 1.1°C occurring in the period A.D. 1650–1680 (Yang et al., 2002; Ge et al., 2003). The coldest times of the Little Ice Age as derived from geochemical studies of sediments from Lake Huguangyan (tropical South China) were A.D. 1470–1550 and A.D. 1670–1730 which are consistent with the indications of the same from phenological data at A.D. 1470–1520 and A.D. 1620–1720 (Chu et al., 2002). A similar study showed cooling during the Little Ice Age with the three times of greatest cold centred on A.D. 1550, A.D. 1650 and A.D. 1750 (Hong et al., 2000).

For Japan, the compilation of Tagami (1996) showed that the Little Ice Age was cooler than the periods either side of it, and that there was distinctly more centennial-scale variability than during the Medieval Warm Period. Summers were anomalously cool A.D. 1730–1760, A.D. 1780–1790, A.D. 1830–1850 and A.D. 1860–1870 while winters were colder than the millennial average A.D. 1680–1700, A.D. 1730–1740 and A.D. 1810–1820. A study of tree-cellulose showed that temperatures in southern Japan A.D. 1600–1700 were ~2°C lower than the long-term average (Kitagawa and Matsumoto, 1995). This could be taken as contradicting the earlier assertion of Sakaguchi (1983) that a warm spell A.D. 1688–1703 interrupted the generally cold conditions of the Little Ice Age.

Precipitation during the Little Ice Age in eastern China varied, supporting the general picture of increased climate variability compared to the constancy of the Medieval Warm Period (Section 4.1.2). A good example comes from Buddha Cave, where conditions were generally wet A.D. 1475–1640 and dry A.D. 1640–1825 (see Fig. 4.4). Another record shows that the first part of the Little Ice Age was generally drier than the second half (see Fig. 5.3B). A synthesis for Japan regarded the Little Ice Age as relatively moist (Sakaguchi, 1983).

For reasons that are unclear, no evidence for the Little Ice Age was found from studies of seasonally laminated sediments at the lakes of the Long Gang Volcanic Field in northeast China (Mingram et al., 2004). It may be that in such areas of monsoonal climate, opposing conditions (warm and wet rather than cool and dry) prevailed during the Little Ice Age (Davis, 1994) or that monsoonal signals here simply dominate long-term interannual signals rendering these invisible in such palaeoclimate records.

The record of agricultural change in the New Guinea highlands – regarded as a proxy for climate change by Brookfield (1989) – shows that the dryland cultivation characteristic of the Medieval Warm Period was replaced during the transition to the Little Ice Age by raised-bed cultivation in wetlands. This change is thought to be an indicator of the replacement of warm climates having low variability by cooler climates punctuated by more frequent droughts and, later in the Little Ice Age, by more frosts.

For eastern Australia, there is evidence from borehole thermometry that temperatures in at least the sixteenth century were significantly lower than in more recent times (Huang et al., 2000). A 420-year record of sea-surface temperatures from the Great Barrier Reef confirms this conclusion, with temperatures A.D. 1565–1700 typically 0.5–1.0°C below the long-term mean (Hendy et al., 2002). The fire history of Worimi Swamp suggests that cooler conditions affected coastal New South Wales A.D. 1570–1720 (Mooney and Maltby, 2006).

The earliest last-millennium palaeotemperature reconstruction for New Zealand (A.D. 1300 to present) concluded that the country

"experienced a climatic deterioration starting around 1300 A.D. [which was] most severe between 1600 and 1800 A.D.". (Salinger, 1976, p. 311)

The study of Wilson et al. (1979) showed the existence of the Little Ice Age with the coldest times approximately A.D. 1600–1700, which agrees with the conclusion of a tree-ring study of the silver pine (*Lagarostrobus colensoi*) by D'Arrigo et al. (1998). The study of a cave stalagmite from Waitomo showed that mean annual temperatures A.D. 1430–1670 were ~0.8°C lower than today (Williams et al., 1999). Speleothem master chronologies for New Zealand show that temperatures started to fall rapidly around A.D. 1350, and then reached a trough about A.D. 1625 (Williams et al., 2004). This picture is confirmed by studies of glacier advance in New Zealand. Many glaciers reached their maximum extent for the last millennium about A.D. 1725–1740, followed by recession until about the late nineteenth century (Winkler, 2004).

A sensitive indicator of periods of both warmer and wetter conditions during the Little Ice Age in New Zealand is provided by records of storm erosion, discussed as a case study below.

6.1.3 Storm erosion in New Zealand

A relationship between climate change and the evidence for sustained erosion attributable to increased storminess was first posited systematically for New Zealand by Grant (1981) who extended his findings – to far from universal approbation – to pre-historic changes in New Zealand society. Most of Grant's conclusions regarding the value of erosion records to broader palaeoclimate reconstructions have now been accepted while his idea that climate was the "prime initiating force" (Grant, 1994, p. 187) in societal change during the Little Ice Age in New Zealand is implicit in the present book.

Apparent New Zealand-wide periods of erosion in pre-European times correspond to warmer intervals. During such times, large storms reach at least northern New Zealand from the tropics and subtropics, something that does not happen during cooler periods. Recent work on Lake Tutira near the east coast of North Island (Page and Trustrum, 1997; Eden and Page, 1998) shows that at least two periods of storminess registered here coincide with those identified in Grant's work and "support the concept of climatically driven periods of increased erosion" (Eden and Page, 1998, p. 53).

The only such period within the Little Ice Age is the Burrell (A.D. 1575–1595) period of Eden and Page (1998) which falls within the Matawhero (A.D. 1510–1620) period of Grant (1994). The Burrell period was a period of comparatively high El Niño activity, as is implied for earlier storm periods (Eden and Page, 1998). Three significant El Niño events occurred within the Burrell storm period (Ortlieb, 2000).

6.1.4 Climates of the Pacific Islands and Ocean

As determined from coral $\delta^{18}O$ data (a proxy for sea-surface temperature), a sustained cool phase of sea-surface temperature occurred A.D. 1550–1895 in the central Pacific (NINO3.4 region – Evans et al., 2002). This approximates to the Little Ice Age and is supported by other coral palaeotemperature records from the tropical Pacific. These include those from Socas Island, Panama (Linsley et al., 1994), and the southern Great Barrier Reef (Druffel and Griffin, 1993). The record from Amédée Lighthouse in New Caledonia shows that mean annual sea-surface temperature here was ~0.3°C lower than the twentieth-century average (Quinn et al., 1998).

Yet coral coring elsewhere in the tropical Pacific appears to show no such clear evidence for the Little Ice Age. A record from Rarotonga Island in the southern Cook Islands shows that sea-surface temperatures during parts of the eighteenth and nineteenth centuries were at least as warm as the twentieth century (Linsley et al., 2000). Coral coring on the Great Barrier Reef revealed a warm period A.D. 1720–1880 during which temperatures were as high as during the early 1980s (Hendy et al., 2002). The long Urvina Bay record (A.D. 1600–1950) from the Galapagos Islands exhibits no readily discernible long-term trend that could be interpreted as Little Ice Age cooling (Dunbar et al., 1994).

It remains possible that not every part of the entire tropical Pacific Ocean and Islands experienced a significant or sustained cooling during the Little Ice Age. If this is correct, then it is possible that the cooling that characterizes the Little Ice Age in most other parts of the world was actually caused by increased poleward transport of water vapour from the tropical Pacific (Hendy et al., 2002). Yet, as noted above, there does seem adequate evidence from the Pacific Rim in the southern hemisphere to dispel any idea that the Little Ice Age did not occur there, even though its effects may have been muted compared to the northern hemisphere.

A coarsely calibrated last-millennium record from the Galapagos Islands suggests that the Little Ice Age (within the period A.D. 1150–1950) was uncommonly dry in this area of the eastern Pacific (Steinitz-Kannan et al., 1997), contradicting the earlier suggestion quoted by Sanchez and Kutzbach (1974) that these islands were wetter about A.D. 1650–1850 (300–100 BP) than either before or after. There is evidence for an increased incidence of drought during the Little Ice Age on Easter Island (Hunter-Anderson, 1998). On the island Nuku Hiva in the Marquesas, there is a suggestion of cooler wetter conditions about A.D. 1450 (500 BP: Sabels, 1966).

Recent research has shown that "conditions in the tropical southwestern Pacific during the Little Ice Age were also consistently more saline" (Hendy et al., 2002, p. 1513) than either before or after. It seems likely that the increased salinity is explicable by reduced precipitation (reduced freshwater input) signifying comparatively dry conditions during the Little Ice Age.

6.1.5 ENSO variability and a two-phase Little Ice Age

If climate variability during the Medieval Warm Period was low, then the situation during the Little Ice Age was quite different. For in common with most reconstructions of ENSO variability spanning the last millennium, it is clear that the Little Ice Age in the Pacific Basin was characterized by a much higher degree of climate variability than the Medieval Warm Period, being associated with a greater incidence of ENSO variations, particularly a higher frequency of El Niño events (see Fig. 5.6).

A key empirical test of this is whether or not periodic droughts were synchronous in diagnostic places along the eastern and western Pacific Rim as well as the Pacific Islands, at least their low-latitude parts, and therefore linked to El Niño. This seems to be true (Haberle and Chepstow-Lusty, 2000). A good example from the Little Ice Age is the El Niño of 1788–1793 during which memorably severe droughts not only extended across both sides of the Pacific but also affected the equatorial Pacific Islands (Ortlieb, 2000).

In the central equatorial Pacific Ocean – another key area for reconstructing palaeo-ENSO frequency – there is evidence suggesting that the frequency of both El Niño and La Niña events was higher during the middle Little Ice Age (A.D. 1650–1680) than at any earlier time during the last millennium (Cobb et al., 2003).

In the tropical southwest Pacific, the pattern of ENSO variations during the later Little Ice Age was indistinguishable from that of modern times (Corrège et al., 2001). Several ENSO data series show no significant variation in ENSO variability within the Little Ice Age. So it can be concluded that whatever happened to cause El Niño events to become more frequent during the A.D. 1300 Event (as shown in Fig. 5.8), something similar did not happen across the transition from the Little Ice Age to the period of Recent Warming in the Pacific Basin.

This conclusion is germane to the question of whether or not a two-phase Little Ice Age occurred in the Pacific Basin, as happened elsewhere in the world (Keigwin, 1996; deMenocal et al., 2000; McDermott et al., 2001). In the Pacific, a few palaeotemperature records hint that this may have been the case (e.g. Fig. 4.2A) but the strongest evidence comes from palaeoprecipitation records along the Pacific Rim. Examples from the eastern Pacific Rim include the Patagonian Andes where A.D. 1600 marks the division between wet and dry stages of the Little Ice

Age (Koch and Kilian, 2005), the Andean *altiplano* where A.D. 1720 marks a similar division (Liu et al., 2005), and Owens Lake in the western United States where A.D. 1550 marks the same (Li et al., 2000). Two studies from China reach similar conclusions, A.D. 1640 marking the point at which the Little Ice Age changed from wetter to drier at Buddha Cave (see Fig. 4.4) and A.D. 1760 the same at Shihua Cave (see Fig. 5.3B). In New Zealand, the change in glacier behaviour A.D. 1725–1740 noted by Winkler (2004) is attributable to a shift from wet to dry conditions.

In summary, there seems good evidence for supposing that, while the Little Ice Age in the Pacific Basin was generally cooler than periods before and after, its earlier part was generally wetter than its later part. This observation can be readily explained by ENSO-linked storminess; in Fig. 5.6D, the frequency of El Niño events is slightly higher during the early part of the Little Ice Age compared to later times. But a two-phase Little Ice Age may also have nothing to do with ENSO variability, rather being a result of cooler temperatures during the early Little Ice Age increasing equator–pole temperature gradients and therefore wind strengths compared to the later part of the Little Ice Age. The paucity and imprecision of the data available render these conclusions somewhat speculative at present.

6.2. ENVIRONMENTS OF THE LITTLE ICE AGE IN THE PACIFIC BASIN

Changes in the environments of the Pacific Basin during the Little Ice Age are considered under three headings – coastal, offshore and inland. For all these environments, considerable change was experienced during the A.D. 1300 Event, largely as a result of sea-level fall (Section 5.3), the legacy of which continued to be felt during the Little Ice Age.

6.2.1 Coastal-environmental change

Most coastal environments affected by sea-level fall during the A.D. 1300 Event continued to experience lower-than-present sea level for most of the Little Ice Age. Supporting data from the Pacific are few, perhaps because much of the evidence is now underwater. One corroborative study comes from the near-coastal Nakaumi Lagoon in Japan, where palaeosalinity was low for most of the Little Ice Age, suggesting that sea level was also low at this time (see Fig. 4.5). Among the many human proxy studies suggesting that sea level remained low during the Little Ice Age is the evidence throughout Japan for reclamation of tidal flats in this period, particularly from about A.D. 1600, following the Edo Regression (Sakaguchi, 1983). Another such study is the spread of fishponds in the Hawaiian Islands (see Fig. 6.6). Both these developments are consistent with the emergence of large areas of what had during the Medieval Warm Period been shallow ocean floor making them available for use by humans.

The other evidence of continued low sea level during the Little Ice Age comes from the apparent lack of change – at least in the early part of the period – to the shorelines exposed by sea-level fall during the A.D. 1300 Event (see Fig. 5.8). This not only allowed their utilization by humans but also their incorporation into terrestrial landscapes. Good examples are Ropera Swamp on Tikopia Island in

Solomon Islands (see Fig. 5.11) and Kawai Nui Marsh on O'ahu Island in Hawaii (see Fig. 6.1B).

Another result of the expansion of new land areas at the start of the Little Ice Age was the accumulation of coastal sand dunes. While aridity and increased windiness are clearly important factors in this process, the sea-level fall that in many places marked the A.D. 1300 Event exposed large areas of sandy sea floor thereby presenting a source as well as targets for windblown sand. Much of this sand was blown inland during the Little Ice Age, sometimes accumulating some distance from the exposed shoreflat, sometimes on it. A good example comes from Kunashir Island (Kurile group, northwest Pacific) where the Little Ice Age was

"characterized by intensive accumulation of aeolian material and input of sand to soil profiles in the coastal area". (Korotky et al., 2000, p. 329)

A similar conclusion was reached for nearby Iturup Island (Razjigaeva et al., 2002). A typical feature of these dunes is the absence of any soil cover, and it seems likely that the same processes that operated here led to the decline of the valley-lowland civilizations of the Moche and Chimú cultures along the Pacific Rim of South America.

Other examples of sand-dune accumulation along the Pacific Rim come from New Zealand, where the influence of natural (rather than human) factors in sand-dune accumulation can be inferred from the synchrony of periods of dune accumulation and stability on mainland New Zealand and the Chatham Islands, 750 km offshore (McFadgen, 1994), for while human impact has been proposed occasionally as a likely contributor to periods of dune instability on the main islands, this is implausible for the remote, sparsely populated Chatham Islands. Thus, in both areas it seems more likely that the effects of sea-level fall perhaps combined with increased storminess led to a dune-building episode approximately A.D. 1500–1550 (450–400 BP).

Another closely argued study of sand-dune accumulation during the Little Ice Age was that of Kumar et al. (2006) in which sand dunes at the mouth of the Sigatoka River in southwest Viti Levu Island, Fiji, were interpreted as a result of sediments being deposited during the early Little Ice Age across a strandflat exposed by lower sea level (Fig. 6.1A). Sand dunes in this location are composed of sediments from the island's interior, released following land clearance associated with initial inland hilltop settlement in the aftermath of the A.D. 1300 Event (see Fig. 6.4).

Another way in which the landscapes of newly emerged shoreflats evolved during the Little Ice Age was in the enclosure of what had earlier been shoreline embayments, typically creating brackish-water lakes or wetlands in the place of saltwater inlets. This situation was described in Section 5.3.3 as part of the coastal-environmental evidence for sea-level fall during the A.D. 1300 Event. But many examples come from observations of how humans interacted with coastal environments in the Pacific during the Little Ice Age. By way of illustration, two examples shown in Fig. 6.1 are discussed further below.

Prior to the Little Ice Age, Kawai Nui Marsh in eastern O'ahu Island in the Hawaii group (Fig. 6.1B) was transformed from being a coastal inlet to a lagoon

progressively blocked by the Kailua sand barrier, and the emerged reef that forms its base, to a brackish-water lake and marsh (J. Allen, 1997). There is some suggestion that infilling of the lake during the early Little Ice Age was associated with increased soil erosion from the surrounding hillslopes, something that may have been a result of their intensified use for agriculture and/or the impacts of higher levels of precipitation, particularly during storms. The existence of a marsh at Kawai Nui during most of the Little Ice Age allowed its use for taro pondfield agriculture throughout this period.

Qaranilaca is a large coastal cave at the southernmost tip of Vanuabalavu Island in northeast Fiji (Fig. 6.1C). During the Medieval Warm Period, its floor was inundated regularly at high tide; there is no evidence for human occupation until around A.D. 1250 (700 BP) when the sea level is likely to have fallen sufficiently to render the inner parts of the cave permanently dry and habitable. The modern cave fill, pictured behind the people sitting in Fig. 6.1C, reaches 1.2 m thick and shows signs of temporary human occupation throughout the Little Ice Age. It is likely that Qaranilaca served during this period as a sheltered and secure place for marine-food processing prior to foods being carried to the hilltop settlements that existed in the vicinity (Thomas et al., 2004).

Finally in this section, it is pertinent to discuss how sea-level fall can have an impact on landscape denudation rates irrespective of human impact (Bull, 1991). The form of graded river channels is controlled by base level, commonly sea level, and, should this change, it is expected that channel form will respond. The typical response to sea-level rise is for river channels to build up and for aggradation to occur on the surrounding lowland floodplains. In this scenario, valley-side slopes would become less steep, reducing gravity-driven movement of sediment downslope. Conversely, when sea level falls, river channels incise, and valley-side slopes steepen, thereby increasing the amount of sediment washed downslope under the influence of gravity.

What this means is that, given that sea level appears to have fallen in many parts of the Pacific Basin prior to the start of the Little Ice Age, it might be expected that terrestrial environments would have been adjusting to the effects of this during the early part of the Little Ice Age. Specifically, it would be expected that increased amounts of terrigenous sediment washed off valley-side slopes in coastal hinterlands would be arriving in coastal areas, perhaps causing shorelines

---

FIGURE 6.1 Examples of coastal-environmental changes marking the beginning of the Little Ice Age in the Pacific Islands. (a) The high sand dunes at the mouth of the Sigatoka River, Viti Levu Island, Fiji, formed during the early part of the Little Ice Age in response to accelerated inland erosion associated with that area's sustained human settlement following the A.D. 1300 event [photo by Patrick Nunn]. (b) Kawai Nui Marsh on O'ahu Island in Hawaii is a wetland that formed during the A.D. 1300 Event. Sea-level fall at this time converted an open bay into a wetland [photo courtesy of Eric Guinther]. (c) Qaranilaca (cave) on Vanuabalavu Island in Fiji is a limestone sea cave that was occupied mostly only during the Little Ice Age, as a place where seafood gathered from the adjacent reef and nearshore areas was processed prior to being carried to hilltop settlements in the vicinity. The people are sitting on the modern shore platform, behind which is exposed a 1.2-m thick fill of largely cultural material [photo by Patrick Nunn].

to extend (prograde) or contributing to the growth of dune fields. It is difficult to tease apart the various potential causes (non-human and human) of coastal-environmental change during the (early) Little Ice Age, but that is no excuse for blithely explaining all such change by human impact.

6.2.2 Offshore environmental change

The effects of sea-level fall during the A.D. 1300 Event on offshore environments (those below mean sea level) were most marked in nearshore areas. This is because at the time of higher sea level during the later Medieval Warm Period, there were – as there are today – many shallow-water environments where marine foods abounded and which were accessible to coastal dwellers.

Outside the tropics, shoreflats exposed at low tide would also have been favoured places for the collection of edible shellfish, which played a major role in the subsistence economy of many people occupying temperate Pacific coasts during the Medieval Warm Period. During the Little Ice Age, as a result of the intervening fall in sea level, the shore ecology of many such offshore areas changed, with a consequent change in the nature and pattern of human predation. Examples include the abrupt changes recorded in the Bass Point midden (east Australia) discussed in Section 5.4.3, and those in New Zealand middens at the same time (Rowland, 1976; Szabó, 2001).

In tropical parts of the Pacific Basin during the Medieval Warm Period, many nearshore environments were lagoonal, bordered by coral reef and containing diverse ecosystems dependent on the regular tidal flushing of lagoons by ocean water pouring across the reef around the time of high tide, and moving largely unimpeded through lagoons. Yet after the sea-level fall that marked the A.D. 1300 Event, many reefs bordering coastal lagoons emerged, presenting barriers to routine tidal flushing, meaning that water exchange between lagoons and the open ocean became confined to narrow reef passes (where these existed). Within the lagoons, many formerly shallow areas – from shoreflats to patch reefs – would also have emerged, slowing water movement within lagoons. The combined effects would have significantly reduced water quality and circulation within many Pacific coastal lagoons, rendering them far less productive for the purposes of human subsistence than they had been previously.

Nowhere within such systems would seafood production have been so affected as on the surfaces of the emerged reefs themselves. Significant amounts of reef foods for human consumption are found within the upper 20–30 cm of reef surfaces where primary productivity levels are highest. Thus, when those reef surfaces emerge (as a result of relative sea-level fall), food production falls sharply.

Some inferences about the health of ecosystems in tropical lagoonal settings during the Little Ice Age can also be made from changes in the pattern of associated human subsistence. On the northwest Pacific islands of Guam and Saipan, for example, insights have been gleaned from looking at the changes in percentage of different shellfish within last-millennium middens (Amesbury, 1999). In particular, the percentage of *Anadara antiquata* has been demonstrated to have declined when sea level fell – as it appears to have done during the A.D. 1300 Event – but also to have risen again when sea level stabilized and mangrove forests developed, as appears to have been the case during the Little Ice Age (see Fig. 6.5C). When sea level falls, large areas of reef would be killed and covered with sand, providing habitats favourable for *Strombus* (Paulay, 1996), numbers of which increase as those of *Anadara* decline during the Little Ice Age (Amesbury, 1999).

A similar situation appears to have existed at the Natia Beach site on Nacula Island in Fiji (Alex Morrison, personal communication, 2006). Here, after some 1,600 years of constant shellfish availability, there was a drastic reduction in the amount of shellfish available around A.D. 1300 (650 BP), something that is most plausibly explained by increased lagoonal sedimentation resulting from sea-level fall.

6.2.3 Inland environmental change

It is difficult to distinguish inland environmental changes caused by climate and sea-level change during the Little Ice Age from those caused by the accompanying societal changes, particularly the first sustained occupation of upland inland areas by people whose ancestors during the Medieval Warm Period had lived along the coast. But it is important to recognize the potential contribution of both natural and human influences on observed changes in particular places.

As noted above, on many Pacific coasts, the Little Ice Age was marked by increased lowland sedimentation and shoreline progradation attributable to increased inputs of sediment from inland areas transported to coastal areas by rivers. While such observations can be explained by increased human impact on coastal hinterlands, perhaps associated with the abandonment of coastal settlements, it is also likely that there were significant contributions from non-human factors in places. Increased storminess during particular times of the Little Ice Age would clearly have had major effects on inland areas. The increased climate variability that characterized the Little Ice Age would have undermined long-established methods of terrestrial food production, perhaps forcing new ways of production or even a return to dependence on largely wild foods, as happened in New Zealand (see Section 6.3.2).

Finally, as noted above, relative sea-level fall during the A.D. 1300 Event is likely to have caused valley-side slopes to steepen and more sediment to be moved downslope and downstream. Although it is challenging to separate such non-human impacts from possible human impacts occurring at the same time, it is clearly unsatisfactory to attribute all upland erosion during the Little Ice Age to the latter merely by default.

One place where it may be possible to isolate non-human causes of landscape change during the Little Ice Age is Yadua Island, a 13.6 km^2 volcanic island in the dry part of northern Fiji. Probably never able, on account of the paucity of fresh water, to support more than the 200 or so people that it does today, it therefore seems likely – as oral traditions suggest – that many coastal flats and their hinterlands were unoccupied during the Little Ice Age. One such coastal flat is at Vagairiki on the island's south coast, where a series of landslides covered the coastal flat A.D. 1430–1670 (Fig. 6.2; Nunn et al., 2005). It is plausible to suppose that these landslides occurred for reasons that had nothing to do with human impact but did so largely because of the steepening of valley-side slopes resulting from sea-level fall during the A.D. 1300 Event.

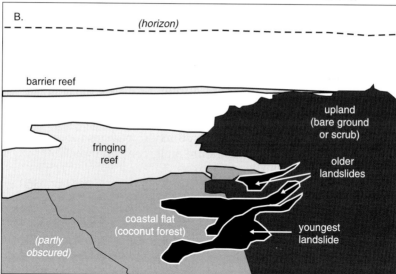

FIGURE 6.2   Evidence of landscape change during the early Little Ice Age possibly associated with steepening of valley-side slopes resulting from relative sea-level fall during the A.D. 1300 Event comes from Yadua Island in northern Fiji (figure after Nunn et al., 2005). (A) Photo of the southwest part of the Vagairiki coastal flat in southern Yadua [photo by Patrick Nunn]. (B) Interpretative sketch of the same area showing the landslides that covered it during the early Little Ice Age. The bareness of the upland is due to the impact of goats, introduced in the mid-twentieth century.

6.3. SOCIETIES OF THE LITTLE ICE AGE IN THE PACIFIC BASIN

In many parts of the Pacific Basin, profound differences can be observed in human societies between the Medieval Warm Period and the Little Ice Age. Many of these differences are attributable to the processes of climate change, environmental change and societal change during the A.D. 1300 Event described in Chapter 5.

The principal factor responsible for societal change between the Medieval Warm Period and the Little Ice Age appears – in many coastal parts of the Pacific Basin – to have been a significant drop in the amount of food available for humans. While the associated food stress is likely to have been the main reason behind the changes in settlement pattern and conflict that marked the early Little Ice Age in many parts of this region – a view that underpins the model discussed in Chapter 7 – this is ultimately unprovable.

Yet there is ample direct evidence of food stress in many Pacific Basin societies during the (early) Little Ice Age manifested by skeletal signs of malnutrition and evidence for unsatisfactory diets. Examples include the sixteenth-century skeletons from Palliser Bay in New Zealand that exhibited Harris lines in limb bones, a sign of near starvation (Leach, 1981). And in Fiji, around A.D. 1400 when many coastal settlements were abandoned, the inhabitants of the Sigatoka River Delta – long a favoured site for human occupation – "were foraging widely, if not desperately, for food resources" (Burley, 2003, p. 313). A rise in interpersonal violence around A.D. 1300 on the Channel Islands off California was marked by a number of non-life-threatening injuries sustained by a high proportion of people at that time (Walker, 1989), likely to have arisen as a result of food shortages during the later part of the Medieval Warm Period (Raab and Larson, 1997). A similar conclusion can be reached for the mainlanders living at nearby Calleguas Creek, ~10% of whom died from arrow wounds (Walker and Lambert, 1989). This may have been part of a more widespread rise in conflict around this time, dates for the construction of fortified settlements among the Anasazi people in the southwest United States being 600–1,100 years ago and in the Great Plains ~500 years ago (Bamforth, 1994). The origins of conflict in the Pacific Basin around this time are discussed further in Section 6.3.7.

Critics will argue that this is very slender evidence from a vast region to support the environmental-determinist view advanced in this book. Yet, as will be seen in the present section, there is a vast array of evidence to show that societies across the Pacific Basin collapsed during the early Little Ice Age. The near simultaneity of this collapse demands a region-wide driver, moderated of course by local conditions, and such a driver cannot therefore be something that was driven from within particular societies, only from outside them all. This argument is extended in Chapter 7.

The present section is followed by overviews of societies during the Little Ice Age along the Pacific Rim (Sections 6.3.1 and 6.3.2) and in the Pacific Islands. Owing to their circumscribed and often comparatively remote nature, those societal changes likely to have been climate-driven were commonly amplified on Pacific Islands compared to the continental rim, where societal stresses could be more readily reduced by dispersal. For this reason, the remainder of the discussion of societal changes during the Little Ice Age in the Pacific Basin focuses on its islands (Sections 6.3.3–6.3.7). A number of other societal responses to resource depletion during the Little Ice Age on Pacific Islands are discussed in Section 6.3.8.

6.3.1 Overview of societal change along the Pacific Rim

As noted above compared to Pacific Islands, the societal effects of food crisis during the early Little Ice Age were muted along most of the Pacific Rim because of the existence of vast hinterlands, often sparsely populated, and a greater number of alternative options for sustenance. This is not to trivialize the problems that the A.D. 1300 Event brought to many Pacific Rim societies, some of which endured great tribulations during the Little Ice Age in consequence.

Examples can be found throughout the eastern Pacific Rim. Most settlements in coastal Ecuador were abandoned after A.D. 1400 (Paulsen, 1976). On the Santa Delta in Peru, the Little Ice Age was coincident with the Late Tambo Real Period (beginning about A.D. 1350) that was characterized by the diminishment of settlements in open coastal areas (Wells, 1992).

As elsewhere in the Pacific Basin, there is evidence from the Pacific Rim in South America that many of the innovations of the Medieval Warm Period that were probably introduced to help societies cope with increasing drought frequency (and intensity?) were abandoned early in the Little Ice Age. This was probably due to the increased climate variability, although evidence for a clear connection has not been found. Examples from Peru come from the Pampa de Chaparrí where irrigation canals and furrowed fields were abandoned ~500 years ago (Nordt et al., 2004) and at Wawakiki where abandonment occurred after A.D. 1400 (Zaro and Alvarez, 2005).

Mayapán and other cities of the Postclassic Maya period in Yucatán were abandoned during severe drought and cold A.D. 1451–1454 (Curtis et al., 1996). The decimation through flooding sometime between A.D. 1100 and A.D. 1300 of the shellfish population in a Mexican estuary led to people abandoning this region completely (McGoodwin, 1992). On Santa Cruz Island in the Channel Islands off southern California, there was widespread abandonment of settlements, populations departed "virtually simultaneously" in the period A.D. 1200–1300 (Arnold, 1992).

In the northeast Pacific, the Little Ice Age was a time when there were abundant Alaskan salmon available in nursery lakes in places such as Kodiak Island in the Gulf of Alaska (Finney et al., 2002). This may reflect changes in ocean–atmosphere circulation in this region that allowed the high levels of salmon productivity achieved during the later Medieval Warm Period to be sustained, but it is also possible that societal changes at this time caused the human exploitation of these salmon resources to fall off significantly, allowing optimal numbers to be maintained despite a deterioration in the conditions for high production levels.

Along the western Pacific Rim, the climate variability of the Little Ice Age brought about often profound societal disruption throughout East Asia. For example, in the 1630s widespread drought in China led to numerous uprisings, harshly put down by the Ming empire, and increasing numbers of Manchu attacks along its northern borders (Fagan, 2000). Climate variability affected even the societies of the fertile Changjiang valley, first drought and then flooding which brought about famines in which millions died. The resultant disruption led to the fall of the Ming Dynasty and its replacement by the Manchus. Similar effects were experienced in Korea and Japan around the same period.

A study of aboriginal occupations of the arid margin of southeast Australia showed that the area, abandoned for more desirable areas during the Medieval Warm Period, was reoccupied shortly after the A.D. 1300 Event (Holdaway et al., 2002). In New Zealand, settled first around A.D. 1250–1300 (Hogg et al., 2003; Wilmshurst and Higham, 2004), there was a wholesale move from open coastal settlements to upland fortified settlements (*pa*) by A.D. 1500 (Schmidt, 1996), perhaps earlier. In terms of societal change during the Little Ice Age, New Zealand is one part of the Pacific Rim that has been particularly well studied. The main reason for this is that it was probably only colonized 100 years or so before the start of the Little Ice Age, possibly even in response to environmental deterioration in the source island groups of the tropical eastern Pacific (see Section 4.3.7). New Zealand is discussed in a case study below.

6.3.2 Crisis in New Zealand

New Zealand, while geologically part of the Pacific Rim and formed of islands that are larger and more temperate than most Pacific Islands inhabited during the early Little Ice Age, was nonetheless colonized by descendants of the same people who colonized most low-latitude Pacific Islands. The changes in settlement pattern that marked the first few hundred years of New Zealand's history are similar to those that took place at the same time on many smaller Pacific Islands that had been settled far longer. The synchronicity of these changes between New Zealand and lower latitude island locations – areas with sharply contrasting settlement histories – reinforces the case for an external driver of these changes.

The earliest settlements in New Zealand were mostly coastal but "around 500 years ago [A.D. 1450], settlement entered a new phase" (McGlone et al., 1994, p. 152). This was marked by

- fewer coastal settlements;
- a sharp decline in the consumption of marine foods (Davidson, 1984), possibly a consequence of coastal-ecological changes associated with cooling and sea-level fall (Nunn, 2000a, 2000b); and
- the replacement of a horticulture-based lifestyle with "a subsistence mode which characterised other island populations only during famine or when pushed into highly marginal lands" (McGlone et al., 1994, p. 156).

Empirical studies in support of this scenario are provided by changes in key edible shellfish species (Rowland, 1976; Szabó, 2001); the latter cites environmental change as the likely cause of the disappearance at around A.D. 1300–1400 (650–550 BP) of the limpet *Cellana denticulata* from North Island sites where it was previously dominant. Modelling on the basis of studies of human mtDNA also suggests a scarcity of food resources early in the Little Ice Age (Whyte et al., 2005).

This "new phase" of New Zealand settlement began early within the Little Ice Age and appears to have been associated with a widespread food crisis. One synthesis of evidence from the southern Wairarapa coast of North Island portrayed this time (A.D. 1450–1550) as a period of "environmental decline" associated with the "loss" of shellfish beds, a marked drop in interaction and conditions of "near starvation" (Leach, 1981).

During the early Little Ice Age, most lowland settlements were abandoned with people moving upslope, upvalley or far inland from coasts. An example of an upslope settlement of the kind established at this time – located for defence and fortified – is shown in Fig. 6.3. A good example of upvalley movement comes from the Coromandel Peninsula in North Island, where discrete valley-side fires in the decades leading up to A.D. 1470 have been linked to intentional forest burning, perhaps to stimulate the growth of edible bracken fern (*Pteridium esculentum*) around valley-side settlements (Byrami et al., 2002). During the Little Ice Age in New Zealand, this wild fern became the staple food for most people, the cooler and more variable climate conditions making agriculture less successful than it appears to have been around the time that New Zealand was colonized (McGlone, 1983).

An influential study of last-millennium settlement-pattern change in New Zealand is that of the Palliser Bay settlements in the south of North Island which

FIGURE 6.3   The *pa* (or "fortified village") at Tolaga in New Zealand, as seen by Herman Spöring in the mid-eighteenth century, and published ~1784 [courtesy of the Alexander Turnbull Library, National Library of New Zealand]. The wooden palisade is typical of many *pa*, as is the construction in a naturally defensible site.

are thought to have started to be abandoned around A.D. 1400 because of climate change (Leach and Leach, 1979); tsunami may be among contributory causes of site abandonment (Goff and McFadgen, 2003). The original climate-linked explanation of Leach and Leach involved increased precipitation during the A.D. 1300 Event, flooding, shoreline erosion, and the disappearance of filter-feeding shellfish from nearshore areas in response to "increased sedimentation from the rivers" (Grant, 1994, p. 186).

There is evidence – as on many tropical Pacific Islands – that the abandonment of the Palliser Bay settlements was a two-stage process. The first people to live in Palliser Bay developed a lifestyle based on marine foods (primarily shellfish), forest-dwelling birds, and horticulture. Yet during the early part of the Little Ice Age, signs of stress appeared, with settlements moving as much as 2 km up the valleys and the first signs of conflict appearing (Leach and Leach, 1979; Grant, 1994). Skeletal evidence for periodic malnutrition or near-starvation date from this time (Leach, 1981) and shortly afterwards the area was abandoned altogether.

The food crisis at this time was also marked by the fact that many defensive positions were established to enclose food-storage pits, both for long-term food preservation and for protection of food resources from competing groups. These upland, often inland, sites (named *pa*) were often fortified and became the main settlements for people throughout New Zealand during the Little Ice Age (Fig. 6.3). While most writers have concluded that *pa* establishment in New Zealand was approximately simultaneous, there is uncertainty about the age of this. The survey by Schmidt (1996) concluded that *pa* occupation had become widespread by around A.D. 1500 but this date was challenged by Kirch (2000a) who suggested that a date of A.D. 1300, originally suggested by Groube (1971), is more accurate.

Not only is the early Little Ice Age in New Zealand marked by settlement movement inland, but it is also marked – as it is throughout the Pacific Islands (see Section 6.3.4) – by the permanent occupation of offshore islands. Life may have been more difficult for many reasons in such places but they had the advantage of being more secure from attack than mainland locations. One example is the permanent occupation of offshore Mayor Island that began about A.D. 1500. This may have been a human response to crisis on the larger islands of New Zealand but may also have been to secure control of the island's obsidian source – one of the few in New Zealand (Empson et al., 2000). A comparable situation occurred off the coast of tropical east Australia, where many smaller islands were permanently occupied for the first time only during the Little Ice Age. These include North and South Keppel islands (Rowland, 1985).

6.3.3 Overview of societal change in the Pacific Islands

The changes that characterized most Pacific Island societies during the early Little Ice Age are explainable by food crisis, amplified compared to the continental Pacific Rim by insularity, and amplified even more on smaller, more remote islands and those with higher population densities. Human responses to crisis were also conditioned by geography, various options being open to inhabitants of larger islands that were comparatively lightly populated which were not available to inhabitants of many smaller islands, including atoll islands. So there is considerable variability in both the manifestations of crisis and people's responses to it, variability that is in turn obscured by the incomplete and often inconclusive evidence available.

For example, in the Kingdom of Tonga, major political changes took place at least twice during the Little Ice Age, first around A.D. 1450 (500 BP) and next around A.D. 1600 (350 BP) (Burley, 1998). It is possible that these changes were unrelated to external (non-human) forcing although it has been suggested that the A.D. 1453 eruption of Kuwae Volcano in Vanuatu – the largest eruption on Earth within the past 700 years (Gao et al., 2006) – brought severe and widespread famine to islands in the region, leading to societal disruption which in Tonga was expressed by the establishment of a new dynasty (Luders, 1996).

Despite such uncertainties, it is clear that four major changes to almost all Pacific Island societies occurred during the Little Ice Age. The first is population dispersal (Section 6.3.4), interpreted largely as a response to the food crisis brought about by cooling and sea-level fall during the A.D. 1300 Event. The second is the change in pattern of resource utilization (Section 6.3.5), ranging from changes in the nature and strategies of food acquisition to changes in the raw materials used for tool manufacture. The third (Section 6.3.6) is that of reduced interaction, both long-distance and subregionally, between island communities that also led to the isolation and abandonment of particularly remote islands. The fourth theme is conflict (Section 6.3.7), as expressed both directly and inferred from the construction of fortified sites on many islands. Some other responses to resource depletion are discussed in Section 6.3.8.

6.3.4 Population dispersal

Most people occupied coastal settlements in the Pacific Islands throughout the Medieval Warm Period – and indeed ever since they were first colonized, pursuing lifestyles based around marine foods. The region-wide collapse of marine-food production during the A.D. 1300 Event was followed – by about A.D. 1500 in most places – by the abandonment of coastal settlements and the occupation of areas that would in earlier times have been considered undesirable, not least for their low food-production potential. Among such marginal areas were:

• inland areas (discussed in Section 6.3.4.1);
• drier areas and areas of low fertility (discussed in Section 6.3.4.2); and
• smaller islands (discussed in Section 6.3.4.3).

It is likely that these types of area were occupied during the Little Ice Age mainly because they were unoccupied (or lightly occupied) by anyone else. In other words, they were refuges where people were able to subsist at less risk of confronting other people (competitors or aggressors) than in the areas where they lived previously. This may also explain why some human groups built and occupied offshore artificial islands at this time (discussed in Section 6.3.4.4).

Each of these refuge situations is discussed below, with examples from a range of Pacific Island groups. A case study is given of population dispersal on atolls (discussed in Section 6.3.4.5).

6.3.4.1 Occupation of inland areas

Although some parts of the larger fertile valleys in the Pacific Islands were occupied at earlier times, these occupations appeared to have been coast-associated in the sense that valley foods were moved downstream and marine foods upstream. But what happened during the early Little Ice Age was quite different, for both coastal and valley-floor settlements appear to have been abandoned in favour of settlements in defensive positions, typically on hilltops or around large caves. Some examples are given below.

Most parts of the 100 or so islands occupied at this time in the Fiji archipelago witnessed a shift in settlement from coasts to inland/upland fortified sites around A.D. 1400 (Kuhlken and Crosby, 1999). Recent work on the large Sigatoka Valley on Viti Levu Island shows that almost all the hillforts and fortified caves were occupied first during the early Little Ice Age (Fig. 6.4). Another example from Lakeba Island in eastern Fiji is illustrated in Fig. 7.7.

A synthesis of the evidence for changes in Hawaiian settlement pattern over the last millennium by Hommon (1986) concluded that Phase II (A.D. 1400–1600) was a critical time, marked by both large-scale inland settlement and wholesale replacement of the system of land ownership; the latter involved the replacement

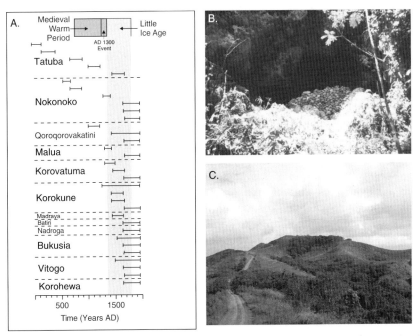

FIGURE 6.4 Societal effects of the A.D. 1300 Event on the Sigatoka Valley, Viti Levu Island, Fiji. (A) Dates for the establishment of various hilltop and cave sites in the Sigatoka Valley, Viti Levu Island, Fiji (after Kumar et al., 2006 with original dates in Field, 2004). Note the concentration of dates within the Little Ice Age. (B) View of the entrance to Tatuba Cave, with the remains of the fortification. (C) The site of Korokune fortified hilltop settlement that was established early in the Little Ice Age. [photos by Patrick Nunn]

of land-holding descent groups (*maka'ainana*) by territorial areas controlled by chiefs (*ahupu'a*). On O'ahu Island, inland movements began in earnest during the period A.D. 1200–1400, with well-dated examples coming from the Maunawili Valley and the Halawa valleys. Throughout the island, there is evidence that "settlement was focused more seaward" during the Medieval Warm Period (Cordy, 1996, p. 597).

During the fifteenth century, on the larger higher islands in the Cook Islands,

> "nucleated coastal villages decline in favour of a more dispersed settlement pattern. Coastal settlements were absent from most islands at the time of contact [A.D. 1778] and instead, the habitations were scattered over the inland [areas]". (Walter, 1996, p. 522)

In the higher islands (not the atolls) of Micronesia, movement inland appears to have occurred at approximately the same time in different places. Examples come from Yap, where "occupation of the inland for residential purposes" occurred around A.D. 1300–1400 (Dodson and Intoh, 1999, p. 18). On Guam, the development during the Little Ice Age of "large interior villages may be … associated with a period of cooler and drier climatic conditions and heightened competition among polities" (Hunter-Anderson and Butler, 1995, p. 66).

Some written accounts of Pacific Islands during the Little Ice Age report details of settlement patterns. For example, in A.D. 1568 in northwest Guadalcanal, Solomon Islands, there were

> "so many villages on the hill-tops that it was marvellous, for more than 30 villages of 10 and 20 houses and more, could be counted within a league and a half of road. And all the slope around the hills was full of huts, clearings and plantations, kept in very good order". (Catoira quoted by Amherst and Thomson, 1901, pp. 308–309)

6.3.4.2 Occupation of drier and/or less fertile areas

For people accustomed to diets that routinely included significant amounts of marine foods – as is true of most Pacific coastal dwellers during the Medieval Warm Period – moving inland and upslope away from sites near the coast was indeed a move to a marginal area. For although some of the people who settled the inland hilltop and cave sites evidently made occasional forays to the coasts to collect marine foods, the change in settlement location required dietary changes that were also required by the outbreak of conflict on many islands. This section discusses the move that characterized the early Little Ice Age on many high islands from more desirable areas to more marginal (typically uninhabited leeward) areas within the same island.

In the Hawaiian Islands, Phase III of Hommon's (1986) settlement-pattern model lasted from A.D. 1600 to A.D. 1778 and was marked by variability in food supply, and an "increasing dependence on marginal, drought-prone and, possibly, exhausted lands" (p. 67). One example from southwest O'ahu Island is a "hot, dry emerged limestone reef characterized by numerous sinkholes" named the 'Ewa Plain', where the earliest human settlement dates from about A.D. 1344 (606 cal BP: Athens et al., 2002, p. 57). The sustained occupation of this marginal area during the Little Ice Age is shown in Fig. 5.12. On Mau'i Island, the

inhospitable water-limited Kahikinui area on the arid southern flank of Hale-
akala Volcano was one of the marginal areas occupied during the Little Ice Age,
beginning around A.D. 1400 with the most intense occupation about A.D. 1500–
1860 (Kirch et al., 2004). It is likely that this area – between 400 and 600 m above
sea level – was occupied because lower areas were too dry at this time and upper
areas too wet. The leeward settlement at Kawela on Moloka'i Island was established about A.D. 1550 (Weisler and Kirch, 1985).

Marginal areas occupied at this time in the Hawaiian Islands not only included leeward locations or those where soils were comparatively poor, but also
upvalley locations. Many of these – less desirable than lower valleys because of
the distance to the sea – became occupied only during the early part of the Little
Ice Age. An example is the Anahulu Valley on O'ahu Island, the upper parts of
which began to be occupied only during the fifteenth century (Dega and Kirch,
2002).

For New Caledonia, the last 700 years "saw a continuous expansion of
settlement into ever more marginal environments" (Spriggs, 1997, p. 219), the
climax being when even the most inhospitable areas were occupied around
A.D. 1400 (550 BP), the date for initial settlement of Kaden in the centre of the
island (Sand, 1995).

On some islands, there is evidence that sandspits – marginal locations by any
yardstick – were occupied for the first time early in the Little Ice Age. For
example, the Tafunsak beach berm in northwest Kosrae was unoccupied until
A.D. 1400–1500 despite having been in existence since ~3,000 BP (~1050 B.C.). The
fragile berm was densely populated for much of the Little Ice Age (Athens, 1995),
probably because of the defensive attributes of the site.

6.3.4.3 Occupation of smaller islands

There is evidence throughout the Pacific Islands region of population dispersal
early within the Little Ice Age, not only within islands but also offshore, particularly to smaller islands that had not been permanently inhabited before. This
is a process that began during the A.D. 1300 Event, and several examples were
described in Section 5.4.4 and illustrated in Fig. 5.9.

A classic example of the movement of larger island peoples offshore during
the Little Ice Age comes from Palau, an island group in which larger higher
volcanic islands are adjoined by smaller lower limestone islands called the Rock
Islands. The volcanic islands were the first to be occupied but then – around
A.D. 1440 – they were abandoned almost entirely in favour of fortified "stonework
villages" on the inhospitable Rock Islands (Masse et al., 2006).

In the Mangareva (Gambier) Archipelago, the third largest island – Akamaru
– was occupied only during the early Little Ice Age, between A.D. 1450 and A.D.
1620, about the same time as up-valley rockshelter occupation began on the larger
islands (Anderson et al., 2003). The smaller island of Kamaka was occupied
permanently A.D. 1600–1750 (Green and Weisler, 2004).

On Easter Island, during the early part of the Little Ice Age, people not only
occupied fortified lava-tube caves on the main island but also some of the offshore islets (Bahn and Flenley, 1992).

The idea of groups of environmental refugees seeking uninhabited islands to colonize early in the Little Ice Age may also be implicated in the wave of east–west (backwards) colonization of uninhabited islands in the tropical western Pacific. Since its first colonization more than 3,000 BP, this region had been settled by waves of people from the island groups to the west, but apparently not all islands were occupied at the start of the Little Ice Age. Some islands, usually those most remote (within archipelagoes) or driest or having the most challenging terrain for settlement and agriculture, were unoccupied, at least permanently. These "Polynesian outliers" became occupied, mostly early in the Little Ice Age, by people from the east who were distinct in terms of both their language and cultural practices from those occupying the generally larger, more resource-rich islands in the area (Kirch, 1984b). One good example is Anuta in the eastern outer Solomon Islands which was first settled during the last millennium "not before the 16th century" (Kirch and Rosendahl, 1976, p. 227), or "sometime before A.D. 1400" (Spriggs, 1997, pp. 206–207). With a land area of 39 ha and the closest inhabited land being remote Tikopia Island (5 km^2 in area) 140 km away, Anuta is indeed comparatively small and remote.

Some Polynesian outliers may have been colonized accidentally, during drift voyages, when perhaps unexpected winds or currents (as might occur during El Niño events) carried sailing groups away from familiar areas. Oral traditions suggest that one such drift voyage resulted in the colonization of Sikaiana Island in Solomon Islands by a group from Vaitapu Island in Tuvalu (Spriggs, 1997), plausibly early in the Little Ice Age.

On a few Polynesian outliers, existing populations were evidently displaced or absorbed into the dominant group of newcomers. A good example is Tikopia in the eastern Solomon Islands where both the oral traditions and material culture suggest the introduction about A.D. 1300–1400 of basalt adzes, a distinctive style of trolling lure, and other things, as a result of

> "the arrival of successive canoe loads of migrants from Western Polynesia who achieved social and political dominance over the original inhabitants". (Spriggs, 1997, p. 203)

A similar situation marked the arrival of migrants from Wallis Island in the Loyalty Islands of New Caledonia, which occurred at a time of warfare. The migrants allied themselves with one side, which won, and were gradually incorporated into the community (Ozanne-Rivierre, 1994).

6.3.4.4 Creation of artificial islands

The existence of artificial islands off larger islands in the tropical Pacific has been known for some time, although no systematic treatment of them has been undertaken. Many of these islands, such as those in the lagoons off high Malaita Island in Solomon Islands (Ivens, 1927) and those at various locations in Fiji (Bau and Serua off Viti Levu Island, Naniubasaga on Moturiki Island, Tavuki on Kadavu Island), were probably created during the past few hundred years, possibly as refuges for elites from the adverse conditions during the early Little Ice Age that made living on larger islands less secure than at earlier times.

The only well-dated artificial island is that of Touhou in the Kapingamarangi Atoll lagoon in Micronesia (Leach and Ward, 1981). Beginning about A.D. 1250 (700 BP), the creation of Touhou apparently involved both the importation of material from elsewhere and the intentional concentration of natural material along the embryonic island's ocean-facing coast by the construction and maintenance of a seawall. It is equally probable in the author's view that sea-level fall exposed an area around a sandbank (Touhou) that was recognized by its occasional visitors as an opportunity for creating a new island. In this scenario, the survival of Touhou through the Little Ice Age was a product of both persisting low sea level and the maintenance of the island mass by construction of seawalls, with perhaps the occasional import of additional material.

The island Lelu (Leluh) off the larger island of Kosrae has been the object of awe for a long time, owing to the belief that its historically inhabited lowland area is a largely artificial creation constructed on a fringing reef surrounding a smaller bedrock island (Athens, 1995; Rainbird, 2004). Most chronologies suggest that Lelu was first occupied by A.D. 1400 with walled compounds being built out across the fringing reef and filled with sand to raise them. It is likely that this process was significantly facilitated – even suggested to the people involved – by the exposure of the reef surface as a result of the sea-level fall that affected this region during the earlier A.D. 1300 Event (see Section 5.4.4). There is evidence of a societal crisis on Kosrae around this time, and it is plausible to suppose that the higher ranking families that settled Lelu did so because it was safer than remaining on the main island. As this crisis deepened and the lowland Lelu settlement was periodically inundated, so its occupants came to appreciate its vulnerability and began to artificially raise the area of reef flat on which they lived and built walled compounds. Possibly sea level had begun rising by the time of René Lesson's visit to Lelu in A.D. 1824 because he was in no doubt that this is why it was "entirely surrounded by a belt of walls" (Ritter and Ritter, 1982, p. 53).

The suggestion by Kirch and Yen (1982) that the emerged reef barrier (Ravenga Tombolo) at the mouth of Lake Te Roto on Tikopia Island in Solomon Islands was an artificial construct is likewise not accepted here, given that its appearance probably also coincided with sea-level fall at this location (see Fig. 5.11).

6.3.4.5 Population dispersal on atolls

Many atolls in the low-latitude Pacific were colonized ~2,000 BP. Although there is obviously a comparatively small range of options for settlement location on such islands, there is evidence from some inhabited atolls of profound changes in settlement pattern consistent with the effects of sea-level fall occurring around the start of the Little Ice Age. Sea-level fall in such places would not only have negatively affected reef resources but would also – by having caused reef surfaces to emerge – have created the foundations on which reef islands (*motu*), made from sand and gravel thrown upon beyond the reach of the waves, could develop. By creating new islands, sea-level fall also provided the inhabitants of atolls with alternative places for settlement, something that allowed population dispersal.

Some of the clearest evidence comes from the Micronesian atolls of Bikini and Kapingamarangi. On Bikini, nucleated settlements on the main *motu* were

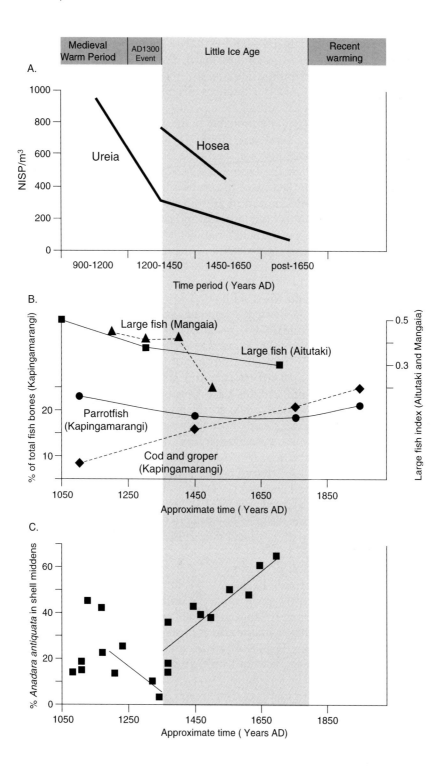

replaced by smaller dispersed settlements around A.D. 1350 (Streck, 1990). On Kapingamarangi, smaller *motu* on the reef were first occupied about A.D. 1250–1650 (700–300 BP) as the nucleated population on the largest *motu* dispersed (Leach and Ward, 1981). Among the new *motu* occupied at this time is Touhou Island, once thought to be a wholly artificial creation, but more plausibly an island that appeared during the early Little Ice Age as a result of earlier sea-level fall which its inhabitants may have preserved by the judicious placement of artificial structures.

Reef islands not only appeared on atoll reefs, but also on fringing and barrier reefs surrounding higher islands. In a similar way, these *motu* became available for occupation early in the Little Ice Age. The *motu* named Motutapu off the east coast of Rarotonga in the Cook Islands appears to have originated – or at least become large enough and stable enough to attract human settlement – about A.D. 1250 (700 cal BP: Okajima, 1999). This is what would be expected as a result of sea-level fall around the same time, for which there is independent evidence from Rarotonga (Moriwaki et al., 2006).

6.3.5 Changing patterns of resource utilization

Population dispersal during the early Little Ice Age also led to changes in human subsistence systems. The nature of available natural resources also changed, with certain marine foods becoming scarcer on many islands after the start of the Little Ice Age. Several examples are illustrated in Fig. 6.5.

Surrounded on its southern side by a large lagoon, the island Aitutaki in the southern Cook Islands is inferred to have experienced severe effects from cooling and sea-level fall during the A.D. 1300 Event (Nunn, 2000a). This is shown by changes in the fish catch, which may be a result of reduced numbers of fish living and breeding in the lagoon, itself a likely result of the increased turbidity and

---

FIGURE 6.5 Changing patterns of marine-food consumption in the Pacific Islands during the last millennium. (A) Changing patterns of fish bone density at two sites excavated on the main island of Aitutaki, southern Cook Islands (after Allen, 2002, with time periods based on Allen, 1994). It is possible that these trends are linked to the lower sea level during the Medieval Warm Period that reduced lagoonal water circulation and negatively impacted lagoonal ecosystems, thereby reducing fish stocks and fish catch. (B) Changes in the percentages of cod/groper (Serranidae) and parrotfish (Scaridae) as percentages of total fish catch from Kapingamarangi Atoll, Federated States of Micronesia (from data in Leach and Ward, 1981). Parrotfish are comparatively easy to catch, usually by nets in shallow water, but their abundance may have dropped as a result of ecological changes during the A.D. 1300 Event, requiring fish catch to be supplemented by deeper water species such as cod/groper from farther offshore. Large fish catches from the Cook Islands show similar declines during the Little Ice Age that are consistent with both ecological changes affecting fish stocks and climate and societal changes rendering the catching of large fish less likely. Aitutaki data are for large-bodied inshore fish (after Allen, 2002) and Mangaia data are for large-bodied fish (after Butler, 2001). (C) Changes in the percentages of *Anadara antiquata* through time in 20 shell middens at Manenggon Hills, Yona, Guam (after Amesbury, 1999, trend lines added). As with earlier sea-level falls, that which appears to have taken place around A.D. 1300 here may have led to a decline in habitats for *Anadara*. The later establishment of mangrove swamps at the lower sea level during the Little Ice Age may have led to an increase in the numbers of this species of *Anadara*, which is typically associated with mangroves.

reduced ecosystem health (Fig. 6.5A). Aitutaki was also the island from which the shell of pearl oysters (*Pinctada margaritifera*) was traded widely within the central Pacific during the Medieval Warm Period (Walter, 1990). Commercial harvesting of Aitutaki pearl shell ceased abruptly A.D. 1400–1500 (550–450 BP: Allen, 1992). Allen suggested that accelerated terrigenous sedimentation in the Aitutaki lagoon caused by gardening along the island's east side may have reduced lagoon health and reduced numbers and production of pearl oyster, but gives no reason for the sudden impact of this process early in the Little Ice Age. More plausible is the idea that cooling and sea-level fall across the A.D. 1300 Event led to increased lagoonal turbidity (by impeding water circulation) that removed the conditions suitable for optimal growth of pearl oysters, a situation that continues today (Dalzell, 1998).

Pearl shell from Aitutaki is believed to have been traded largely because of its suitability for manufacturing strong fishhooks. The decline in the amount of Aitutaki pearl shell during the early part of the Little Ice Age was so severe that, even on Aitutaki itself, there was insufficient to make fishhooks. This saw inferior materials – principally the turban shell (*Turbo setosus*) being used for fishhook manufacture. The weaker hooks made from turban shell dominated in Aitutaki after A.D. 1400–1500 (550–450 BP) and may explain the changes in the most common fish species caught after this time; these are smaller fish, of a size that would not break the inelastic turban-shell fishhooks (Allen, 1992).

Such inferior fishhooks also dominated in the Little Ice Age sequence on other islands in the region. On Mangaia Island, also in the southern Cooks, where pearl shell does not occur naturally, its import for fishhook manufacture ended about A.D. 1300 and turban-shell fishhooks were manufactured subsequently (Kirch, 1997b). The decline by almost 50% in large-bodied inshore fish on Aitutaki between the Medieval Warm Period and the Little Ice Age, and the even larger decline in large fish from Mangaia Island are shown in Fig. 6.5B.

On the same graph are shown changes in the relative catch of parrotfish and cod/groper on Kapingamarangi. The decline in parrotfish – a lagoon dweller – probably reflects the reduced health of the Kapingamarangi lagoon after sea level fell during the A.D. 1300 Event. The concomitant rise in cod and groper – caught farther offshore – probably points to the development of new deepwater fishing strategies in response to the reduction in lagoon-species yields.

Changes in the nature of shellfish collected by coastal foragers at various times during the last millennium may also reflect climate changes. For example, on Guam, the *Strombus–Anadara* ratio is thought to be a key indicator of sea-level fall. The increased numbers of *Anadara* being consumed during the Little Ice Age (Fig. 6.5C) suggest a lower sea level in which shoaling allowed the spread of mangroves, a favoured habitat for *A. antiquata*. Comparable examples come from Fiji and New Zealand (see Sections 6.2.2 and 6.3.2).

In most parts of the higher Pacific Islands, the movement of people inland during the early part of the Little Ice Age was accompanied by a decreased role for marine foods in human subsistence. Typically on smaller islands, some communities made continued use of marine foods, particularly shellfish that could be rapidly gathered. An example comes from Qaranilaca (cave) in northeast Fiji (Thomas et al., 2004), pictured in Fig. 6.1C.

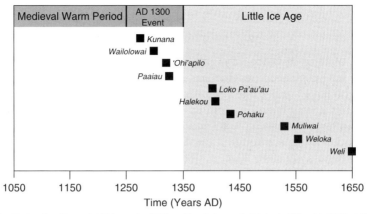

FIGURE 6.6 Dates for the establishment of Hawaiian fishponds (data in Kikuchi, 1976; Athens, 2001).

On some islands, the environmental crisis associated with the early Little Ice Age may have been less severe than on others allowing, somewhat exceptionally, continued use and development of coastal areas. This appears true of most larger islands in the Hawaii group where the construction of fishponds during the early Little Ice Age points not only to the changed coastal environment (that made such constructions comparatively easy) but also to the presumably unfettered access that fishpond users had to the coast. Most of the 449 fishponds recorded by Kirch (1984a) in Hawaii were built after A.D. 1400 (Fig. 6.6). Initial development of fishponds is seen to have occurred during the A.D. 1300 Event, with a hiatus of ~100 years before the next phase of fishpond development. If this hiatus is real (rather than just an artefact of data collection), then it may be that it represents a period of societal disruption – the immediate food crisis that followed sea-level fall – before a degree of readjustment. It has been suggested that fish were raised in these ponds to ensure a constant supply of particular species to elite groups (Kikuchi, 1976).

Supplies of other resources, particularly non-living ones, did not suffer as a result of climate change across the A.D. 1300 Event, but their dispersal was uniformly reduced during the early Little Ice Age because of the reduction in interaction between people on different islands. This is discussed in the following section.

6.3.6 Reduced interaction

During the Medieval Warm Period, networks of cross-ocean exchange existed throughout the Pacific Islands region. Valued commodities from one island were exchanged with those from others. They included fine-grained igneous rocks for making long-lasting stone tools, uncommonly strong shells for carving hooks capable of catching large fish, and even marriage partners (Weisler, 2002; Weisler and Green, 2001). The abrupt discontinuation of these interactions during the Little Ice Age is well known. The collapse of interaction networks during the early Little Ice Age is discussed below under two headings:

- the end of long-distance voyaging (discussed in Section 6.3.6.1), including both interarchipelagic and intra-archipelagic voyaging, and

- the abandonment of remote, resource-limited islands (discussed in Section 6.3.6.2).

6.3.6.1 End of long-distance voyaging

The end of long-distance voyaging has puzzled many archaeologists and anthropologists who have studied the last-millennium history of the Pacific Islands, most assuming that it was linked to conflict on and among particular island communities. Only a few scientists have acknowledged that the end of voyaging represents "simultaneous changes through time in the overseas transport of functionally different materials" (Rolett, 2002, p. 185), fewer still that its cause should lie in "general patterns of regional divergence" (Walter, 1996, p. 523). The author has suggested that the end of voyaging was approximately simultaneous across the Pacific Islands region, not merely subregionally, and that it therefore demands a regional or global explanation (Nunn, 2000a, 2003b). Some of the evidence for changes in interaction spheres during the early Little Ice Age is presented in Fig. 6.7.

In his monumental survey of Pacific Islander oral traditions, Fornander (1969) concluded that most long-distance interaction ceased abruptly by A.D. 1400. This included contact between New Zealand and the tropical islands of eastern Polynesia from which its first settlers originated, and journeys within the central and eastern tropical Pacific Ocean including those that saw Pacific Islanders reach the Pacific Rim in central and/or South America.

While the Hawaiian Islands remained in contact with various island groups in French Polynesia during the Medieval Warm Period, long-distance voyaging

"ceased after about A.D. 1300, after which the Hawaiian Islands became completely isolated from the rest of Polynesia". (Kirch, 2000a, p. 291)

Long-distance interaction in eastern Polynesia linked the Mangareva group with the Cook Islands, Austral Islands, Marquesas Islands and Easter Island is attested to by similarities in fishhooks, harpoons and adzes in the period A.D. 1220–1450, after which time it stopped (Green and Weisler, 2002). Contacts between the southern Cook Islands group and others such as Samoa, Tonga and the Society Islands of French Polynesia involving the movement of lithic material were most intensive A.D. 1000–1500 (Walter, 1990; Allen and Johnson, 1997).

Cross-ocean interaction also ceased in a subregional sense around the same time. A well-constrained example of the evidence by which an end to long-distance voyaging has been obtained comes from the Marquesas Islands of French Polynesia (Rolett, 2002). On Tahuata Island, the best-formed adzes were made of imported basalt from Eiao Island but the import of these declines rapidly around A.D. 1450. Similarly, import of phonolite to Tahuata from Ua Pou Island for making flake tools "disappears entirely" after this time. By the time of European arrival in the Marquesas in A.D. 1595,

"Marquesans apparently lacked the means for regular communication with islands lying beyond their own". (Rolett et al., 1997, p. 134)

The declines in interaction that these data represent are coeval with the end of Marquesan lithic exports to the resource-poor Line Islands (mostly eastern

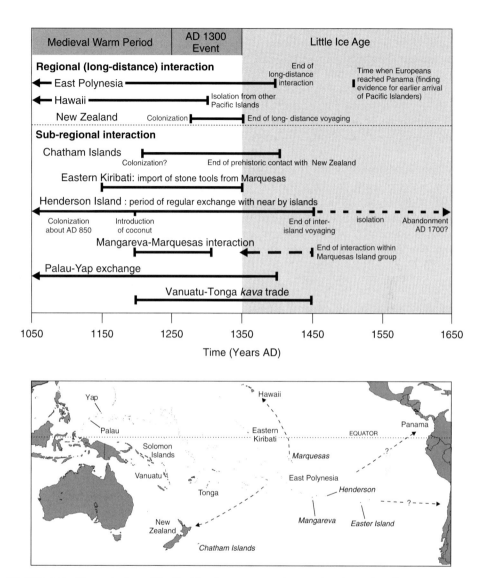

FIGURE 6.7   Chronologies of Pacific Island cross-ocean interaction (upper) with a map to show locations of places mentioned (lower). Chronologies from East Polynesia (Fornander, 1969), arrival of Europeans in Panama A.D. 1513 encountering evidence for Pacific Islander settlement (Ward and Brookfield, 1992), Hawaii (Kirch, 2000a), New Zealand (Buck, 1949; Anderson, 1991), Chatham Islands (interaction with New Zealand; Sutton, 1980), eastern Kiribati (Kiritimati, Manra and Tabuaeran islands – Di Piazza and Pearthree, 2001b), Henderson Island (Weisler, 1995, 1996a), Mangareva–Marquesas (Weisler and Green, 2001), within-Marquesas interaction determined by basalt and phonolite exchange (Rolett, 2002), Palau (Berg, 1992) and Vanuatu-Tonga (Luders, 1996). Use of the recently revised date for colonization of New Zealand (Hogg et al., 2003) implies that shown for the Chatham Islands needs revision.

Kiribati) and with general patterns of exchange in the Mangareva–Pitcairn–Henderson region to the southeast (Weisler, 2002). The Mangareva (Gambier) Island group is located east of the eastern end of the Tuamotu Island group with only more remote islands (such as the Pitcairn group) farther east. Mangareva acted as an important cog in the exchange spheres of eastern Polynesia, exporting a range of stone tools, plants and animals, and pearl shell to smaller, more isolated islands such as Pitcairn and Henderson (Green and Weisler, 2002). This interaction sphere was active A.D. 1000–1450 after which time contacts effectively ceased.

Many adzes made from the prized hawaiite on Tutuila Island in American Samoa were carried around the Pacific, probably beginning about A.D. 1050 (900 BP: Best et al., 1992). The industrial quarries on Tutuila continued to be operated during the Little Ice Age but its long-distance export around the Pacific, which had been such a dominant feature of earlier times, was apparently reduced during the early Little Ice Age. For example, finds of Samoan adzes in the southern Cook Islands suggest that their import ceased about A.D. 1400–1470 (Bellwood, 1978; Walter, 1990). Quarry chronologies are poorly known (Clark et al., 1997).

Another conspicuous item moved around the central Pacific at this time was pearl-oyster shell – valued for fishhook manufacture. The export of pearl shell from Aitutaki in the southern Cook Islands to islands such as Mangaia and Ma'uke, where it did not occur naturally, ceased around A.D. 1500 (Weisler, 1993), around the same time (approximately A.D. 1450) as exports of pearl shell from Mangareva to Henderson ceased (Weisler, 1995).

6.3.6.2 Abandonment of remote islands
Not only were settlements on larger islands moved inland and offshore during the early part of the Little Ice Age across the Pacific Islands region, but also the settlements on more remote islands

> "were forced back on the larger more resource-rich islands, where settlement became crowded ... dates for these abandonments ... point shakily towards the middle of the second millennium A.D.". (Irwin, 1992, p. 180)

A good example is the raised-limestone Henderson Island in the southeast Pacific, occupied approximately A.D. 900–1500 (Weisler, 1996a). For most of this period, cultural deposits excavated on the island contain imported materials, notably basalt for adze manufacture and pearl shell for fishhooks, but these disappear abruptly about A.D. 1450. Thereafter inferior, locally available materials were used; adzes are made from fossilized clam shells, and fishhooks from flat-oyster shell. The basalts and other volcanic rocks in the Henderson deposits came from the islands of Mangareva (~1,300 km away) and Pitcairn (400 km away) and it is likely that the Henderson colonists were in regular contact with people on these islands during the Medieval Warm Period but that these contacts ceased about A.D. 1450 (Weisler, 1996a, 1996b). Like many other "mystery islands" of Polynesia, the abandonment of Henderson is ascribed to the difficulties of life in such marginal locations but, given that this time also marked the end of the era of long-distance voyaging (see Fig. 6.7), it may be that the onset of adverse conditions during the Little Ice Age led to survival in such places becoming increasingly difficult to the point where abandonment was the only practical option.

Norfolk Island, 1,500 km east of Australia, also appears to have been abandoned by its pre-European inhabitants about A.D. 1600 (Anderson and White, 2001). The available information about the isolated smaller islands Necker and Nihoa in the Hawaii group suggests that they were occupied by people A.D. 1000–1500 but abandoned after that (Emory, 1928; Cleghorn, 1987).

Similar arguments apply to the mystery islands of the central Pacific islands that were uninhabited when first visited by Europeans yet showed abundant signs of former occupation. The commonest explanations for mystery-island abandonment have involved the lack of sufficient food resources to sustain the population (Di Piazza and Pearthree, 2001a; Weisler, 2002). Explanations for island abandonment involving climate extremes and deforestation were found wanting by Anderson (2002) who points out that many of the islands found to have been abandoned during the early Little Ice Age had clearly not become uninhabitable; he suggests that

> "isolation, operating through the hazards of long ocean passages, probably induced social pressures to abandon very isolated islands, especially once the easy resources were severely depleted". (p. 384)

In a similar vein, Di Piazza and Pearthree (2004) regarded abandonment of the Phoenix Islands (eastern Kiribati) as a strategic decision explainable in cost–benefit terms by settlers who knew of islands within sailing distance that contained more abundant resources. These explanations are consistent with that proposed in Section 7.7.6.

6.3.7 Conflict

The reasons for pre-historic conflict (or warfare) have long been a subject of considerable debate among anthropologists. One view is that such conflict arose only at times of resource (food or land) stress (Ferguson, 1998); an opposing view is that the assertion of status and prestige were equally as important (Chagnon, 1990). Given the lack of relevant material and skeletal data for most parts of the Pacific Basin during the last millennium, no generalization is readily apparent for the origins of the conflict that evidently gripped the region – particularly its islands – during the early part of the Little Ice Age. Yet, as might be expected in circumscribed land areas like islands, the widespread evidence of food stress during the A.D. 1300 Event and its aftermath provide a cogent region-wide reason for the abrupt start of conflict at this time in innumerable places simultaneously and, more widely, during the early Little Ice Age.

For example, warfare in Fiji – as in many other island groups and similarly circumscribed areas – appears to have been largely driven by there being insufficient resources to sustain existing populations (Kuhlken, 1999). Explanations that involve this situation being reached only through population growth beyond island carrying capacity (Rechtman, 1992) are unverifiable and fail to

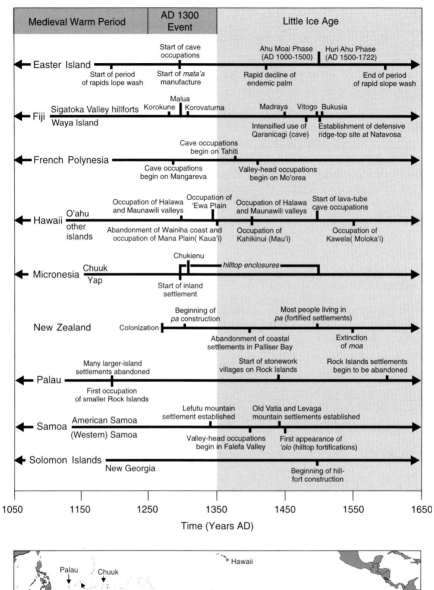

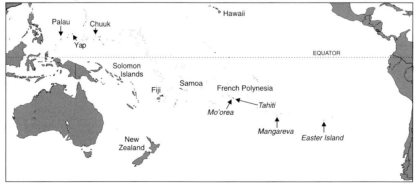

acknowledge the effects of environmental change, as do more elaborate explanations of warfare involving speculations about the political evolutionary process (Thomas, 1986; Ladefoged, 1995).

It is likely that increased climate variability during the Little Ice Age compared to the Medieval Warm Period plus the effects of rapid climate and sea-level change during the A.D. 1300 Event led to a widespread food crisis, most acute on the smaller (most circumscribed) remoter, and most resource-poor Pacific Islands. "People were starving, and engaging in warfare was a culturally appropriate way of dealing with the problem" (Campbell, 1985, p. 91) is how the situation has been explained on just such an island: Tongareva (Penrhyn) Atoll in the northern Cook Islands. This view agrees with that of most island-focused anthropologists and archaeologists, who have linked warfare to resource shortages or at least the unequal distribution of resources. The shortcoming of many of their models – in the present author's view – is to link the shortages of resources to population pressure, a connection that is conjectural and today, when so much is known about contemporaneous environmental variability, somewhat naive.

This section focuses on the most widespread evidence for conflict during the early Little Ice Age in the Pacific Islands, beginning with the construction of defensive fortifications (discussed in Section 6.3.7.1), and then looking at the evidence of cannibalism (anthropophagy) (discussed in Section 6.3.7.2). A case study of the Fiji Islands is also given (Section 6.3.7.3).

6.3.7.1 Fortifications in the Pacific Islands

The appearance of fortified sites on many Pacific Islands during the Little Ice Age "illustrate the late occurrence of endemic warfare on a supra-community scale" (Green, 2002, p. 146). A selection of dates for the development of fortified sites in the Pacific Islands – a proxy for the outbreak of conflict – is shown in Fig. 6.8 and some of the better documented situations described below.

FIGURE 6.8   Chronologies of the establishment of fortified and upland sites in the Pacific Islands (upper) with a map to show locations of places mentioned (lower). Data in the upper part of the figure as follows: the commencement of cave occupations and weapon manufacture on Easter Island (Bahn and Flenley, 1992) with chronological divisions from McCoy (1979), slope wash dates from Mann et al. (2003), endemic palm (*Paschalococos dispersa*) decline from Orliac (2003); the earliest dates for establishment of new hilltop/fortified sites in the Sigatoka Valley, Fiji (Field, 2004; Kumar et al., 2006), Natavosa date (Hunt et al., 1999), use of Qaranicagi (cave) on Waya Island (Cochrane, 2004); initial cave and valley-head occupations on islands in French Polynesia (Orliac, 1997; Anderson et al., 2003; Kahn, 2003); data for hilltop sites from Chuuk (Takayama and Intoh, 1978), inland settlement on Yap (Dodson and Intoh, 1999); chronology of establishment of *pa* sites and *moa* extinction in New Zealand (Groube, 1971; Leach and Leach, 1979; Anderson, 1989; Schmidt, 1996; Kirch, 2000a) and date of island colonization from Hogg et al. (2003); dates for abandonment of larger island settlements on Palau and occupation of the Rock Islands (Masse et al., 1984; Parmentier, 1987) with start of stonework village era on Rock Islands (Masse et al., 2006); mean dates for mountain settlement establishment in American Samoa from Pearl (2004, errata); dates for up-valley and *'olo* settlement in (Western) Samoa (Green, 2002); hillfort establishment dates for part of New Georgia, Solomon Islands (Sheppard et al., 2000).

Throughout the Aleutian Islands in the northernmost Pacific, after about A.D. 1100, there is evidence for a "monumental increase" in the evidence for warfare: evidence that is both skeletal and material (Maschner and Reedy-Maschner, 1998). In the Kodiak Islands, fortified settlements were established on offshore, cliff-faced islands in the period A.D. 1400–1600 (Fitzhugh, 2002) and there is a proliferation in the appearance of bone arrows after about A.D. 1400 (Knecht, 1995). Intertribal warfare became a feature of Kodiak society during the Little Ice Age (mid-Koniag period, about A.D. 1400–1600 – Fitzhugh, 2002).

On the Ryukyu Islands of southern Japan, the period A.D. 1200–1500 is known as the Gusuku Period; *gusuku* means castle in the Okinawan dialect. The naming of this period reflects the spate of castle building that occurred on these islands at the time, the construction of more than 200 castles having been documented (Ladefoged and Pearson, 2000; Takamiya, 2004). This appears to have been a response to increased conflict, which spread throughout Japan early in the Little Ice Age (Mass, 1997).

In the Chuuk archipelago of Micronesia, mountain tops were occupied for the first time in the period A.D. 1300–1500 with considerable evidence, both archaeological and ethnographic, that these locations were selected for their defensive attributes (Rainbird, 2004). In neighbouring Palau, the Little Ice Age was the time when fortified "stonework villages" were constructed on the more readily defendable Rock Islands (Masse et al., 2006).

The earliest written accounts of the islands of (Western) Samoa, dating from around the 1830s and 1840s, state that settlement was exclusively inland, with many fortified sites named '*olo* present on mountain tops (Moyle, 1984; Meleisea, 1987). Residential occupation of inland valleys, while piecemeal in earlier times, appears to have been concentrated between about A.D. 1400 and A.D. 1800 while '*olo* appeared only within the past 500 years (Davidson, 1969; Green, 2002). Wilkes (1845) described '*olo* as being

> "usually on the top of some high rock, or almost inaccessible mountain, where a small force could protect itself from a larger one". (p. 151)

A similar situation has been found in American Samoa where the earliest fortified mountain-top settlements date from the A.D. 1300 Event (Pearl, 2004).

Fortifications begin to appear on the limestone islands of Tonga (South Pacific) around A.D. 1450 (500 BP: Burley, 1998). This was also the time when warfare became endemic in Tonga with the paramount chief, the 24th Tu'i Tonga, Kau'ulufonuafekai (*fekai* means savage), extending his power by force throughout nearby islands (Burley, 1998).

In Hawaii, during the Little Ice Age, "archaeological evidence of warfare increases dramatically" (Culliney, 1988, p. 333). A parallel development saw many lava-tube caves in these islands used as defensive retreats after about A.D. 1500 (Schilt, 1984).

On most of the high islands in French Polynesia, there is evidence for a proliferation of fortified ridgetop sites during the early Little Ice Age (Kirch, 2000a; Weisler, 2002). In the atolls of the Tuamotu Archipelago, warfare dominated interisland relations for several "centuries prior to and subsequent to western contact" (Weisler, 2002, p. 268) with one early report stating that the Tuamotuans were almost constantly at war (Moerenhout, 1837).

Probably nowhere in the Pacific Islands has the onset of conflict during the early part of the Little Ice Age received more publicity than on Easter Island (Bahn and Flenley, 1992). In stark contrast to the cultural efflorescence marked by *moai* (statue) construction during the Ahu Moai Period (A.D. 1100–1500), the succeeding Huri Moai Period (A.D. 1500–1722) saw the "rise of endemic warfare" (Kirch, 2000a, p. 273). Another feature of this period on Easter Island was the pulling down of *moai* and the dismemberment of the worked basalt slabs from *ahu* and chiefly dwellings and their use to fortify cave entrances (Bahn and Flenley, 1992).

Among the other signs of conflict that are found in the Pacific Islands around this time are the manufacture of weapons of war. A key chronometer of warfare on Easter Island is the manufacture of spearheads named *mata'a* from the island's obsidian quarries, which were mined from A.D. 1300 to A.D. 1650 (Bahn and Flenley, 1992).

6.3.7.2 Cannibalism

Cannibalism developed as "a symptom of tensions" within Pacific societies during the Little Ice Age (Kirch, 1984a, p. 159). Like conflict, cannibalism is interpreted to have been largely a response to enduring food crisis resulting from the A.D. 1300 Event, for while cannibalism may have occurred at earlier times, it is only during the Little Ice Age that it appears to have become widespread in the Pacific Islands (including New Zealand) (Kirch, 2000a). On Mangaia Island in the southern Cook Islands, a well-dated site containing dismembered and cooked human remains is evidence of ritual cannibalism A.D. 1390–1470 (Steadman et al., 2000); similar evidence occurs slightly later on Mangareva (Green and Weisler, 2004).

The question of whether or not adjuncts to conflict such as cannibalism and headhunting were truly endemic among Pacific Island people around the time of the earliest European settlement in the early middle nineteenth century is a legitimate one. It has been argued that it was in the interests of proselytizing missionaries to have their potential converts in the Pacific Islands feel as ashamed as possible about their past so that they would be ripe for conversion. The reliance of opinion in Europe (and later elsewhere outside the Pacific Islands) on the accounts of a few observers regarding the warlike habits of Pacific Islanders may also have led to false impressions. For example, it has been argued that the influential 1813 account by Peter Dillon of a cannibal feast in Fiji was entirely fabricated, largely for reasons of self-aggrandizement (Obeyesekere, 2001). But so ingrained was the belief in conflict having been endemic in the Pacific Islands prior to the arrival of Europeans that they often represented themselves as having a calming influence on the indigenous people (Howe, 1984). The debate between Sahlins and Obeyesekere and their adherents, as in the 2003 issues of *Anthropology Today*, provides examples of both views.

6.3.7.3 Warfare in Fiji

As expressed by its popular nineteenth-century sobriquet "The Cannibal Is-lands", Fiji had a reputation for warfare during the later part of the Little Ice Age, the time at which the earliest accounts were written. In the judgement of one missionary, "Fijians, as a people, are addicted to war" (Waterhouse, 1866, p. 315). It is likely that warfare was also pronounced in the earlier part of the Little Ice Age. It is worth noting, however, that indications of conflict and related practices are almost wholly absent from the earlier 2,400 years or so that these islands have been occupied. Warfare began to increase only around A.D. 1400 (Kuhlken and Crosby, 1999).

In the Fiji archaeological record, it is the Vuda (Vunda) Phase that marks the societal breakdown characteristic of the Little Ice Age. The Vuda Phase is that in which warfare commenced, coastal settlements were abandoned and upland settlements – commonly on mountain tops or in caves – were established. The start date of the Vuda Phase is uncertain, a recent synthesis regarding it as marked by

> "the heaviest phase of fortification [that] did not begin for another few centuries [after A.D. 950] with widespread warfare continuing well into the latter half of the 19th century". (Burley and Clark, 2003, p. 244)

At a key site in Fiji, on the Sigatoka Sand Dunes, the Vuda Phase is dated to A.D. 1450–1750 (500–200 BP: Burley, 2003). A coincidence between the Vuda Phase and the Little Ice Age seems inescapable.

Fortifications are found throughout the largest island, Viti Levu, one recent study of the Sigatoka Valley showing that a large number of fortified sites were established early in the Little Ice Age (see Fig. 6.4). The situation at this time and throughout the Little Ice Age involved close links between defence and agriculture (Kuhlken, 1999). Writing in the 1860s, a British Colonial official found that

> "Jealousy that made every village distrustful of its neighbours compelled the inhab-itants to fortify themselves on the most inaccessible heights, and prevented them from cultivating any land beyond the few feet around each man's dwelling; if more were required, the cultivator, afraid to descend into the plain discovered some spot in the recesses of the mountains where he might plant his yams secure from molestation". (COC, 1864)

In the swampy Rewa Delta in southeast Viti Levu, open sites were replaced by fortified ring-ditch sites after about A.D. 1450 (500 BP: Marshall et al., 2000). Else-where along the coasts of Viti Levu and other islands, this may have been the time when substantially artificial islands were created as refuges for high-ranking persons (see above).

Among the smaller islands of the Fiji group, examples abound of fortified sites created early in the Little Ice Age. On the island of Beqa, for instance, inland fortified sites dating from the last 500 years "were found all over the island, on every available peak" (Crosby, 1988, p. 260). Similar examples come from Wakaya and Waya islands (Rechtman, 1992; Cochrane, 2004). On Taveuni Island, the period of building hilltop fortifications has been dated to A.D. 1200–1400

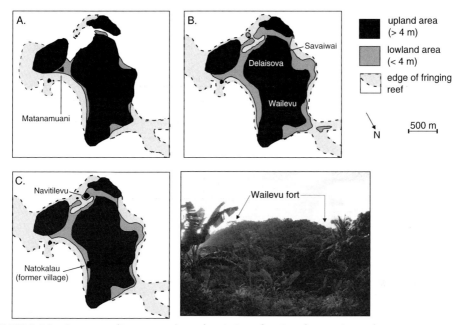

FIGURE 6.9 Summary diagram to show the timing of major changes in settlement pattern on Naigani Island, central Fiji (information from Ramoli and Nunn, 2001). (A) Location of the Matanamuani (Lapita) settlement 800–900 B.C. (B) Reported situation from oral traditions during early Little Ice Age involving abandonment of the three coastal settlements in favour of ones on adjoining mountain tops. (C) Early twentieth century saw abandonment of mountain-top settlements in favour of lowland Natokalau and later modern Navitilevu.

(Frost, 1979). Examples also come from the smaller, generally more resource-poor islands of Lau in eastern Fiji, where the period A.D. 1450–1750 (400–200 BP) marked the wholesale movement of people inland and upslope (see Fig. 7.7). An 1840 account of Totoya Island in the same area states that the inhabitants

"are said to be constantly at war, and are obliged to reside on the highest and most inaccessible peaks, to prevent surprise and massacre". (Wilkes, 1845, p. 145)

On the island of Naigani in central Fiji, archaeological research and oral traditions have allowed the history of settlement to be reconstructed in outline (Fig. 6.9). It appears that permanent coastal settlement was abandoned during the Little Ice Age in favour of fortified hilltop settlement and that only around the end of the Little Ice Age was coastal-lowland settlement re-established.

In the Fiji Islands, although cannibalism may have been an occasional practice in earlier times, there is much evidence for an upsurge during the last millennium, dating from perhaps A.D. 1100–1600 at Vuda, Viti Levu Island (Gifford, 1951), A.D. 1300–1825 on Wakaya Island (Rechtman, 1992) and A.D. 1400–1700 on Waya Island (Cochrane et al., 2004).

This agrees with oral traditions, although the interpretation of these is somewhat subjective. Early recorders of oral traditions in Fiji were told that "fighting" was unknown in the distant past and that "a profound peace prevailed" (Brewster, 1922). The same author was told by the people of Ba and Nasoqo that famine was the cause of cannibalism.

6.3.8 Other societal responses to resource depletion

There are other, less widespread, societal responses to resource depletion during the Little Ice Age that are worth describing briefly. Those discussed in this section are the end of pottery making, the appearance of new belief systems, megalith construction, and pig extirpation.

A widespread symptom of societal change, plausibly connected to resource depletion and associated conflict and inland settlement, was the discontinuation of pottery manufacture on many Pacific Islands. This is a conspicuous time marker for the start of the Little Ice Age in the islands of central Vanuatu, and on Tikopia and Vanikoro islands in the eastern Solomon Islands (Spriggs, 1997). Elsewhere in the islands of the tropical western Pacific, a region that had been characterized by pottery manufacture since first settlement, the start of the Little Ice Age was marked by often radical changes in the design of the pottery being made. Examples come from Fiji and Yap (Dodson and Intoh, 1999; Burley and Clark, 2003).

In many parts of the Pacific Islands, a response to the food crisis that was obtained in the early part of the Little Ice Age may have been the development of new belief systems, often involving the replacement of those that had served the population in the past. A good example comes from Easter Island, where a ritual involving the collection of the first Sooty Tern egg of the season from islands off Orongo led to the apotheosis of the bird man (*tangata manu*). In explanation, it is unsurprising that the people of Easter Island,

> "faced with an ecological crisis that threatened their entire social order, if not their very lives, should take the initiative to fashion a new religion more suited to their precarious times. Instead of a king for life, they opted for one elected by ordeal. Instead of many ancestors who descended from many ancestors, they propounded a single god, whose image they carved in deserted rocks, on the bases of feud-toppled figures, and even around water-holes". (McCall, 1980, pp. 38–39)

A similar situation may have occurred in Samoa with the construction of star mounds early in the Little Ice Age. As pigeon-snaring sites, these star mounds likely represent the importance of personal achievement over hereditary rank (Herdrich and Clark, 1993).

Monumental structures (megaliths) became commonplace on many Pacific islands A.D. 1200–1450 (750–500 BP: Burley and Clark, 2003). Monument construction during the Little Ice Age may have been a manifestation, not of cultural efflorescence and societal complexification, but of cultural and societal stress, involving increased supplication to the gods for relief from the threats of famine and drought (McCall, 1994). In the Northern Marianas Islands, the appearance of latte stones is interpreted as a consequence of the need to demonstrate land ownership at a time of increased resource competition (Hunter-Anderson and Butler, 1995). In Samoa and Tonga, appropriation of labour is implicit in the suddenness of the start of monument construction during the Little Ice Age (Burley, 1998; Green, 2002).

Dates of pig extirpation on many Pacific islands cluster within the early part of the Little Ice Age and "require an [non-local] effect such as the onset of the Little Ice Age" (Masse et al., 2006, p. 121). These authors were discussing the simultaneous extirpation of pigs in the Micronesian island groups of Palau and Yap after A.D. 1450, but this time also marks the point when pigs were evidently extirpated on Mangareva (Green and Weisler, 2004). On Mangaia in the southern Cook Islands, pigs disappeared about A.D. 1450 (Kirch, 2000b) and on Tikopia in the eastern outer Solomon Islands, pigs disappeared after A.D. 1300 (Kirch and Yen, 1982). These islands are likely to be representative of a far greater number for which reliable data about pre-historic pig populations are currently lacking. For these islands, it is suggested that pigs were a casualty of intertribal conflict following the A.D. 1300 Event but it is also possible that they were considered a luxury food item that could no longer be afforded following the decline in food resources available (particularly given pigs' ability to destroy food gardens) and the increased climatic variability.

6.4. THE END OF THE LITTLE ICE AGE

Compared to the A.D. 1300 Event, the terminal Little Ice Age appears to have been marked by a slow rise in temperature in many parts of the Pacific Basin, making its transition to the following period of Recent Warming somewhat difficult to pinpoint in many records. For example, a coral-based palaeotemperature record from Rarotonga in the southern Cook Islands (Linsley et al., 2000) shows that high temperatures persisted A.D. 1730–1760 but then fell again thereafter. In contrast, most such records show that the period A.D. 1750–1850 was generally a time of rising temperature (see Section 6.1.4). What is clear is that there is no abrupt region-wide synchronous change such as the A.D. 1300 Event marking the end of the Little Ice Age in the Pacific Basin.

In a few places, the end is abrupt. In the Quelccaya ice core (Andean Peru), for example, the end of the Little Ice Age about A.D. 1880 is "one of the most abrupt events" recorded (Thompson and Mosley-Thompson, 1987). But in most other parts of the Pacific Basin, the end of the Little Ice Age is diffuse. For example, it is generally not clear in palaeoclimate records from western North America (Luckman et al., 1993, 1997). Likewise sea level rose at the end of the Little Ice Age but, unlike that during the A.D. 1300 Event, this did not occur simultaneously across the Pacific Basin.

In societal terms, changes at the end of the Little Ice Age in many parts of the Pacific Basin that might be attributable to climatic amelioration are obscured by the overprint of European settlement and introductions. The latter ranged from infectious diseases to exotic animals and plantation crops, all of which brought about profound transformations in environments and societies in the Pacific, particularly in the more fragile of these on Pacific islands (McNeill, 1994, 1999; see also Chapter 8).

One key societal change that had begun in parts of the Pacific Ocean towards the end of the Little Ice Age before European influences were felt was the re-commencement of subregional interaction. For example, there is evidence that Fiji–Tonga contact recommenced around A.D. 1600 (Reid, 1977) while within the Bismarck Archipelago of Papua New Guinea, trade across the Vitiaz Strait re-sumed around A.D. 1600 and pottery manufacture increased at Madang and Sio in response to distant demand (Lilley, 1988). Another compelling sign of decreased societal tension in the Pacific Islands region towards the end of the Little Ice Age was the lessening of real conflict in favour of a more conventionalized form. This has not been documented in detail before, so is explained in the remainder of this section.

There is abundant evidence that conflict during the later part of the Little Ice Age became largely conventionalized, plausibly the cultural legacy of far more serious warfare at a time (during the earlier Little Ice Age) when resources were scarcer and fiercely competed for, being played out in a time (during the later Little Ice Age) when there was generally no immediate need for competition because resources were once more abundant (Nunn and Britton, 2001; Nunn, 2003a). Several examples describing the conventionalized nature of warfare around the end of the Little Ice Age follow to emphasize the point that it was probably a Pacific-wide development which, as such, demands a regional ex-planation rather than a series of local ones.

In the Marianas Islands of the northwest Pacific, warfare was a

> "sort of game in which rival groups would test their strength against one another. There was a great show of bravado, but as soon as one side had lost two or three men, it would send a turtle shell to the enemy as a sign of submission". (Thompson, 1945, p. 19)

In southern Vanuatu, "the Tannese … live in a state of perpetual war" (Turner, 1861, p. 82) but "their fighting is principally bush skirmishing; they rarely come to close hand-to-hand fighting" (p. 83).

In Fiji, despite large armies taking the field, there were no pitched battles and only a few casualties (Routledge, 1985); war was regarded as a "social perform-ance" (Tippett, 1954). A more detailed account was given by Derrick (1953):

> "Relatively large armies took the field … but there were no pitched battles. … When opposing parties of Fijians met and fought, little damage was done beyond a few flesh wounds until some lucky shot or cunning blow should bring down one of the warriors. Should one or two men fall in this way, the skirmish was over; the friends of the slain would run, if possible, taking the bodies with them. But if the victors could secure one or more of the bodies their honour and joy were un-bounded". (p. 142)

On Niue Island, traditional

> "warfare was … a game inasmuch as it was carried out under certain fixed con-ventions and with a definite technique to which both sides conformed". (Loeb, 1926, p. 128)

On the atoll of Tongareva in the northern Cook Islands,

> "warfare … was ritualized. Rather than involving gratuitous violence, it was governed by a body of implicit rules. Killing was frequently avoided, and even after hostilities were under way, standardized patterns of behavior provided opportunities for bringing about a peaceful settlement". (Campbell, 1985, p. 84)

On the island of Mangareva (Gambier Islands) prior to European contact in A.D. 1834, oral traditions tell that warfare "was relatively harmless, involving a great deal of invective and few casualties" (Crealock, 1955, p. 122).

A parallel argument can be made for the progressive ritualization of cannibalism during the later part of the Little Ice Age. Many accounts of cannibalism emphasize the rhetoric and convention involved but not the fact (Bensa and Goromido, 1997; Stewart and Strathern, 1999).

Many people have pondered the reasons for the outbreak of conflict on Pacific Islands early in the Little Ice Age but hardly any have considered why it evidently became largely conventionalized towards its end. The reasons plausibly lie in an erosion of the original reasons for warfare. If we assume that warfare broke out because of resource depletion (associated with the A.D. 1300 Event), then perhaps conventionalization developed because the resource base recovered, at least relative to depleted levels. Recovery in this sense does not necessarily mean that the original resource base (that had sustained Pacific Islanders during the Medieval Warm Period) was restored to full productivity, but that islanders adapted to changed environmental conditions and were – by the end of the Little Ice Age – producing enough food in most places to sustain themselves. In this scenario, warfare – which developed its own rituals when it was real – became conventionalized, with many of the original rituals still followed in later times when two groups came together to resolve a disagreement. Many ceremonies in Pacific Island cultures appear to have evolved in much the same way, their original meanings lost or modified to suit new conditions (Ravuvu, 1987).

KEY POINTS

1. In the Pacific Basin, the Little Ice Age was generally marked by cool dry conditions with considerably more interannual climate variability than during the Medieval Warm Period.
2. It is likely that sea level remained comparatively low in most parts of the Pacific Basin during the Little Ice Age, bringing about changes to coastal environments and their utilization by humans. Coastal hinterlands in many places were subject to increased denudation (erosion) as a result of sea-level fall, increased storminess and/or sustained inland settlement.
3. The Little Ice Age was a time when most people in the Pacific Basin, especially the Pacific Islands, experienced food stress that resulted in many cases in societal breakdown. The principal manifestations of this were population dispersal, changed patterns of resource utilization, reduced interaction, and conflict.

CHAPTER 7

The A.D. 1300 Event: An Explanatory Model for Societal Response to Climate Change during the Last Millennium in the Pacific Basin

"... the problem of long-term climatic change in the Pacific Islands deserves more attention than it has hitherto received". (Kirch, 1984a, p. 127)

"... the categorical rejection of environment as a potential cause of cultural change will lead to unsuccessful if not naïve characterizations of prehistoric human behavior". (Jones et al., 1999, p. 137)

"[in the Pacific Islands] ... climatic instability could have increased during the mid-second millennium AD with the onset of cooler, often drier and windier conditions that culminated in the Little Ice Age ... It is plausibly that instability which accentuated the anthropogenic impact on prehistoric East Polynesian forests and erosional regimes generally". (Anderson, 2002, p. 385)

In many parts of the world, explanations of cultural change as the outcome of intrasocietal forces (rather than being externally driven) have been in vogue for decades, and many archaeologists and related scientists are reluctant to admit the possibility that this assumption – and the often stupefyingly complex models that have been formulated in its support – might be wrong. Yet in many parts of the world, this is exactly what is becoming clear, with some archaeologists and other scientists researching pre-historic nature–society interactions now advocating cultural change and particularly cultural collapse as consequences of externally driven climate changes.

The marriage of societal and climate changes has been solemnized by high-resolution data from proxy climate records, especially ice-core data and annually laminated sediment analyses from both of which precise changes in interannual or decadal climate can be obtained. Classic examples from the past two millennia include the use of GISP2 ice-core data to explain the abandonment of Norse settlements in Greenland (Barlow et al., 1997), the use of Quelccaya ice-core palaeotemperature data to calibrate the end of the Tiwanaku civilization on the Andean *altiplano* (Binford et al., 1997) and the study of titanium inputs – as palaeoprecipitation proxies – into the Cariaco Basin of Venezuela to explain the cycles of collapse and recovery shown by the Maya of central America (Haug et al., 2003).

Most of the earliest suggestions that climate change may have been important in the human history of the Pacific Basin focused on the ways in which it may have alternately facilitated and inhibited the movements of people within various parts of the region. Much discussion continues to focus on the times during the Last Glaciation when the Bering Strait was dry land and would have permitted access on foot from East Asia to North America (Elias et al., 1997). A more recent suggestion is that people crossed the entire Pacific Ocean from west to east during the Pleistocene making use of now-submerged island stepping stones (Wyatt, 2004) but this seems quite implausible. More widely accepted is the idea that the first (post-Pleistocene) cross-ocean voyaging by people in the Pacific Ocean was facilitated by optimal climate conditions (Bridgman, 1983; Finney, 1985), an idea recently sanctioned by Anderson et al. (2006) who argued that bursts of eastward (against-the-wind) ocean voyaging took place during El Niño events when the usual wind patterns were disrupted.

Most explanations of post-settlement societal evolution in the Pacific Basin have involved intrasocietal forces, such as population growth, invasion, or the outcome of particular societies attaining particular socio-political developmental thresholds. All these explanations suffer from being undemonstrable. Often they are dependent solely on the premise, making them examples of circular arguments acceptable only, it seems, by virtue of their supposedly intuitive correctness. Only a few authors have worked with last-millennium palaeoclimate records in the Pacific Basin to produce independently verifiable explanations of societal evolution involving climate forcing (Binford et al., 1997; Raab and Larson, 1997; Nunn, 1999; Anderson et al., 2006).

It seems beyond doubt that climate change – in its many manifestations – has the ability to bring about fundamental change in human societies (see Chapter 3). What is presented below is a model of what the author has termed "the A.D. 1300 Event", identified as a period separate from those before and after to emphasize its importance as a driver of environmental and cultural changes in the Pacific Basin. To be explicit, it is contended that it was not principally the Medieval Warm Period or the Little Ice Age that drove environmental or cultural change in the Pacific, but rather the transition from one to the other (the A.D. 1300 Event).

The present chapter is devoted to the model of the A.D. 1300 Event in acknowledgement of the fact that some aspects of it are currently undemonstrable and certainly contentious. Nevertheless, since this model was first suggested it has gathered considerable empirical support from archaeological studies and has been applauded in some quarters. Shorter syntheses have been published – on the basis of complimentary reviews – in some leading academic journals including *Geoarchaeology* (Nunn, 2000a), *Environment and History* (Nunn and Britton, 2001), the *Geographical Review* (Nunn, 2007c) and *Human Ecology* (Nunn et al., 2007).

This chapter begins with an outline of the model of the A.D. 1300 Event (Section 7.1) followed by discussions of alternative models (Section 7.2). There follows an account of key (cause–effect) linkages for climate change (Section 7.3), environmental change (Section 7.4) and societal change (Section 7.5). A series of case studies is given in Sections 7.6 and 7.7 to illustrate the proposed effects of the A.D. 1300 Event on a range of situations in the Pacific Basin. Section 7.8 elaborates several key issues, and is followed by the author's suggestions for future verification of the A.D. 1300 Event model (Section 7.9).

7.1. OUTLINE OF THE MODEL OF THE A.D. 1300 EVENT

The A.D. 1300 Event model (Fig. 7.1) assumes that climate change was the major driver of environmental and societal changes in the Pacific Basin during the last millennium, especially in and around the period A.D. 1250–1350. The primary causes of this climate change are considered only in Chapter 10, because it is regarded as important that their validity – necessarily somewhat speculative – does not influence evaluation of the model itself.

For the purposes of the discussion in this chapter, climate changes are regarded as the "primary effects" of the primary causes. Climate change in this sense refers to temperature fall (cooling), which is regarded as having been the main cause of at least localized sea-level fall, and an increased El Niño frequency, considered to have been responsible for increases in climate variability, storminess and possibly annual precipitation.

Under the heading of environmental (secondary) effects, temperature fall is regarded as having stressed ecosystems, something also caused by water-table fall, particularly in coastal and valley-floor lowlands, and increased climate variability. Water-table fall is one main outcome of sea-level fall, the other being coastline and reef emergence that led to the emergence of shore platforms on which habitable islands could form, particularly *motu* built from wave-deposited sand and gravel. Coastline and reef emergence also killed off the most productive parts of coral reefs and inhibited water circulation within lagoons which in turn depleted nearshore food resources.

Increased lowland sedimentation would also have been caused by sea-level fall, which would – by lowering stream-valley base levels – have caused stream channels to incise, in turn steepening valley-side slopes and increasing the amount of material moved downslope by mass movements. Much of this material would have ended up being carried by rivers down to their lowermost parts and being deposited on floodplains or around river mouths. Lowland sedimentation would also have been increased by the increased upland erosion at this time.

Finally, the lower part of Fig. 7.1 shows how environmental change led to societal change. Ecosystem stress contributed to the abandonment of many coastal and valley-floor settlements (and associated subsistence systems) that had existed during the Medieval Warm Period, something that is also explained by the depletion of nearshore resources and increased lowland sedimentation. Environmental change created new places for settlement, including formerly swampy areas now drained by water-table fall and lowland sedimentation, and islands (commonly *motu*) newly formed following coastline and reef emergence. Other places that already existed became occupied, at least in large numbers and for sustained periods, only after the A.D. 1300 Event. These included inland (including marginal) areas of particular islands together with previously uninhabited offshore islands.

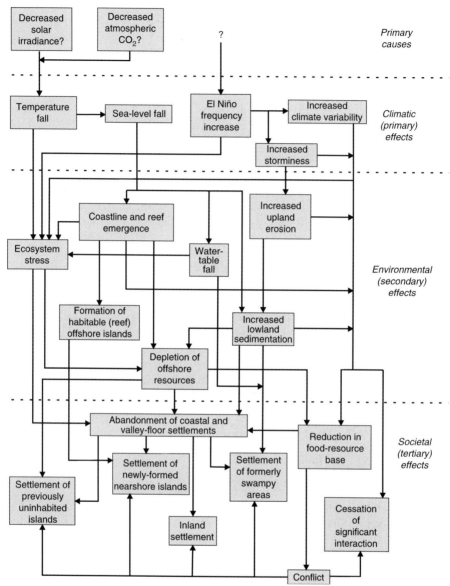

FIGURE 7.1   Model of the A.D. 1300 Event showing plausible connections between climate and climate-associated drivers and environmental and societal effects in the Pacific Basin. The primary causes identified are tentative, and derive from the work of Perry and Hsu (2000) and Weber et al. (2004) on solar irradiance and Kouwenberg et al. (2005) on atmospheric CO_2.

Increased storminess and increased climate variability are envisaged as the principal reasons for the end of cross-ocean interaction following the A.D. 1300 Event. These would also have reduced the food-resource base (along with other factors) in many places, leading to conflict which, as shown in Fig. 7.1, amplified most other societal changes.

Figure 7.1 is an outline of the A.D. 1300 Event model, but does not claim to identify all the cause–effect pathways, only the principal ones for most of which there is empirical evidence from throughout the Pacific Basin. It should be noted that, for reasons of simplicity, many feedbacks are not shown in Fig. 7.1.

7.2. ALTERNATIVE MODELS

Many aspects of societal evolution during the last millennium in the Pacific Basin have been modelled, including:

- the onset of conflict;
- changes in subsistence strategies;
- changes in settlement patterns; and
- the development of chiefdoms (complex hierarchical societies).

All of these can be satisfactorily explained by the A.D. 1300 Event model outlined in the preceding section. But most anthropologists and archaeologists working in the Pacific Basin would not accept that most (if any) such societal changes were brought about by climate and environmental changes, as envisioned here. So it is worth briefly reviewing alternative explanatory models and pointing out their deficiencies compared to that of the A.D. 1300 Event model.

In general (not just in the Pacific Basin), the causes of such societal changes have generally been pondered only by "human" scientists, typically anthropologists, archaeologists and sociologists, so that a predilection for "human-driven" explanations of societal change is understandable, albeit unfortunate. Therefore, most explanations of societal changes, such as those bulleted above, have ignored environmental change or dismissed it as trivial. The most commonly cited driver of societal change is population growth, particularly the ways that it affected resource availability relative to local-area carrying capacity.

A typical (albeit simple) model of this kind is shown in Fig. 7.2A where local-area (island) carrying capacity is shown as effectively constant. In this model, a crisis is reached only when population growth exceeds carrying capacity. Such models have been proposed to explain societal evolution in many parts of the Pacific Basin (e.g. Fiji – Rechtman, 1992; Hawaii – Kirch, 1984a; California – Chartkoff and Chartkoff, 1984) and are central to more complex theories such as those involving socio-political evolution of interdependent societies.

A more realistic model is shown in Fig. 7.2B where carrying capacity is shown as highly variable, creating food crises at several points against a background of constant population growth – itself an improbable scenario, but one that facilitates comparison with Fig. 7.2A. Variations in carrying capacity of the kind illustrated in Fig. 7.2B would be caused largely by non-societal factors such as climate and environmental changes. Such a scenario is integral to models such as the A.D. 1300 Event model as applied to the Pacific Basin (Jones et al., 1999; Schimmelmann et al., 2003; Nunn, 2007c; Nunn et al., 2007) and globally, especially in hunter-gatherer communities (Shnirelman, 1992).

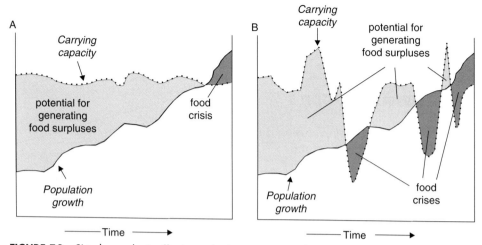

FIGURE 7.2   Simple graphs to illustrate the basic contrasts between explanations of societal change driven by (A) population growth and (B) climate change. In the former, local-area (island) carrying capacity is regarded as essentially constant while in the latter it is regarded as highly changeable, especially during periods such as the A.D. 1300 Event.

In much of the Pacific Basin, especially in the Pacific Islands, earlier ideas about last-millennium societal evolution were driven largely by models based on control of food production (e.g. Sahlins, 1972; Earle, 1980), but as more data have become available to allow palaeodemographic calculations, so population growth became more popular as the main driver (e.g. Hommon, 1986; Kirch, 1990; Cordy, 1996). The basic problems with both these types of model are threefold.

First, the assumptions of these models are unverifiable, and most are self-validating. In other words, acceptance of the model depends on the user's acceptance of its assumptions. For example, it may seem self-evident that population growth fuelled pre-historic societal change, but a connection cannot be proved. So to believe in a model of societal change driven by population growth, it is necessary to believe that such a model is intrinsically viable, and that there are no competing models that are more so. The acceptance of such cultural-determinist models to explain pre-historic societal change in the Pacific Basin (and elsewhere) has become so widespread for so long that they have formed an orthodoxy for many scientists and their students, few of whom apparently ever pause to ponder the fragility of the underlying assumptions of these models. In contrast, because of the existence of independently derived palaeoclimatic and palaeoenvironmental data series, environmental-determinist models such as that of the A.D. 1300 Event can be verified.

Second, cultural-determinist models focus almost exclusively on human-associated changes, and regard the "environment" as a passive backdrop to the human drama. Such an assumption has its roots in the heady days of cultural ecology in the first half of the twentieth century when information about palaeoclimates and palaeoenvironments was sparse and imprecise. But today, when such information is more widespread, commonly serial, and often very precise, there seems little reason for assuming "background" environmental change to have been a trivial force for change in any pre-historic society. Some models, such as that of Kirch (1990) on Hawaii, have gone some way to acknowledging human-caused environmental change as having affected societal evolution, but these still fail to adequately acknowledge the significance of non-human causes of environmental change.

Third, most cultural-determinist models overlook the regional synchrony of societal change. This synchrony is a conspicuous clue to the ultimate causes of societal change for, were these solely internal to a particular society, there is no reason for pre-historic societal changes across a wide region to have been synchronous; in fact, asynchrony would be expected. Yet, in many instances from different regions of the world, it is clear that synchronous societal changes have occurred, and could have done so only as a result of external forcing acting on the entire region simultaneously. Examples include societal collapse throughout the western Asia–Egypt region 2200 B.C. and in central America A.D. 750–950 (see also Section 3.3). Within the last decade, it has been realized that cultural changes throughout the western United States during the Medieval Warm Period were approximately synchronous and therefore driven largely by climate change (Jones et al., 1999), a similar situation to that in the Pacific Islands (Nunn, 2000a; Nunn et al., 2007).

It is worth reiterating that the A.D. 1300 Event model, as outlined here, is necessarily simplified, and that it is focused on the principal drivers of environmental and societal changes, not the proximate causes of particular changes in particular places. The choice is not between the A.D. 1300 Event and any of the other explanatory models of societal change because many (not all) of these identify plausible cause–effect linkages that probably operated within the former. For example, as Sahlins (1972) averred, the evolution of Pacific Island chiefdoms may indeed have been linked to changes in the available amounts of key foods, changed agricultural production and an emerging imperative for redistribution. But this did not happen, as Sahlins envisaged (in the spirit of the times), against the passive backdrop of an unchanging environment. The A.D. 1300 Event model shows that not only was that environment dynamic, but also that the degree to which it changed during the fourteenth century was sufficient to profoundly upset the dynamics of many Pacific Basin societies.

The rest of this chapter proceeds on the basis that the A.D. 1300 Event model is correct as an explanation for at least the crude changes in human societies in the Pacific Basin during the last millennium (before about A.D. 1850).

7.3. CAUSE–EFFECT LINKAGES FOR CLIMATE CHANGE IN THE A.D. 1300 EVENT MODEL

The ultimate causes of the climate changes that took place across the A.D. 1300 Event are shown in Fig. 7.1 as primary causes. It is important that the validity of models such as that of the A.D. 1300 Event proposed here should not depend on the correct identification of primary causes but rather on observations. For this reason, primary causes are discussed in Chapter 10, not here. The present section focuses on the linkages between the principal climatic effects shown in Fig. 7.1 – temperature fall as it is linked to sea-level fall (Section 7.3.1), and increased El Niño frequency as it is linked to increased climate variability, increased storminess and increased precipitation (Section 7.3.2).

7.3.1 Linking temperature change to sea-level change

On many timescales, sea-level change is driven by temperature change at both the ground surface and the sea surface. The least controversial example is of glacial-eustatic (sea-level) changes that occurred throughout much of the latest part of the Cenozoic Era, most conspicuously as regularly spaced oscillations during the Quaternary Period (Haq et al., 1987; Pugh, 2004). On shorter millennium-century timescales, most sea-level changes appear to have followed closely the record of temperature changes, and most scientists have assumed the latter to have been the principal driver of the former (Tanner, 1992; van de Plassche et al., 1998; Korotky et al., 2000; Nunn, 2001; Pugh, 2004). And finally, many decadal changes in sea level have been attributed largely to temperature change, including the net sea-level rise that has been monitored in most parts of the world for 100 years or so and which is expected to continue in the future (Warrick and Oerlemans, 1990; Wyrtki, 1990; IPCC WGI, 2001; Nunn, 2004). Temperature change is regarded as being able to influence sea level in two major ways.

First, temperature changes can alter the amount of ice on the land. Warming decreases the amount of land ice through melting, the meltwater entering the ocean and raising its level. Conversely, cooling increases the amount of land ice by causing precipitation to fall in solid forms and accumulate on land (rather than as liquid and be eventually returned to the ocean).

Second, water expands when it is heated and contracts when it is cooled. So when sea-surface temperatures rise, the upper few metres of the ocean expand and sea level rises. Conversely, when sea-surface temperatures fall, the upper ocean contracts and sea level drops.

It is important to note that there are other causes of non-tidal sea-level changes, ranging from the effects of global-tectonic changes that affect the capacities of the ocean basins to the effects of El Niño. Yet on a millennium-century timescale, such as that underpinning the A.D. 1300 Event model, it is most likely that interannual sea-level changes were driven largely by temperature changes. This is impossible to prove, as will be seen when causes are discussed in Chapter 10, and can be judged only by comparing chronologies of temperature and sea-level changes. Unfortunately, many data series are insufficiently precise – as would seem to be the case with those in Fig. 5.8 (where sea-level fall precedes temperature fall rather than the other way around) – to provide ready support for this argument.

Sea level does not respond instantly to temperature forcing. There is a lag time that differs depending on the magnitude of the temperature change and the location of the particular site. In the Pacific Basin during the last two millennia, lag times seem to have been decadal-century in duration. Examples come from the compilation for the Japanese islands by Sakaguchi (1983), work on the Kurile and Komandar island groups of the northwest Pacific by Razjigaeva et al. (2002, 2004) (see also Fig. 1.3A) and suggested interpretations of Pacific Island (including New Zealand) data by the present author (Nunn, 2000b, 2003b). Beyond the Pacific, the work of van de Plassche et al. (1998) made an explicit connection between global cooling and sea-level fall during the A.D. 1300 Event (see also Fig. 10.4B and C).

7.3.2 Climatic effects of increased El Niño frequency

As can be seen in Fig. 5.6B–D, there was a sharp and sustained increase in the frequency of El Niño events across the A.D. 1300 Event. Both theoretically and empirically, it is clear that any sustained increase in the frequency of El Niño events in the Pacific Basin would lead to increased climate variability (more interannual variation). Most of the Pacific Basin is affected more by El Niño (El Niño-Southern Oscillation (ENSO)-negative) than by La Niña (ENSO-positive) events. El Niño causes the warm pool of the western equatorial Pacific to shift east, often for another warm pool (in which tropical cyclones can form) to develop farther east. On the land surrounding the warm pool in the western Pacific, cooling, aridity, and decreased monsoon strength are typical of El Niño events.

Another important issue is the effect of El Niño on the positions of atmospheric convergence zones such as the Inter-Tropical Convergence Zone (ITCZ) and South Pacific Convergence Zone (SPCZ). Parts of the SPCZ move northeast during El Niño events and southwest during La Niña events (Trenberth, 1976; Folland et al., 2002), bringing unusual weather to land areas in the southwest Pacific Basin including eastern Australia (Power et al., 1999) and the Pacific Islands (Salinger et al., 1995, 2001).

So it seems reasonable to link increased El Niño frequency to increased climate variability, especially interannual variability, which has major implications for those people whose livelihoods depend directly on subsistence from agriculture and/or particularly vulnerable sources of wild foods. This would have been the case for everyone living in the Pacific Basin during the A.D. 1300 Event.

In the central part of the Pacific Basin, the tradewind circulation is weakened during El Niño events, allowing tropical storms both to form within a wider area and to penetrate farther eastwards than they could do normally. The warm pool that often forms in the central eastern Pacific during El Niño events allows tropical cyclones and other unstable phenomena to affect this part of the region, unlike at other times. Both effects seem likely to be applicable to the A.D. 1300 Event with this period mainly a time when tropical Pacific storminess increased and affected larger areas than it did during the Medieval Warm Period. Storminess outside the tropics is also likely to have increased across the A.D. 1300 Event as a result of increased equator–pole temperature gradients. In addition, some changes in storminess, particularly along the Pacific Rim, could be explained by shifts in storm belts, especially those associated with mid-latitude westerlies along the eastern Pacific Rim.

7.4. CAUSE–EFFECT LINKAGES FOR ENVIRONMENTAL CHANGE IN THE A.D. 1300 EVENT MODEL

The effects of climate (and climate-linked) changes on Pacific Basin environments are summarized in Fig. 7.1, and are considered here under the various boxes in the central part of that figure (environmental effects). Only the principal linkages are discussed here, namely ecosystem stress, increased upland erosion, coastline and reef emergence, and depletion of offshore resources.

Ecosystem stress is seen in Fig. 7.1 to be a result of temperature fall, sea-level fall and increased climate variability. Direct stress on terrestrial and nearshore ecosystems as a result of cooling during the A.D. 1300 Event is difficult to demonstrate because of the abundance of competing causes producing the likely same effects. Yet it is plausible to suppose that, as at other times, certain crops and wild foods – both terrestrial and marine – became less able to grow or less common under cooler conditions.

Sea-level fall would have impacted marine ecosystems by exposing areas of shallow sea floor and changing their ecology. In particular, the upper parts of most coral reefs – growing during the Medieval Warm Period to around low-tide level – would have emerged, killing off their most productive parts and reducing water circulation in the associated lagoons. The abrupt decline in pearl-shell production in the Aitutaki Island (Cook Islands) lagoon, inferred from the decline in its regional trade (Section 6.3.5), is an example of how reduced water circulation and increased turbidity impacted lagoonal productivity. In a similar fashion, water-table fall would have deprived coastal-lowland plants, many of which became essential components of human food-production systems during the Medieval Warm Period, of sufficient groundwater to grow properly.

Increased climate variability could have stressed terrestrial and marine ecosystems in a variety of ways. Periods of drought can deprive living things from the moisture essential to their growth and survival, in addition to aridifying environments and making them more vulnerable to erosion by wind and, subsequently, by rain. Conversely, excess precipitation – whether delivered in short intense bursts or over more prolonged periods – can drown terrestrial habitats, killing their living occupants, or even wash them away entirely preventing their subsequent recovery. The juxtaposition of periods of uncommon dryness and wetness can affect landscapes in ways that dislodge them from their evolutionary trajectories. For example, it has been argued that increased twentieth-century tropical-cyclone frequency in Fiji destabilized the landscapes of many high-island interiors, rendering impacts of tropical cyclones more severe than in earlier times (Nunn, 1998).

Related to this is the effect of increased upland erosion resulting from increased annual precipitation and increased storminess. Many records described in Sections 5.2.5 and 5.2.6 recall the environmental consequences of assumed increased storminess during the A.D. 1300 Event, including upland erosion and deforestation, and associated downslope valley and coastal-lowland sedimentation.

Coastlines that emerged as a result of sea-level fall during the A.D. 1300 Event were often transformed, typically because ocean embayments were converted to brackish lakes or wetlands (see Sections 5.3.3 and 5.3.4). On atoll reefs, islands formed early in the Little Ice Age in many parts of the Pacific as a result of reef emergence (Section 6.3.4.5).

The last major environmental effect is the depletion of offshore resources used particularly in human subsistence. As shown in Fig. 7.1, this effect was contributed to by ecosystem stress, coastline and reef emergence, and increased lowland sedimentation. Ecosystem stress – anything that alters the regular and potential growth of particular organisms – will lead to a reduction in offshore resources, be they food crops or trees for boat-making. Coastline emergence (relative sea-level fall) will deprive many coastal and valley-lowland plants of groundwater, as it may cause accustomed sources of surface water for animal (including human) consumption to dry up. Reef emergence will, as seen above, result in widespread death of reef organisms through exposure and others in lagoons because of increased water turbidity and reduced circulation. Both effects will result in the depletion of offshore resources. Increased lowland sedimentation may bury existing production systems, whether through alluviation or dune encroachment, and will also adversely affect shallow-water ecosystems by burial and increased water turbidity.

7.5. CAUSE–EFFECT LINKAGES FOR SOCIETAL CHANGE DURING AND AFTER THE A.D. 1300 EVENT

In Fig. 7.1 (bottom section – societal effects) are shown the principal societal effects regarded as having arisen primarily from climate and environmental changes during the A.D. 1300 Event in the Pacific Basin, especially the Pacific Islands. What cannot be shown in Fig. 7.1 is the near synchrony of these societal effects, something noted only occasionally by earlier commentators. For example, referring to the Peninsular Alaska and Aleutian Islands region, Maschner and Reedy-Maschner (1998) asked

"why these [societal] changes should occur in both regions simultaneously in the context of completely different languages, social organizations, or ceremonialism". (p. 28)

As in Chapter 6, the societal changes resulting from the climatic and environmental changes described above are divided into four groups – population dispersal (Section 7.5.1), changing patterns of resource utilization (Section 7.5.2), reduced interaction (Section 7.5.3) and conflict (Section 7.5.4).

7.5.1 Population dispersal

Settlement-pattern changes during and after the A.D. 1300 Event in the Pacific Basin were both necessitated by the environmental and societal changes that occurred and encouraged in places by the appearance of new environments. In other words, there were push factors and pull factors. Although changing climate – perhaps largely through increased storminess – probably had significant effects on particular settlement locations occupied during the Medieval Warm Period, most regional drivers of settlement-pattern change are considered – as shown in

Fig. 7.1 – to have been driven proximately by environmental changes. These drivers (the push factors) are seen to have been the food crisis for coastal dwellers and conflict, while attractants (the pull factors) are the presence of newly formed islands or unoccupied (pre-existing) coastal hinterlands or smaller islands.

First, the push factors will be discussed. The reduction in food resources that occurred along Pacific coasts during and after the A.D. 1300 Event provided the most pragmatic reason for their inhabitants to move elsewhere – there was simply insufficient to eat. To this can be added the other environmental stresses on coastal living that appeared at this time, including perhaps increased coastal flooding (associated with increased storminess) and reduced supplies of surface water for drinking. Conflict among people where none (or little) had existed in earlier times would have caused them to abandon settlements located in places that could not be easily defended in favour of places that could. Just as people today move away from conflict spheres, so there is an expectation that this would have occurred following the A.D. 1300 Event.

But it was not solely such push factors that led to the reorganization of settlement pattern early in the Little Ice Age in many parts of the Pacific Basin. There were also pull factors. Sea-level fall during the A.D. 1300 Event created new areas of land for people to settle, both offshore and along coasts, particularly around river mouths. Of these, only newly created offshore islands clearly appear to have been targeted for occupation, the new areas along coasts perhaps being regarded as too exposed. In most cases it seems reasonable to suppose that the decision to occupy these new lands was not driven solely by their appearance but also by the undesirability of continuing to occupy existing settlement locations. Nevertheless, the appearance of new lands would have encouraged their occupation, and we see a move to such places early in the Little Ice Age (see Section 6.3.4).

Not only were newly appeared lands occupied at this time, but also lands that had hitherto been unoccupied (or sparsely occupied). These include coastal hinterlands but also often smaller islands which people had not occupied (permanently) before, and which had similar defensible attributes to coastal hinterlands. Many examples of such occupations after the A.D. 1300 Event were given in Section 6.3.4.

Figure 7.3 summarizes last-millennium settlement-pattern changes on Pacific Islands, and at least some parts of the Pacific Rim (such as New Zealand). The early part of the Medieval Warm Period was characterized by smaller nucleated coastal settlements (Fig. 7.3A) that in many places aggregated into fewer larger ones by the end of the Medieval Warm Period (Fig. 7.3B), perhaps in response to increasing aridity (Section 4.3.6). During the early part of the Little Ice Age, settlement pattern in many parts of the region, particularly on the tropical islands (see Section 6.3.4), involved smaller inland settlements, the earliest sustained occupation of smaller offshore islands, and out-migration – all a likely response to the food crisis and conflict resulting from the A.D. 1300 Event.

Not all coastal-environmental changes in the Pacific Basin during the A.D. 1300 Event repelled rather than attracted people. In one of the clearest examples of the likely influence of sea-level fall on coastal environments at this time, various

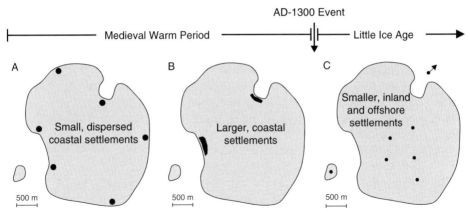

FIGURE 7.3 Changes in settlement pattern during the last 1,200 years on many high Pacific Islands (after Nunn, 2003a). Map A shows the situation at the start of the Medieval Warm Period, perhaps about A.D. 900. Map B shows the situation at the end of the Medieval Warm Period, just prior to the A.D. 1300 Event about A.D. 1250. Map C shows the situation during the Little Ice Age, after the end of the A.D. 1300 Event about A.D. 1500.

changes occurred in different parts of Kaua'i Island in Hawaii at this time (Nunn et al., 2007; see also Section 7.7.4). The common effect is illustrated by the Wainiha Valley where permanent coastal settlement characterized the Medieval Warm Period but where sea-level fall during the A.D. 1300 Event led to people abandoning these settlements and establishing others inland at higher elevations during the Little Ice Age (see Fig. 7.9A and C). The opposite effect is seen on Mana Plain where a marshy coastline proved unattractive for humans during the Medieval Warm Period but where – plausibly as a result of sea-level fall across the A.D. 1300 Event – sustained coastal settlement characterized the subsequent Little Ice Age (see Fig. 7.9B and D). A comparable situation affected parts of the coast of eastern Australia, with people being drawn to the coast, probably as a result of ecosystem changes that reversed the attractiveness for people of the hinterland, where most people lived during the Medieval Warm Period, and the coast, where most lived during the Little Ice Age (see Section 5.4.2).

7.5.2 Changing patterns of resource utilization

For most people inhabiting the Pacific Basin during the Medieval Warm Period, the offshore (shoreflat, lagoon, reefal) environment was an important source of food. On Pacific Islands, there is ample evidence that it was the most important source of wild food for people at that time (see Chapter 4).

Various climatic and environmental changes during the A.D. 1300 Event contributed to a general reduction of marine-food supply (see Fig. 7.1). These included offshore ecosystem changes resulting from cooling and sea-level fall. Evidence of offshore ecosystem changes during the A.D. 1300 Event is widespread throughout the Pacific Basin, marked especially well among communities depending on shellfish and lagoon-fish species (see Fig. 6.5 for examples).

There is also evidence that coastal-terrestrial food sources were reduced during the A.D. 1300 Event although most such evidence is inferential, only a few studies of coastal dwellers at the time having recorded direct effects of food stress; examples come from the western United States (Walker, 1989), New Zealand (Leach, 1981) and Fiji (Burley, 2003). Indirect evidence for food stress among coastal dwellers during the A.D. 1300 Event comes from the widespread population dispersal (Section 7.5.1) and conflict (Section 7.5.4) that ensued.

It may be that falling sea-surface temperature during the A.D. 1300 Event also affected the growth of seaweeds in the nearshore areas of the Pacific. This has been suggested for coastal communities of southern California, where kelp-bed species of fish, mammals and birds were important for subsistence, particularly for the inhabitants of the offshore islands (Arnold, 1992). During warm periods, such as those that obtained at the end of the Medieval Warm Period, when sea-surface temperatures rose above 20°C, it has been suggested that the deleterious effects of this on the kelp beds led to "environmental disruption" which in turn led to "rapid social evolution" expressed by the appearance of the first chiefdom on Santa Cruz Island (Arnold, 1992, p. 66). For a variety of reasons, seaweed may also have been important to Pacific Islanders during the last millennium (Abbott, 1991), its declining productivity around A.D. 1300 contributing to food stress and societal breakdown.

When food resources become scarcer or their supply less predictable, then preservation and storage become important (Goland, 1991). A tropical cyclone can wipe out an island's food crops for 6 months, so that methods of preserving key food crops (out of those that lend themselves to preservation) have become widely developed in the Pacific Basin, especially in the more remote Pacific Islands. The most common method of long-term preservation is pit fermentation (of various root crops, particularly breadfruit) which has been recorded ethnographically for the Marquesas and the eastern outer Solomon Islands. It has been suggested that such technologies developed about A.D. 1300 on Tikopia Island, Solomon Islands (Kirch and Yen, 1982), and increased in importance in the Marquesas group of French Polynesia about the same time (Kirch, 1984a) in response to reduction in the food-resource base.

7.5.3 Reduced interaction

For the purposes of this account, interaction refers to the routine contact between adjacent communities, whether occupying contiguous land areas within a few kilometres of each other or on islands separated by ocean gaps – perhaps hundreds of kilometres wide – that were regularly traversed in earlier times. Some examples of changing interaction across the A.D. 1300 Event are illustrated in Fig. 6.7.

While the breakdown of interaction early in the Little Ice Age in many parts of the Pacific Basin was probably driven in part by climate change, as shown in Fig. 7.1, interaction networks also undoubtedly collapsed as a result of conflict and the resulting abandonment of coastal and marine-associated lifestyles. Interaction became dangerous, abruptly subordinate to issues of individual survival, and irrelevant in the changed Pacific world that emerged after the A.D. 1300 Event.

Also it is envisaged that increased storminess in particular would have inhibited long-distance interaction, particularly across the ocean, by creating hazards that were significantly fewer or encountered less commonly during earlier times. Storms at sea may have driven vessels off course, increased climate variability may have led to unexpected changes in wind direction, increased cloudiness may have adversely affected stellar navigation, while even sea-level fall during the A.D. 1300 Event would have created shoals and made nearshore navigation more difficult than in earlier times.

7.5.4 Conflict

Much conflict (warfare) in pre-historic times is well established as being an adaptive response to scarcity of resources, particularly food (equated with land) resources, although of course many other factors contributed to particular conflicts (Vayda, 1976; Durham, 1976). The study of warfare in 186, mostly pre-industrial, societies led Ember and Ember (1992) to conclude that warfare arose mostly from people's fear of nature, perhaps a natural disaster or from the fear that a natural disaster could affect them through abruptly diminished food supply – with potentially drastic consequences – at some point in the future. In this sense, it seems that in the Pacific Basin during the Medieval Warm Period, the constancy of the climate was one reason why conflict appears hardly to have occurred, whereas the onset of a period of rapidly changing and unpredictable climate during the A.D. 1300 Event sowed the seeds for conflict, much of which appears to have begun during the early part of the Little Ice Age (see Section 6.3.7).

In Fig. 7.1, conflict is pictured as arising solely from food stress. While it is acknowledged that conflict, particularly locally, may have been driven by other climate-linked forces, it is considered that food stress was the most widespread driver and that which either directly drove conflict or immediately underpinned its proximate causes.

It takes little imagination to picture scenarios of how food stress could engender conflict, especially among populations at levels close to the carrying capacity of rigidly circumscribed areas such as remote islands. As food stress continued, and communities came to realize that their condition would not be relieved in the foreseeable future and that short-term strategies were not working, it is likely that the social structures that defined the community would have begun to fall apart, and conflict would have ensued. Examples of conflict arising from food stress associated with prolonged drought are known from more recent times, especially from islands whose circumscribed nature often means that critical thresholds in supply–demand are reached sooner and more frequently than on larger landmasses. These examples are instructive because they deal with societies that can be assumed to be similar in many ways to those that were affected by food crisis during the A.D. 1300 Event in the Pacific Basin.

During the years 1925–1963, there were several periods of moderate to severe drought in northern Australia that brought conditions of "food deprivation" to the Kaiadilt people of Bentinck Island in the Gulf of Carpentaria (Cawte, 1978). At times, food deprivation was so severe that conflict broke out between different kin groups on the island, sometimes expressed as cannibalism. In a survey of famine in the Pacific, Currey (1980) concluded that at the "peak" of famine-induced food crises "Pacific Island societies will also break down" (p. 452). On the remote Pacific island of Tikopia ($5\,km^2$), the effects of food shortages following tropical cyclones led to societal breakdown, as shown by:

- groups of kinsfolk becoming isolated from each other because of the shortage of food, which was normally exchanged in rituals, and
- the widespread stealing by people who, fearing that another tropical cyclone would reduce them to starvation, sought to "accumulate a margin of safety" (Firth, 1959, p. 65).

Beyond the Pacific Basin, during the period of slave labour on smaller Caribbean Islands, the effects of drought on already underfed people led to various "modes of competitive behavior" including violence (Dirks, 1978, p. 123).

While conflict need not be the immediate consequence of prolonged food shortage, it could be regarded as inevitable in such places after other adaptive strategies were tried and failed. Among these in Pacific Islands might have been the use of alternative foods, the introduction of novel forms of food storage and preservation, migration and even searches for imagined lands of plenty. Each of these is discussed briefly below.

Many Pacific Islands have designated "famine foods", ranging from wild yams and arrowroot in the Cook Islands, through the flesh of pandanus and the fetid fruit of *noni* (*Morinda citrifolia*) in Hawaii, to dried breadfruit and wild taro leaves on atolls such as Kapingamarangi (Currey, 1980). The implication of the existence of such foods, not consumed under normal circumstances, is that their consumption began at times of food crisis and continued during periods of increased climate variability.

Although Kirch and Yen (1982) found that pit ensilage – a form of long-term food storage – began about A.D. 1300 on isolated Tikopia Island, the times when this and other forms of storage originated elsewhere in the Pacific Islands are generally unknown. Yet it is plausible to suppose that some forms of storage began or were introduced as strategies for creating surpluses to be eaten in times of hardship, conceivably at times of climatically induced food crisis. The converse strategy is that of reducing consumption of luxury foods ("feast foods") at such times. In this context, the extirpation of pigs from many remoter Pacific Islands, apparently by the early Little Ice Age (Section 6.3.8), is germane.

Migrations of people during famines are well documented during the recorded history of the Pacific. For Guadalcanal Island in Solomon Islands, 35% of the migrations of entire settlements that occurred between 1850 and 1972 were because of the lack of accessible food (Bennett, 1974). During the 1941 famine in the New Guinea highlands, migration was considered an insurance mechanism (Waddell, 1975). It is reasonable to assume that recorded migrations in the Pacific Basin during the A.D. 1300 Event and early Little Ice Age, principally abandonment of mystery islands (see Section 7.7.6), were responses to food crisis.

During recorded history on more remote islands, migration was evidently entertained as an option during famines, even though it had no appreciable chance of success. Sometimes "suicide voyages" were intentionally undertaken, sometimes canoes full of people set off in search of imagined lands of plenty. According to Firth (1959), *foraus* or suicide voyages were planned in response to starvation on remote Tikopia Island. Records of more than 800 people setting off on fruitless *he fenua imi* (land-seeking) voyages from the Marquesas Islands during times of famine were obtained by Handy (1930).

7.6. CASE STUDIES OF THE A.D. 1300 EVENT ALONG THE PACIFIC RIM

To fully appreciate the case for the A.D. 1300 Event model, three case studies from the Pacific Rim are given below. The first (from Peru) looks at examples of societal collapse and survival under similar conditions of climate forcing (Section 7.6.1). The second contrasts the effect of the A.D. 1300 Event on hunter-gatherer communities of the western United States, some on the mainland, some on islands offshore (Section 7.6.2). The third looks at the contrasting effects of sea-level fall during the A.D. 1300 Event on two parts of the southwest Pacific Rim (Section 7.6.3).

7.6.1 Case study: upland and lowland Peru, east Pacific Rim

The changes that took place in Peru (and adjacent parts of Bolivia) during the A.D. 1300 Event illustrate not only its contrasting effects on human societies in the same region but also the likely effects of local climate variations on modifying regional forcing (Fig. 7.4).

One example comes from the *altiplano*, where the Tiwanaku civilization collapsed at this time, beginning with the abandonment of cities and the farms on which they depended A.D. 1150–1200 (Binford et al., 1997). It is generally agreed that protracted severe droughts in this area were responsible for this. In the Quelccaya ice-core palaeoclimate record, the period of severest drought is A.D. 1245–1310 (Thompson et al., 1985), a time range that is supported by analyses of sediment cores from Lake Titicaca (Binford et al., 1997).

But it is legitimate to ask why the Tiwanaku civilization did collapse rather than recover in this instance. Why, for instance, were the fertile agricultural lands around Lake Titicaca that had sustained the civilization before about A.D. 1150 not reoccupied after A.D. 1310? It may be that new lifestyles had by then evolved and people had no inclination to return to earlier, memorably more risky, food-production systems. But it may also be that the climate did not suggest this option. As seen in Fig. 7.5A, cooling affected this area during the A.D. 1300 Event, and it is also likely that an increased frequency of El Niño led to greater climate variability. This may have been manifested by the increasing juxtaposition of drought and periods of prolonged storminess associated with shifts in storm belts: good reason for continuing the less vulnerable lifestyles (pastoralism and opportunistic dry farming – Kolata and Ortloff, 1996) that developed after the abandonment of the Tiwanaku area.

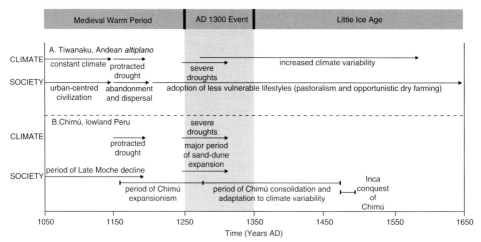

FIGURE 7.4   Climate–society changes A.D. 1050–1650 in upland and lowland Peru. References in text. (A) Changes on the *altiplano* of Peru–Bolivia, where a comparatively constant climate during the Medieval Warm Period allowed the development of the complex Tiwanaku civilization that began to collapse during a period of prolonged drought A.D. 1150–1200. Tiwanaku never recovered, perhaps because of a later period of severe drought during the A.D. 1300 Event. (B) Changes in lowland Peru, with detailed examples from the Jequetepeque Valley. Following earlier drought-linked stresses, Moche society continued to decline during the Medieval Warm Period and was taken over by Chimú culture. Shortly afterwards, a period of severe drought during the A.D. 1300 Event was marked by stress and adaptation of Chimú society to more variable climate conditions, a preoccupation that rendered them vulnerable to conquest by the Inca about A.D. 1470.

A similar explanation is tenable for the Chimú culture of lowland Peru at this time. The spread of sand dunes into cities and across previously fertile agricultural lands in places such as the Jequetepeque Valley was associated with the time of severest drought recorded at Quelccaya (Section 4.3.2). Yet while this led to a manifest downturn in the rise of Chimú culture, it did not spell its end for, as the threat of drought receded and such areas became well watered once again, the Chimú returned, adapting their food-production strategies to the more variable climate.

Ironically, the ability of the Chimú culture to adapt successfully to the increasingly variable climatic conditions in the aftermath of the A.D. 1300 Event proved its undoing. As Dillehay and Kolata (2004) explain,

> "overinvestment in a highly complex, productive, but vulnerable agricultural infrastructure may explain why the Chimú were unable to respond effectively to the challenge of Inca military expansionism and succumbed with relative ease to imperial ambitions of the Inca kings" (p. 4230)

around AD 1470.

In summary, the A.D. 1300 Event in this area would have exacerbated the effects of drought late in the Medieval Warm Period, ensuring that the displaced Tiwanaku people never reconstructed their Andean civilization and laying the

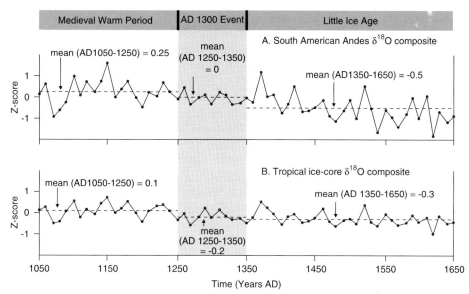

FIGURE 7.5 Ice-core palaeoclimate records (decadal averages) for the period A.D. 1050–1650 (after Thompson et al., 2002). Graphs drawn using the supporting information on the PNAS website with means for various periods calculated by the author. Note the evidence for climate change represented by falling levels and lower variability of $\delta^{18}O$ during the A.D. 1300 Event, likely to be primarily a result of cooling. (A) Composite record for three Andean cores (shown separately in Fig. 4.1). (B) Combined records from low-latitude high-elevation ice fields.

ground for Chimú exposure to Inca aggression. It is possible that sea-level fall during the A.D. 1300 Event further destabilized the economic base of the Chimú by lowering valley-floor water tables and exposing areas of formerly productive shallow ocean floor.

Similar forcing probably occurred throughout the tropical Pacific Rim (Fig. 7.5B), explaining why a number of complex societies had difficulties weathering the A.D. 1300 Event.

7.6.2 Case study: southern California, east Pacific Rim

In southern California, both on the mainland and on islands offshore, increasing social complexity among the Chumash people during the late Holocene has been attributed to a bountiful environment, which enabled ample food to be collected by hunting, gathering and fishing (King, 1990). According to this model, the distribution of food and other resources through an intervillage network that permitted accumulation and loss led to culture change among the Chumash, which has also been used to explain the appearance of chiefdoms among the peoples of the northwest United States (Ames, 1995).

The foundations of a more realistic – at least more readily verifiable – model of changes in Chumash society were laid with the work of Arnold (1992) who argued that rising sea-surface temperatures and increasing numbers of severe droughts during the later Medieval Warm Period (A.D. 1150–1300) led to chiefs on

Santa Cruz Island gaining power through control of resources. More than this, the chiefs came to control the manufacture of shell beads that were exchanged for food with mainland Chumash during times of food stress. Recent work, utilizing more detailed palaeoclimate records than those available to Arnold, has argued that the Chumash were not responding to local changes in resource availability attributable to high sea-surface temperatures alone but to "late Holocene climatic perturbations whose effects were felt far beyond southern California" (Raab and Larson, 1997, p. 322). The principal of these are the protracted and severe droughts that affected the southwest United States during the period A.D. 800–1400, particularly A.D. 1150–1300. Within this period, various studies have shown the incidence of harsh drought A.D. 1120–1150 along the California coast (Raab and Larson, 1997) and A.D. 1209–1350 in the Sierra Nevada (Stine, 1990). Increased moisture characterized the early Little Ice Age beginning in A.D. 1400.

Many human changes – among them internal societal changes (such as the emergence of hierarchies), physiological changes (signs of disease or violence) and changes in settlement pattern – have been attributed to climate change at this time but these are commonly poorly dated, so the association remains suggestive. Yet the abandonment of Santa Cruz Island A.D. 1250–1300 (Arnold, 1992) does provide data that support this association, as does the more general conclusion of Jones et al. (1999) that along the central California coast, sites occupied earlier than A.D. 1200 show signs of abandonment in the following century. Changes in weaponry and obsidian production are also suggestive of abrupt cultural changes about A.D. 1200–1300.

It is argued that here the A.D. 1300 Event came at the end of a century-long period of increased drought which had already brought about profound changes in human societies along and off the California coast. Cooling and sea-level fall in this area, together with continued drought, would have further stressed its human inhabitants, leading to the conflict and out-migration that apparently characterized much of the early Little Ice Age here.

7.6.3 Case study: east Australia and New Zealand, southwest Pacific Rim

Many studies along the east coast of Australia and New Zealand suggest that shallow-water marine ecosystems changed in response to sea-level fall during the A.D. 1300 Event (Fig. 7.6). The scenario envisaged is one that simply involved a fall of sea level exposing areas of formerly inundated shoreflat, causing its ecology to change, and resulting in changes in the predation targets of coastal-tethered humans. In the cases where the coastline was heavily embayed at the time of higher sea level during the late Medieval Warm Period, the sea-level fall may also have drastically affected water circulation offshore with associated effects on ecosystem composition and human predation.

Other changes in eastern Australia and New Zealand also suggest a near-simultaneous response to changes in nearshore marine-food-resource supply across the A.D. 1300 Event. Principal among these are changes in settlement pattern, with examples of parts of the Australian coast being occupied for the first time around A.D. 1300 – along the Curtis Coast of Queensland (Ulm, 2004) and along the arid margin of southeast Australia (Holdaway et al., 2002) – and

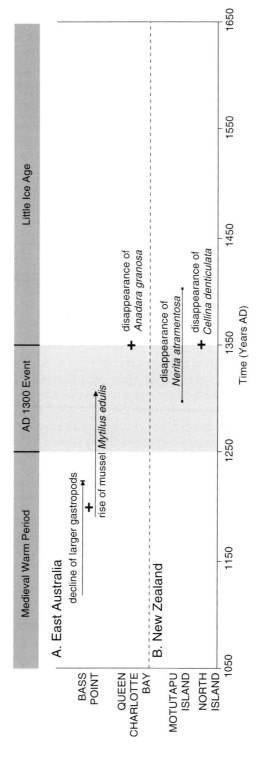

FIGURE 7.6 Changes in nearshore human predation inferred from abrupt changes in shell middens in east Australia and New Zealand. Data from Bass Point (Bowdler, 1976), Queen Charlotte Bay (Hiscock and Kershaw, 1992), Motutapu Island (Szabó, 2001) and North Island (Rowland, 1976).

conversely in New Zealand a mass abandonment of coastal settlements in favour of upland fortified *pa* settlements as early as A.D. 1300 (Kirch, 2000a).

It is argued here that all these changes can be explained by sea-level fall changing the nature and quantity of shallow-water marine foods during the A.D. 1300 Event. The point is also clear that climate forcing does not always evoke identical responses for much depends on the antecedent relationship between humans and the environments they occupied. In New Zealand, for instance, the earliest settlers followed marine-associated lifestyles, whereas those in the two parts of eastern Australia mentioned did not. In New Zealand, climate and sea-level changes during the A.D. 1300 Event are envisaged as having rendered near-shore areas significantly less bountiful and upland/inland areas safer from the conflict attendant on resource depletion. Conversely, in parts of eastern Australia, these climate and sea-level changes rendered inland areas (and nomadic life-styles) generally less sustainable than they had been in earlier times, in contrast to coastal areas where new opportunities for subsistence appeared around the same time.

7.7. CASE STUDIES OF THE A.D. 1300 EVENT IN THE PACIFIC ISLANDS

Various case studies are given below to illustrate how the A.D. 1300 Event model is consistent with empirical data for the Pacific Islands region. Each case study highlights a particular issue.

The case study of Nuku Hiva and Tahuata islands in the Marquesas (Section 7.7.1) shows how lowland communities were displaced upslope during the A.D. 1300 Event, and how associated ecosystem changes altered patterns of food consumption. The Lakeba case study (Section 7.7.2 uses detailed information about last-millennium settlement-pattern changes to illustrate the impact of the A.D. 1300 Event. The Palau case study (Section 7.7.3) shows the effects of climate and sea-level change on an isolated island group, where people responded by building fortified settlements in both open and protected locations. The Kaua'i case study (Section 7.7.4) explains how opposing settlement-pattern changes occurred on the same island, and emphasizes the importance of antecedent environmental conditions. The Easter Island case study (Section 7.7.5) shows the dramatic effects of increased El Niño frequency on a remote island where coasts were never targeted for settlement. Finally, the case study of the mystery islands of the central eastern Pacific (Section 7.7.6) shows how remoteness and comparative resource impoverishment forced island abandonment in the aftermath of the A.D. 1300 Event.

7.7.1 Case study: Nuku Hiva and Tahuata Islands, Marquesas Islands, French Polynesia

Nuku Hiva and Tahuata, like most other islands in the Marquesas group, have little flat land along the coast and during the Medieval Warm Period most people lived in open (unfortified) communities along the floors of the island's "verdant valleys" (Suggs, 1961, p. 182) following agriculture based mainly on lowland field cropping (Kirch, 1984a). The A.D. 1300 Event (beginning about A.D. 1200 on

Nuku Hiva – the start of the Expansion Period according to Kirch, 1984a) saw the "sudden" movement of people out of these communities and scattered elsewhere on the island, including its inhospitable coasts and upland areas, where fortified sites were constructed. This is exactly what would be expected if there was an abrupt reduction in water tables that reduced yields of valley-floor crops, as would have occurred during the sea-level fall of the A.D. 1300 Event, and would have led to the "food stress" that characterized Marquesan society at this time (Kirch, 2000a, p. 261). Other factors such as increased storminess, expected as a result of increased El Niño incidence in this region, may have contributed to these settlement-pattern changes.

By the start of the Classic Period on Nuku Hiva (A.D. 1400 according to Suggs, 1961), settlement distribution "remained widespread and dispersed" (Kirch, 2000a, p. 261), but with the vulnerable coastal sites being abandoned;

> "the entire population seemed to gravitate inland, moving up high into the valley heads for protection". (Suggs, 1961, p. 185)

Suggs explains the abandonment of the coastal (beach) sites by fear of increased "raiding from the sea" but it is more likely that this was a result of both their vulnerable locations, particularly to raiding from the land, and the depletion of offshore food resources associated with cooling and sea-level fall.

On Tahuata Island, valley-floor, near-coastal settlements were abandoned during the A.D. 1300 Event in favour of upland, often ridgetop locations where natural defendability was easily enhanced by artificial defences (Rolett, 1998). Changes in vertebrate consumption revealed by excavations in the Hanamiai Valley permit insights into the likely effects of climate change on human subsistence (Table 7.1). Numbers of commensal species such as dog, rat and chicken, all of which were probably introduced deliberately by humans to the Marquesas, rise during the Medieval Warm Period, perhaps manifesting human control, but

TABLE 7.1 Selected vertebrate remains from the Hanamiai excavation, Tahuata Island, Marquesas (after Rolett, 1998).

	Phase I	Phase II	Phase III	Phase IV
	(A.D. 1000–1250) Medieval Warm Period		(A.D. 1250–1400) A.D. 1300 Event	(A.D. 1400–1800) Little Ice Age
Dog (*Canis familiaris*)	2	26	20	50
Rat (*Rattus exulans*)	60	269	165	506
Medium whale (Odontoceti)	10	4	0	2
Human (*Homo sapiens*)	0	1	3	2
Native seabirds	448	88	13	14
Native landbirds	64	5	8	6
Chicken (*Gallus gallus*)	3	9	3	6
Turtle	20	10	2	4
Fish	2292	2216	2001	4401
All species	3299	3059	3197	7542

fall abruptly during the A.D. 1300 Event, perhaps manifesting the breakdown of human control, before recovering somewhat thereafter. Vertebrates not subject to human management – including whales, native seabirds and turtles – all show declines during the course of the Medieval Warm Period that are readily attributable to their unsustainable exploitation, but this decline accelerates during the A.D. 1300 Event – probably because of climate-driven ecosystem change – and reverses slightly during the Little Ice Age. The fall in fish numbers is likewise dramatic during the A.D. 1300 Event, probably because of offshore ecosystem deterioration; the subsequent rise during the Little Ice Age may indicate the recovery of these ecosystems off the reef-free coasts of Tahuata. The increases in native landbird consumption during the A.D. 1300 Event probably reflect increased human occupation of the environments where such birds commonly lived. The increase in human remains may signify cannibalism associated with the outbreak of conflict.

The Marquesas provides a superb example of the effects of the A.D. 1300 Event on a Pacific Island society. While there are some questions about timing, it seems that the Medieval Warm Period was characterized by peaceful lowland agriculture-based communities while the Little Ice Age saw widely scattered communities experiencing food shortages and in conflict with each other. In this scenario, it is clearly a mistake to regard the sudden movement of people out of valley-floor settlements as a result of rapid population growth rather than an abandonment of undefendable valley-floor settlements in favour of more defendable locations where smaller groups established themselves.

7.7.2 Case study: Lakeba Island, Fiji

The island Lakeba is one of the largest in the Lau group of eastern Fiji, an important zone of interaction between Fiji and Tonga in pre-historic times. Research by Best (1984) allows an unusually high degree of understanding of changes in settlement pattern during the last millennium (Fig. 7.7).

Although there were movements of people inland prior to the start of the Medieval Warm Period, all the settlements during this period (Period III) were coastal. It is conceivable that the aridity of the Medieval Warm Period led to the temporary abandonment of Lakeba by all but a small number of people, although it is also possible that some of the evidence of coastal occupation during this period has been washed away. By the time of the A.D. 1300 Event (Period IV), most settlements were within a few hundred metres of the coast but during the Little Ice Age (Period Va) most settlements were inland and upland; at this time coastal sites may have been occupied only temporarily. Following European contact with Lakeba, people left their upland and inland settlements and once more established coastal settlements.

The Lakeba settlement data fit the A.D. 1300 Event model well. Coastal settlements existed during the Medieval Warm Period – as they had in earlier times – but began moving inland and upslope during the A.D. 1300 Event in response to falling resource availability along previously productive reef-fringed coasts primarily as a result of sea-level fall. This situation was amplified during the Little Ice Age as coastal food supply fell markedly and conflict broke out. Inland,

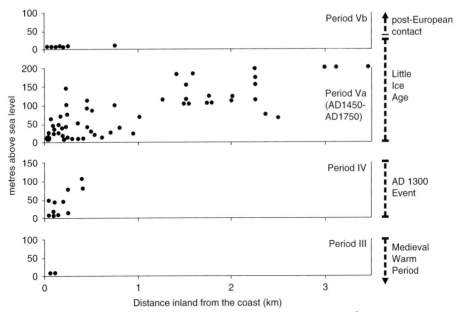

FIGURE 7.7 Last-millennium settlement history for Lakeba Island (56 km^2), eastern Fiji, as shown by locations and numbers of settlements in different periods (distinguished by pottery style) (after Best, 2002). Note that during the Medieval Warm Period, settlements are all near-coastal; their low number might be because they were large and/or have low archaeological visibility. Alternatively, aridity during the Medieval Warm Period may have led most people to abandon islands such as Lakeba. During the A.D. 1300 Event, settlements begin moving inland and upslope, a trend that peaked during the Little Ice Age.

upland settlements were commonly fortified, sometimes linked with coastal sites occupied perhaps during times when conflict abated temporarily or as transitory marine-food processing centres.

Similar conclusions can be drawn from the study of Totoya Island, also in eastern Fiji, although changes in settlement pattern are less well dated (Clark et al., 1999). Elsewhere in the Lau group, the author has seen the remains of upland fortified settlements that locally recounted oral traditions (unpublished) suggest are similar in age to those on Lakeba; examples come from Mago, Moce, Nayau, Vanuabalavu and Yacata islands. In the Sigatoka Valley on Viti Levu – the largest island in the Fiji group – dates from fortified inland upland settlements show that almost all were established early in the Little Ice Age (see Fig. 6.4) as was the case on Lakeba.

7.7.3 Case study: Palau Islands, Micronesia

The islands of the Palau (Belau) Archipelago in the equatorial western Pacific comprise larger volcanic islands (such as Babeldaob) and numerous smaller high limestone islands (such as the Rock Islands). Research for this case study was presented by Masse et al. (2006).

Drier conditions prevailed in Palau for much of the Medieval Warm Period and were succeeded by wetter conditions during the Little Ice Age (Fig. 7.8). Droughts may have increased societal stress during the later part of the Medieval Warm Period leading to the establishment of fortified settlements (stonework villages) on the Rock Islands and the outbreak of warfare throughout Palau about A.D. 1250.

Important evidence for sea-level fall comes from the sustained decline in level of Lake Ngerdok on Babeldaob. A hint of the effects of sea-level fall on marine resources comes from the unmistakable decrease in size of marine fauna recorded in excavations at Uchularois Cave in the Rock Islands. Sea-level fall is also suggested by subsequent formation of new coastal wetlands which attracted people to the coast on Babeldaob where they established fortified villages; this is interpreted as a compromise between access to wetlands for agriculture and the need for enhanced security in the open setting.

An increase in numbers of fortified settlements in the more readily defendable Rock Islands around A.D. 1450 coincided with pig extirpation, both observations interpreted as demonstrating a deepening of the resource crisis affecting Palau at this time. The subsequent abandonment of Rock Island fortified settlements may represent the easing of this crisis as climate ameliorated.

In terms of the A.D. 1300 Event model, societal changes in Palau across the A.D. 1300 Event can be explained largely by climate forcing. Dry conditions stressed Palauan society during the later Medieval Warm Period but the situation worsened significantly as sea level fell during the A.D. 1300 Event. This led to resource impoverishment and conflict manifested by increased construction of fortified settlements, including some near newly formed coastal wetlands. Despite the onset of wetter conditions early in the Little Ice Age, the effects of sea-level fall during the A.D. 1300 Event continued to mean that fewer food resources were available to Palauans at this time and that conflict continued for at least a few hundred years.

7.7.4 Case study: Kaua'i Island, Hawaii group

On the island Kaua'i there is evidence of opposing human responses to the A.D. 1300 Event that can be satisfactorily explained by sea-level fall, and which are likely to have been the same as on other Pacific Islands. This research was reported by Carson (2003, 2004, 2006).

Along some parts of the Kaua'i coast, such as that around the mouth of the Wainiha Valley, people during the Medieval Warm Period followed lifestyles that were centred around marine-food consumption (Fig. 7.9A). Temporary uses were made of the hinterland for horticulture and wild-food gathering. Elsewhere along the Kaua'i coast, as on the Mana Plain, the land was marshy and only higher spots were temporarily occupied by people (Fig. 7.9B).

Sea-level fall during the A.D. 1300 Event saw coastal settlements such as those around the mouth of the Wainiha Valley abandoned and people moving inland, occupying more defensive locations and cultivating the immediate vicinity (Fig. 7.9C). Most coastal areas were used only occasionally. Yet on the Mana Plain, human responses changed in quite different ways (Fig. 7.9D). Here sea-level fall

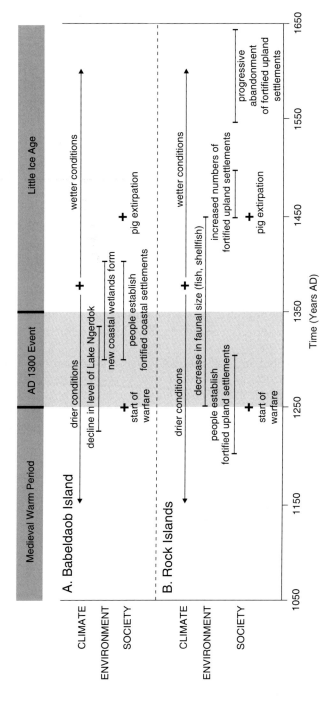

FIGURE 7.8 Changes in climate and environment in Palau and the possible societal responses (all data from Masse et al., 2006).

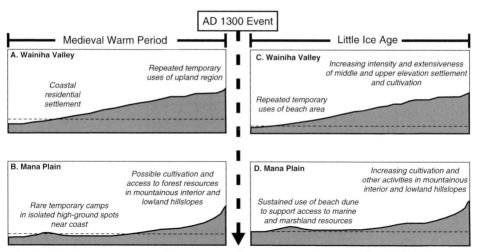

FIGURE 7.9   Contrasts in coastal-environmental response to relative sea-level fall. Schematic view of sea level and potential land use areas on the island Kaua'i (Hawaii) during the Medieval Warm Period (A: Wainiha Valley; B: Mana Plain) and Little Ice Age (C: Wainiha Valley; D: Mana Plain) (after Nunn et al., 2007, from work by Carson, 2003, 2004).

drained many swamps, facilitating access to marine resources, and exposed areas of formerly shallow-water shoreflat rendering them available for accumulation of windblown sand.

Similar situations to that in the Wainiha Valley are known throughout the Pacific but it appears uncommon to find the converse, as illustrated by Mana Plain, within the same island. Likely analogous situations are found on Kosrae (Athens, 1995), Palau (Masse et al., 2006) and maybe elsewhere in Micronesia where, for a variety of reasons, the societal crisis at the start of the Little Ice Age may not have been as severe as it was elsewhere. And in Hawaii, the effects of the A.D. 1300 Event on human societies were evidently milder than elsewhere, the Hawaii Islands lacking the unmitigated endemic conflict that occurred during most of the Little Ice Age in the tropical Pacific Islands.

In summary, it is envisaged that higher sea level during the (later) Medieval Warm Period simultaneously allowed coastal settlement in some parts of the Kaua'i coast yet prevented it in others. During the A.D. 1300 Event, sea-level fall caused a resource crisis that led to many formerly coastal communities moving inland yet elsewhere along the Kaua'i coast, because sea-level fall had transformed coastal environments, people settled for the first time.

7.7.5 Case study: Easter Island

A fine example of societal complexification during the Medieval Warm Period comes from remote Easter Island (Rapanui) in the southeast Pacific, famed for its giant statues called *moai*. It has long been known that the pre-European history of Easter Island could be neatly divided into two parts. The earlier Ahu Moai (or *moai* construction) period (*ca.* A.D. 1100–1500) was a time of plenty, when most settlements were in open locations on flatter parts of the island. In contrast, the latter Huri Moai (or *moai* toppling) period (*ca.* A.D. 1500–1722) was a time of societal stress, marked by abandonment of open settlements in favour of caves and smaller offshore islands, and the outbreak of conflict, conspicuous for the first use of the island's obsidian to manufacture spearheads (*mata'a*) about A.D. 1300 (Bahn and Flenley, 1992).

Easter Island's coasts were never settled by a significant number of people because they are mostly cliffed and difficult to access, and because the island's interior was particularly inviting and contained areas of fertile soils. Easter Island's early settlers developed an extraordinary culture, popularly manifested by *moai*, and one based on probably unsustainable resource exploitation during the times of plenty (Ahu Moai). The times of less (Huri Moai) were prefaced by ecological impoverishment – likely to have been associated with the A.D. 1300 Event – and the toppling and destruction of many *moai*. When Roggeveen reached Easter Island in 1772, the islanders were mostly living in caves, apparently practising cannibalism, and occupying a conspicuously resource-depleted environment – a distant cry from their ancestors' prosperity.

The basic thesis of Bahn and Flenley (1992) and Flenley and Bahn (2003) is that the people of Easter Island caused the conspicuous collapse of their society early in the Little Ice Age by unsustainably exploiting the island's tree resources, a scenario that has been taken up by those who study societal collapse (Diamond, 2005) and those who see the Easter Island story as a parable for our own time: an example of a society that committed "ecological suicide by destroying their own resources" (Diamond, 1999, p. 411). The collapse of Easter Island is regarded as "a story with an urgent and sobering message for our times" (Flenley and Bahn, 2003, p. vii).

It does not weaken the imperative for global sustainable development to question the scientific interpretation of the data used to make the case for human-driven collapse of Easter Island society, and several writers have drawn attention to alternative explanations involving last-millennium climate change (McCall, 1993, 1994; Hunter-Anderson, 1998; Haberle and Chepstow-Lusty, 2000). The most persuasive of these alternative explanations involve:

- the correspondence between the Ahu Moai Period and a period of climate stability corresponding to the Medieval Warm Period and characterized, critically, by low ENSO activity and consequently a degree of climate constancy that allowed optimal food production which led to societal complexification;
- the increase in both El Niño frequency and the incidence of large-scale volcanic eruptions that increased both climate variability and drought incidence during the Huri Moai Period, which corresponds to the Little Ice Age, thereby making agriculture a less-dependable source of food; and
- across the A.D. 1300 Event, both sea-level fall and increased precipitation may have changed offshore environments and impacted offshore ecosystems (including the birds that also were also sustained by them), resulting in a drop in wild foods available.

It is worth emphasizing that during the Medieval Warm Period on Easter Island, as elsewhere, low climate variability led to the time of plenty and allowed the cultural efflorescence conspicuously represented by *moai* construction. As Haberle and Chepstow-Lusty (2000) noted,

> "there is a striking correspondence between the period of statue building, the most resource demanding phase of human occupation, and the phase of low ENSO activity". (p. 362)

The A.D. 1300 Event, marked by a sharp increase in ENSO activity, led to a degree of climate variability to which Easter Island's complex society was unaccustomed. In particular, the droughts associated with El Niño, perhaps increased storminess, all contributed to a decline in food production. The cooler temperatures of the Little Ice Age would have rendered this decline long term;

> "the cause of the ultimate collapse of the society that made these statues … appears to be closely linked to a period of intense ENSO activity". (Haberle and Chepstow-Lusty, 2000, p. 362)

The combination of these climatic effects with what with hindsight can be seen to have been unwise choices by the islanders – such as the continued, even increased, production of *moai* at a time when energies could have been more profitably directed towards new strategies of food production – are more likely to have brought about the eventual, spectacular collapse of Easter Island society.

7.7.6 Case study: the mystery islands of Polynesia

The mystery islands of Polynesia are so called because, although no one was living there when they were first visited by Europeans, typically in the nineteenth century, there is abundant evidence that they were occupied for many generations, particularly during the Medieval Warm Period. Examples of these mystery islands include those of the Pitcairn group (especially Henderson) and the northern Line Islands (including Kiritimati and Tabuaeran), today politically part of Kiribati.

Henderson Island had a permanent population between about A.D. 900 and A.D. 1500, during most of which time its inhabitants participated in an exchange network with Pitcairn and Mangareva islands. Turtles and red feathers were the main exports of resource-impoverished Henderson Island, while diverse imports of edible plants and animals, basalt, pearl shell, even marriage partners came in from Pitcairn and Mangareva (Weisler, 2002). The end about A.D. 1450 of long-range contact – the shrinkage of this interaction sphere – made life more difficult on Henderson, which is believed to have been abandoned early in the Little Ice Age, probably because its population could not survive without imports (Weisler, 1995).

The story in the northern Line Islands is similar, with low islands being colonized or at least regularly visited for prolonged periods during the Medieval Warm Period but being abandoned early in the Little Ice Age (Di Piazza and Pearthree, 2001a, 2001b, 2004). Shrinkage of interaction spheres – in this case, notably with the high volcanic islands of the Marquesas – is implicated in abandonment, although there are other possible contributors to this. Kiritimati (Christmas) and Tabuaeran (Fanning) are both atolls, and it has been convincingly argued that their occupation during the Medieval Warm Period was maintained because of the islands' importance for birds and gardening respectively (Di Piazza and Pearthree, 2001b). Yet if it is assumed that sea level fell in this part of the Pacific during the A.D. 1300 Event, as seems likely, then the inhabitants of these resource-poor atolls would have had to cope with environmental changes that reduced productivity, which may have compelled abandonment.

In the case of Kiritimati – the "island for birds" – most ancient settlement sites are located along the shores of the former lagoon, probably that which existed near the end of the Medieval Warm Period when sea level was higher than today. Sea-level fall during the A.D. 1300 Event would not only have dried up much of this lagoon, but in doing so would have changed the lagoonal ecosystem, reducing its area, slowing water circulation and in consequence making it far less attractive to birds.

Tabuaeran was the site of a single large village during the Medieval Warm Period, built on a terrace of phosphate rock. Its inhabitants were agriculturalists, focused largely on taro cultivation in the freshwater marsh. It is hypothesized that, as sea level fell during the A.D. 1300 Event, this freshwater marsh dried up and was no longer available for taro cropping. Faced with the loss of this essential ecosystem, people abandoned Tabuaeran. Today the island has been resettled and, because of the higher sea level, its people are once more able to grow taro in the freshwater marsh.

The low mystery islands of the central eastern Pacific all appear to have been occupied during the Medieval Warm Period and abandoned by the early Little Ice Age. There is no consensus as to the reasons for abandonment (Weisler, 1996a; Di Piazza and Pearthree, 2001a, 2004), although they appear to relate to both

- discontinued contacts with parent islands necessitating use of local (often inferior) materials for artifact manufacture, and
- declines in food resources, plausibly linked to sea-level fall especially on atoll islands.

In summary, climate change across the A.D. 1300 Event can plausibly be seen to have contributed to societal stress on mystery islands resulting in their abandonment. This was in two ways. First, as recognized by many authors, the shrinkage of interaction spheres – a likely result of deteriorated conditions for successful long-distance voyaging and conflict on parent islands – deprived mystery islands of essential resources. Second, as suggested for the first time here, sea-level fall along the coasts of mystery islands – especially atoll islands – would have reduced food resources. Given the unproductive nature of island interiors, the only long-term solution must have been for people to abandon these islands.

7.8. KEY ISSUES CONCERNING CAUSE AND EFFECT

The point was made above that the model illustrated in Fig. 7.1 was simple, unable to show all the cause–effect pathways that were undoubtedly involved in the interaction of climate, environment and society during and following the A.D. 1300 Event. Yet several issues are worth highlighting, particularly those involving the response times of particular systems to particular forcing.

This section looks at four key issues associated with the theme of system response to external forcing during the A.D. 1300 Event. In Section 7.8.1, there is a discussion of the relative predictability of environmental and societal systems to climate-linked forcing during the A.D. 1300 Event. Then in Section 7.8.2, the question of system lags is discussed. In Section 7.8.3, there is a discussion of how spatial (geographical) variations in climate forcing may have caused spatial variations in the environmental and societal responses. In Section 7.8.4 there is a discussion of why both environments and societies on Pacific islands were generally more sensitive to climate-associated forcing than those along the Pacific Rim.

7.8.1 Predictability of system responses

In considering the response of the environment (including sea level) to climate forcing, we must consider that natural systems – while often responding complexly – also ultimately respond predictably. A natural system – think of a hill-slope or a river channel – will change in response to external forcing, and that change is predictable. A mid-valley hillslope will steepen if base level falls for a prolonged period, and has the potential to enter a completely new equilibrium condition if certain thresholds are crossed (Bull, 1991). A river channel may downcut or aggrade depending on a variety of external forcings, such as sea-level change, catchment-precipitation change and so on.

Yet when we deal with societal responses to either climate or environmental forcing, it is clear that these responses would not necessarily be predictable, at least not within similar narrow margins as climate–environment interactions. The reason for this is that, although many human societies appear to have responded similarly – at least at a crude level – to climate forcing during the A.D. 1300 Event in the Pacific Basin, human societal responses are basically unpredictable because they are experiential. Landscape has no memory that allows it to respond unpredictably to external forcing, but human societies – because their evolution is based on decisions by human individuals whose experiential knowledge varies (and may even be ignored) – can respond unpredictably. A tribal leader confronted by persistent food shortage may decide to move his people elsewhere, eliminate half of them, confront the neighbouring tribe, or supplicate the gods by constructing megalithic structures. The decision, assuming it to be a rational one, will be one based largely on experience of which solution is likely to be most effective – and all of those mentioned were clearly tried in the Pacific Basin shortly after the A.D. 1300 Event.

7.8.2 Lag times

There is evidence throughout the Pacific Basin that, during the A.D. 1300 Event, climate forced environmental changes that were predictable and near simultaneous. Small apparent disparities in time and space are largely attributable to

- lags in the responses of the various systems to external forcing, discussed in this section, and
- the variable magnitude of forcing in particular parts of the region, discussed in the following section.

In contrast, the evidence of societal response – especially initial societal response – to the climate and environmental changes of the A.D. 1300 Event is generally less predictable and sometimes conspicuously diachronous. This is the outcome of both factors bulleted above, together with what might be best called the "human factor". Yet despite all these sources of potential misfit, there was still – at a crude level – a number of simultaneous societal responses to external forcing during the A.D. 1300 Event. Conflict became widespread, probably endemic in the most circumscribed parts of the region. Vulnerable locations for settlement were abandoned. Subsistence strategies changed.

Lags in system responses to external forcing sometimes obscure the nature of the relationship between two variables. Yet most natural scientists would accept that such a relationship that is *a priori* probable could be validated with less empirical support than a more debatable one. Especially when considering natural systems, lag times associated with the attainment of system thresholds that evoke response are well documented (Bull, 1991). For example, for sea level to change in response to a change in solar insolation requires first that the latter be registered in the upper levels of the ocean over prolonged periods of time, that the increased heat is not dissipated but builds up to a point where the water column expands, producing sea-level rise. Such responses are not instantaneous. For the last millennium, a lag time of 125 years was calculated for sea-level response to solar-insolation change along the eastern seaboard of the United States (see Fig. 10.4B), similar to that of 90 years suggested for the Pacific (see Fig. 10.4C).

Owing to the human factor, societal responses to climate and environmental forcing are likely to be far more variable. Much will depend on the antecedent conditions – the degree to which the society was already utilizing the available resource base before the food crisis (see Fig. 7.2) – much on the appropriateness of the response(s) selected to deal with the crisis.

7.8.3 Spatial variations in climate forcing as an explanation for variations in the timing of human response

Apparent lags in environmental and (especially) societal responses to external forcing may be a product of spatial variations in the magnitude of that forcing. Just as temperature and precipitation vary today within the Pacific Basin, so too did they during the A.D. 1300 Event and – in doing so – obviously produced varying effects on environments and societies. Yet, for many human communities, the A.D. 1300 Event did not simply involve climate change but

- the start of a fundamentally different climate regime characterized in every part of the region by increased climate variability, and
- sea-level fall.

So it is these processes that provided a region-wide driver of societal change which was modulated locally by other effects.

Good examples come from studies of the modern effects of El Niño, which increased in frequency across the A.D. 1300 Event and into the Little Ice Age, on the western Pacific. This is particularly well illustrated by the movement of the SPCZ which shifts northeast during El Niño events; warmer and wetter conditions occur to its north, cooler and drier conditions to its south (Power et al., 1999; Salinger et al., 2001). So during periods of high El Niño incidence, climates will become increasingly variable and unpredictable year to year in those parts of the western Pacific (such as northeast Australia) crossed by the migrating SPCZ which is itself a zone of unstable, stormy weather.

Another example is provided by the effects of climate changes in central America during the first millennium A.D., discussed by Neff et al. (2006). In seeking to explain the differing human response, marked by various indicators of the collapse of Maya civilization, these authors point to the differences in conditions during the Medieval Warm Period in this region. Across much of the region, as indicated by proxy data from the Cariaco Basin (Haug et al., 2003), the start of the Medieval Warm Period that witnessed the collapse of the Classic Maya civilization (see Fig. 3.1) was marked by periods of extreme and prolonged drought but followed by a return to more humid conditions ~1,000 cal BP (A.D. 950) that allowed population recovery. In contrast, the Pacific coast of central America was subject to persistent dry conditions after A.D. 950 that kept populations there to a minimum until the end of the Little Ice Age (Neff et al., 2006).

7.8.4 Why island environments and societies are more sensitive than those of continents

During the last millennium in the Pacific Basin, island environments and societies appear more sensitive to external (climate-associated) forcing – at least when regarded with hindsight from the twenty-first century – than those of continents.

In terms of environments, the changes that occurred as a result of the A.D. 1300 Event are less obscured on islands than on continents. Consider the widespread development of coastal wetlands early in the Little Ice Age (Section 6.2.1) or the evidence for decreased water circulation in lagoons around the same time (Section 6.2.2), both attributable to sea-level fall. Similar evidence exists along continental shores but is not nearly so readily identifiable nor so apparently ubiquitous. The reason is largely the magnitude of post-A.D. 1400 geomorphological processes. On islands, where rivers are generally smaller than on continents, fluvial sedimentation has been insufficient to obscure or remove the evidence for environmental changes during the A.D. 1300 Event. Yet along Pacific continental coasts, where river drainage basins are generally orders of magnitude larger than on islands, and where – largely because of the mountain ranges fringing the continental coasts of the Pacific Basin – precipitation levels are often greater and/or droughts more severe, there is a greater chance that the evidence of a 700-year-old landscape change will have been buried or obliterated.

Studies of the A.D. 1300 Event in the Pacific Basin also show that its inferred societal effects are often more visible on islands than along the contiguous continental rim. The reasons have to do with geography and resource availability. A good comparative example is provided by California where droughts during the later part of the Medieval Warm Period had more visible effects – settlement abandonment and the emergence of a chiefly elite – on the offshore Channel Islands than on the mainland (Raab and Larson, 1997). Precipitation is considered to be the key factor in the cultural changes at this time, mainlanders responding by shifting close to perennial streams (True, 1990) but islanders – occupying places where stream courses were short and precipitation less – being forced to more drastic coping measures (Arnold, 1992).

The contrast can also be seen at a regional scale, by noting the paucity of evidence for a range of societal responses to external forcing along continental coasts compared to the plethora on islands. Conflict sparks on most Pacific Islands (including New Zealand) early in the Little Ice Age, but is apparently only localized along the Pacific continental rim. Clearly, there are non-confrontational ways of dealing with food stress on large land areas such as continents that are not readily available to people occupying islands.

A similar contrast can be noted between larger islands, where societal responses to food crisis might lag compared to smaller islands on which a condition of food crisis may have developed more quickly because of the absence of any unutilized food resource. There is evidence that smaller, very isolated, Pacific Islands were abandoned shortly after the A.D. 1300 Event, perhaps because the food crisis developed so rapidly there (see Section 7.7.6).

7.9. VERIFYING THE A.D. 1300 EVENT MODEL

Many anthropologists and archaeologists in particular are uncomfortable *per se* with models that explain culture change by external causes, such as climate and sea-level change, and will be ill at ease with the model of the A.D. 1300 Event suggested above. Two crucial tests of this model are suggested in the paragraphs below. These tests are regarded as crucial in the sense that if they fail, then the fundamental premises of the A.D. 1300 Event model might need to be rethought. Yet, if both tests are successful, then this should be interpreted as adding significant credibility to the model.

One way of evaluating the A.D. 1300 Event model, at least partly, is by trialling it in a place where there is currently no (or very little) relevant information. The reef-fringed Ryukyu (Nansei) Islands in southern Japan are such a place. There is very little information on the age of castle-building and settlement-pattern changes in these islands, but they would be expected to have had a similar last-millennium history to that of other reef-fringed Pacific Islands.

The other way of testing the A.D. 1300 Event model is to find a data series, in this case most persuasively a history of last-millennium societal change from the Pacific Islands, that manifestly contradicts the model. If the A.D. 1300 Event model for explaining (in large part) last-millennium societal change in the Pacific Basin is not valid, then there should be data that clearly contradict it. Given that the proposed response to climate and sea-level forcing of societal change in the Pacific Basin is somewhat subdued in many parts of the continental Pacific Rim, for reasons explained in Section 7.8.4, the site should be among the Pacific Islands.

If the A.D. 1300 Event model is invalid, and the societal changes during the last millennium in the Pacific Islands region were due solely to internal forcing mechanisms such as population pressure and socio-political evolution, then it would be expected – given the greatly differing times of colonization of Pacific Island groups (see Fig. 2.1) – that some island communities would have declined during the Medieval Warm Period and flourished during the Little Ice Age. Put another way, if climate forcing is insignificant, or if the subdivisions of the last millennium are inapplicable to the Pacific Islands region, then this is exactly what would be expected. The author knows of no such island (group) in the Pacific but, if research does indeed demonstrate beyond doubt that such a place exists, then the A.D. 1300 Event model would seem to be inapplicable there.

KEY POINTS

1. The A.D. 1300 Event model is proposed as an explanation, involving mainly environment-filtered climate change, for the profound Pacific-wide societal changes that occurred in the fourteenth and fifteenth centuries.
2. The A.D. 1300 Event model is explicitly environmental determinist and is superior to cultural-determinist alternatives because its chronology and its proposed cause–effect pathways are independently verifiable.
3. Verification of the A.D. 1300 Event model will be accomplished if it is found to be applicable to places for which no relevant data currently exist and if no chronology of culture change in the Pacific Basin is found to clearly contradict it.

CHAPTER 8

Recent Warming and Sea-Level Rise
(since A.D. 1800) in the Pacific Basin

The time since A.D. 1800 in the Pacific Basin has witnessed developments of unprecedented complexity in human societies that have driven changes in environment and even – most scientists would agree – changes in the Earth's climate. It is not the intention of this chapter to give a comprehensive review of these changes and their multifarious interactions – a daunting task – but to continue to identify this era's principal climate changes, and link these to environmental and societal changes to demonstrate that such linkages have remained significant up until the present day. The difference is that for this recent period, in some parts of the Pacific Basin (not all), there were for the first time in its post-settlement history significant counter-influences by human societies on environment and climate.

To decide which observed changes are climatogenic (climate-caused) and which are anthropogenic (human-caused) is no easy task, and inevitably the interpretations in this chapter will attract criticism. Many observed changes are probably a result of both. A good example is global warming: warming began before the Industrial Revolution in many parts of the Pacific, suggesting that the recent warming trend is probably an amalgam of a natural and an anthropogenic one. It actually matters only slightly whether or not causes within this period of Recent Warming are correctly attributed, because there are numerous undeniable examples of climate change causing environmental change that in turn caused societal change. Thus, the basic argument of this book – that societal changes are driven by external changes – is sustained into modern times and even into the future.

This chapter begins with a description of the climate changes that affected the Pacific Basin during the period of Recent Warming (Section 8.1), followed by an account of those environmental changes that appear likely to have been caused by climate change (Section 8.2), and finally an outline of those societal changes that are likewise attributable to the climate and environmental changes described (Section 8.3). Then there is a brief section that considers the probable effects of future climate change on Pacific environments and societies (Section 8.4).

8.1. CLIMATE CHANGE DURING THE PERIOD OF RECENT WARMING IN THE PACIFIC BASIN

In a general sense, the transition from the Little Ice Age to the period of Recent Warming marks a change from cooler-than-present conditions throughout the

Pacific Basin to warming conditions, a situation that continues in the early twenty-first century. For the purposes of this chapter, it makes no sense to try and separate the warming that is likely to be natural from the warming that may be anthropogenic, so the warming described below is discussed as though it was – as in earlier times – driven wholly by natural processes.

One important difference for understanding climate change within this Recent Warming period compared to other periods discussed in this book is that sources of data change in both their nature and their resolution. For, typically towards the end of the nineteenth century or early in the twentieth century in many parts of the Pacific Basin, meteorological stations were established and data began to be collected on a subdaily basis. Selected temperature time series are shown in Fig. 8.1.

It is important to understand the significance of this change because it contrasts so sharply with the methods available for reconstructing the climate earlier in the past millennium. One consideration is not to attribute undue significance to the more recent high-resolution directly recorded data series just because they are more dependable than earlier data series. Another consideration is that we should not regard trends in particular climate variables as commencing only at the time when direct monitoring began simply because of the contrast in resolution between these data and the lower resolution less-dependable data available for earlier times.

It follows that to understand clearly the nature of climate change during the entire period of Recent Warming, it is best to look at palaeoclimate data series that extend throughout the entire period with the same degree of resolution, and then to use the higher resolution series available through monitoring in the second half of the period to fill in the details. This is the approach taken in this section, each of the regional Sections 8.1.1–8.1.3 focused first on palaeoclimate data series that cover the entire period, and second on what monitored data series can tell us about the variations within that time. Since we have more precise information about El Niño-Southern Oscillation (ENSO) for the last century than for earlier times, Section 8.1.4 is devoted to examining how ENSO variations affected Pacific Basin climates during this time. For similar reasons, there is a separate discussion of tropical cyclones (hurricanes) in Section 8.1.5.

8.1.1 Climates of the eastern Pacific Rim (including Antarctica)

On James Ross Island, off the tip of the Antarctic Peninsula, the warmest time of the last 400 years was A.D. 1750–1850; from 1850 to 1980 cooling of ~2°C occurred here (Aristarain et al., 1990). At Talos Dome in Antarctica, a cold period endured A.D. 1680–1820 with subsequent warming pulses occurring about A.D. 1930 and A.D. 1970 (Stenni et al., 2002).

Along the Pacific Rim in South America, ^{18}O enrichment of ice, indicating warming, began in the high Andes at A.D. 1830 and continued into the twentieth century, something that is consistent with the recession of glacier fronts associated with these ice caps (Thompson et al., 2006). Work on lake sediments from the Pacific Rim in Chile showed that the period A.D. 1850–1998 was a time of greatly increased frequency of flood events suggesting that the Recent Warming

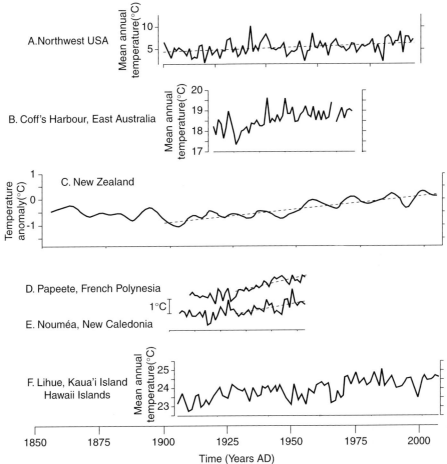

FIGURE 8.1 Selected temperature records from the Pacific Basin in which long-term trends can be identified. All trends are from the sources indicated. (a) Temperature record 1900–1998 for the northwest United States (Pacific northwest) (after Mote et al., 1999). (b) Temperature record 1953–2006 for Coff's Harbour, eastern Australia (after Goddard Institute for Space Studies, http://data.giss.nasa.gov/gistemp/, accessed in November 2006). (c) Mean annual temperature over New Zealand 1855–2005 (after National Institute for Water and Atmospheric Research, www.niwascience.co.nz/ncc/clivar/pastclimate, accessed in October 2006). (d) Temperature records 1959–2002 for Papeete, Tahiti Island, French Polynesia (eastern tropical Pacific), showing trend lines for 1976–2002 (after Meteo France, supplied by Jean-Francois Royer, October 2006). (e) Temperature records 1953–2002 for Nouméa, New Caledonia (western tropical Pacific) showing trend lines for 1976–2002 (after Meteo France, supplied by Jean-Francois Royer, October 2006). (f) Temperature record 1905–2006 for Lihue, Kaua'i Island, Hawaii (after Goddard Institute for Space Studies, http://data.giss.nasa.gov/gistemp/, accessed in November 2006).

period was significantly wetter here than was the second half of the Little Ice Age (Jenny et al., 2002).

For North America, a cogent palaeotemperature record from the northern Coast Mountains of Pacific Canada shows that warming occurred during the entire period of Recent Warming (Clague et al., 2004). During the twentieth century, temperatures rose 1–2°C, associated with a rise in treeline, the northward spread of tree cover in the subalpine zone, and increased biological productivity of ponds. Studies of Farewell Lake in southern Alaska show that surface-water temperatures increased by ~1.75°C from A.D. 1700 to the present (Hu et al., 2001). With regard to palaeoprecipitation, studies of Dog Lake in British Columbia where the early decades of the nineteenth century are uncommonly dry and marked by an almost unprecedented peak in forest fires (unrelated to humans in this location – Hallett et al., 2003) suggest that the start of the period of Recent Warming was not only very dry but was also marked by a rapid temperature increase here. Over the next 50 years of the nineteenth century, wetter conditions reduced the incidence of forest fires although the period about A.D. 1880–1930 was a conspicuous dry period with several large fires.

Many meteorological stations on the Pacific side of the Antarctic Peninsula have reported rapid warming over the past 30–50 years (King and Harangozo, 1998). Yet decadal-century temperature trends from Antarctica are notoriously variable in both rate and often in sign (rising or falling), suggesting that local conditions exert a more enduring influence on long-term trends than elsewhere, where 30–40 years are usually sufficient to identify a long-term trend within the short-term "noise". To resolve the issue of the variability of recent Antarctic temperature change, Comiso (2000) analysed data from 21 stations and found an average warming of 0.012°C/year for the last 45 years.

Evidence for a temperature rise of ~0.1°C per decade since 1939 in the tropical Andes was described by Vuille and Bradley (2000), who also found that the rate of warming had tripled over the past 25 years. It seems likely that the conspicuous recession of Andean glaciers since at least the early twentieth century has been largely driven by warming (Thompson et al., 2003; Fig. 8.2). These authors also make the important point that, given that intra-annual temperature variations in the tropics are commonly small, their data have shown that "a large and unusual warming … is underway at high elevations in the tropics" (p. 150).

Warming of the Pacific Rim in North America is also represented by the time series for the northwest United States shown in Fig. 8.1A. Despite considerable variation since recording began in A.D. 1900, an overall warming trend is clear.

8.1.2 Climates of the western Pacific Rim

Studies of glacier advances in Kamchatka suggest that a minor advance about A.D. 1860–1910 was probably largely a response to regional cooling (Savoskul, 1999). The same cold period was widespread throughout China and Japan (Sakaguchi, 1983).

Yet throughout the Pacific Rim in East Asia (including Japan), there has been overall warming during the period of Recent Warming. For example,

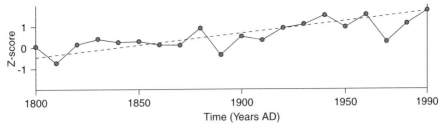

FIGURE 8.2 Recent warming of the tropics A.D. 1800–1990 as shown by changes in the $\delta^{18}O$ content of low-latitude high-altitude ice cores (after Thompson et al., 2002, drawn from supporting information on the PNAS website). Trend line drawn by author. The interpretation of this graph as primarily a warming one is supported by the observation that it coincides with rapid melting of glaciers around the ice fields sampled. In the tropics, temperature rise (not reduced precipitation) is the main cause of glacier recession (Thompson et al., 2002).

phenological data for eastern China show that temperatures rose rapidly (~1°C in 60 years) from ~1870 to 1930 (Ge et al., 2003). Rapid temperature rise has been recorded since ~1981, temperatures having increased 0.5°C between 1981 and 1999, almost as high as the warmest time of the past 2,000 years that occurred at the end of the Medieval Warm Period. Tropical cyclones in the northwest Pacific reached their maximum frequency for the period A.D. 1470–1931 in the early to middle nineteenth century (Chan and Shi, 2000). Based on records from 1491 to 1931, it is clear that during the early to middle nineteenth century, there was a greater than usual number of tropical cyclones (typhoons) that came on land in Guandong Province in China.

A 420-year coral-core record from Australia's Great Barrier Reef suggested that the end of the Little Ice Age was marked here by an abrupt decrease in tropical Pacific sea-surface salinity about A.D. 1870 (Hendy et al., 2002). But the later part of the Little Ice Age here was actually marked by warmer-than-present conditions A.D. 1700–1870, so the salinity decrease around A.D. 1870 represents a cool interval that lasted for some 50 years. This anomalous condition seems confined to the tropical (west and central?) Pacific Ocean and was discussed elsewhere (see Section 6.4). Of more interest here is the cause of the abrupt salinity decrease ~1870. This is explainable by an abrupt increase in precipitation compared to most of the Little Ice Age (during which salinities here were comparatively high) suggesting lower precipitation on the adjacent continent. Yet Hendy et al. (2002) also point to the likely contribution of decreasing evaporation across this transition, resulting from the effects of a weakened tradewind circulation after 1870.

Using records of glacier-front recession as proxies for warming, Salinger (1976) was able to show that the period of Recent Warming in New Zealand began about A.D. 1800, since which time "a slow warming has occurred, culminating in the rapid 1°C warming of the past 40 yr" (p. 311).

Monitored data show evidence also for warming along the western Pacific Rim; time series for eastern Australia (Coff's Harbour) and New Zealand (composite) are shown in Fig. 8.1B and C. In New Zealand, the present warming trend may have begun only around A.D. 1900.

8.1.3 Climates of the Pacific Islands and Ocean

Some of the most precise palaeotemperature records for the period of Recent Warming come from coral-core proxies ($\delta^{18}O$) of Pacific sea-surface temperature. Like the results from the Great Barrier Reef discussed in the preceding section, coral-core palaeoclimate records extending back through the period of Recent Warming in the tropical Pacific also show that temperatures were generally as high (or higher) than present during the eighteenth and nineteenth centuries. Examples come from Socas Island in Panama (Linsley et al., 1994), Espiritu Santo Island in Vanuatu (Quinn et al., 1996) and Amédée Lighthouse in New Caledonia (Quinn et al., 1998). Shorter term coral-core palaeotemperature records, typically extending back to around A.D. 1900, exhibit similar evidence for warming. Examples include records from Tarawa Atoll in Kiribati (Cole et al., 1993), Cebu in the Philippines (Patzold, 1986) and Palmyra Atoll in the central Pacific (Cobb et al., 2003).

Results of coral coring from Maiana Atoll, Kiribati (Fig. 8.3), show "a trend towards warmer, wetter conditions from 1840 to 1995" (Urban et al., 2000, p. 990). These changes are represented by decreasing coral $\delta^{18}O$ which probably represents a combination of rising temperature and falling salinity, the latter probably largely a result of increased rainfall. Temperatures monitored at Maiana since 1976 show a warming of 0.6°C, about twice the rate of warming farther east, and indicative of the expansion of the western Pacific warm pool (see Section 8.1.5).

In their review of coral-core proxies for sea-surface temperature change in the southwest Pacific, Quinn et al. (1998) highlighted the apparent common trends in many records of warm–cold–warm–cold for the period A.D. 1805–1840. This may mark an oceanic response to the transition from the Little Ice Age to the period of Recent Warming. Another period of note in western Pacific records is the warming episode A.D. 1890–1900.

Climate data series from routine monitoring in many parts of the Pacific Basin show temperature increases comparable to those along the Pacific Rim. Two series from French island territories and one from Hawaii are shown in Fig. 8.1 as examples.

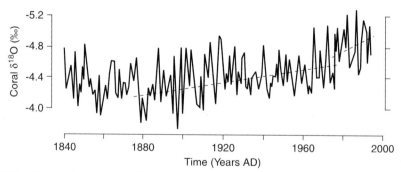

FIGURE 8.3 Oxygen-isotope data from a coral core at Maiana Atoll, Kiribati, show a 155-year trend towards warmer and wetter conditions (after Urban et al., 2000). The pre-1976 part of the record shows a gradual temperature rise, the post-1976 part shows this rise accelerating. Trend lines added.

There is ample evidence that the world's oceans have warmed during the past 60 years or so (Levitus et al., 2000). Most sea-surface temperature data series for the Pacific show warming over the past 100 years or less (Hoegh-Guldberg, 1999b). The warming of the Pacific has been significantly less than for the other oceans. For the period 1955–1996, the upper 3,000 m of the Pacific warmed 30%, significantly more in the North Pacific (37%) than in the South Pacific (16%). By comparison, the Atlantic warmed 89% and the Indian Ocean 57%.

8.1.4 ENSO, PDO/IPO and Pacific Basin climate change

El Niño (ENSO-negative) events affect the Pacific every 3–5 years and last as long as 18 months; they are the principal source of short-term interannual climate variability. For the tropical Pacific, the early nineteenth century was characterized by enhanced ENSO activity, similar to that which occurred A.D. 1982–2002, which may signal rapid warming (Evans et al., 2002). Some records show a significant rise in ENSO variability around A.D. 1870, beginning a 50-year period of strong regular ENSO (Stahle et al., 1998). In contrast, the 1920–1960 period was associated with a weak, irregular ENSO while the period since then has been characterized as one of the few periods of strong, regular ENSO within the entire late Holocene (Tudhope et al., 2001).

Within the last decade, the Pacific Decadal Oscillation (PDO) and the Interdecadal Pacific Oscillation (IPO) – climatic fluctuations similar to ENSO but longer term – have been recognized in the Pacific (Mantua et al., 1997; Zhang et al., 1997; Salinger et al., 2001). The PDO usually refers to the North Pacific and the IPO (incorporating the PDO) to the entire Pacific. They represent the same interannual climate effect, moderated by subregional climatic and topographic factors. Like ENSO, the PDO/IPO has warm and cool phases that refer to sea-surface temperatures in the tropical eastern Pacific but – unlike ENSO – these phases last 20–30 years. Strong El Niño events are more likely to occur during positive phases of the PDO/IPO (and La Niña events during their negative phases).

Within the period of Recent Warming, there were cool phases of the PDO/IPO from 1890 to 1924 and 1947 to 1976 and warm phases from 1925 to 1946 and 1977 to 1998 that had impacts on fisheries (Mantua et al., 1997; Lehodey et al., 2006). In the Pacific Ocean, the clearest effects of the PDO are in higher latitudes, probably affecting tropical-cyclone frequency in the North Pacific (Chu and Clark, 1999) and precipitation and shoreline erosion in Hawaii (Mantua et al., 1997; Rooney and Fletcher, 2005) for example. The effects of the IPO in the South Pacific are most marked in its influence on the position of the South Pacific Convergence Zone (SPCZ). During the cool phase of the IPO 1958–1977, the SPCZ was farther south compared to its position during the warm phase of the IPO 1978–1998 (Salinger et al., 2001).

8.1.5 Tropical cyclones (hurricanes)

The formation of tropical cyclones in the South Pacific is confined to places where the sea-surface temperature is greater than 27°C, usually only in the warm pool in the tropical western Pacific. As would be expected as a result of global warming, the warm pool has expanded, particularly in an eastward direction, over the past few decades. Palaeotemperature data obtained from Maiana Atoll at the eastern edge of the warm pool demonstrate its recent expansion (Urban et al., 2000).

The expansion of the warm pool under normal (not El Niño) conditions is likely to be the principal reason why the frequency of tropical cyclones has increased in the Pacific in recent decades (Nunn, 1994). Another consequence of the eastward expansion of the warm pool has been that tropical cyclones are moving farther east than they were traditionally regarded as doing, sometimes in the past two decades reaching South Pacific island groups such as Samoa and Niue and the Cook Islands. Another effect of the expansion of the warm pool is that its contraction in winter in the southern hemisphere, for example, is delayed and that tropical cyclones therefore develop increasingly outside what is regarded as the normal cyclone season.

It is well known that tropical cyclones in the South Pacific are more likely to develop during El Niño events rather at other times. But it is not simply an issue of frequency, for the warm pool in the western Pacific not only extends east at such times but also another warm pool (in which sea-surface temperatures exceed 27°C) generally develops in the central tropical Pacific during El Niño events. In this eastern warm pool, tropical cyclones can develop that affect islands of this region, such as those in French Polynesia that were traditionally regarded as immune from these climatic phenomena. A similar situation obtains in the North Pacific where tropical cyclones are more likely to form farther east during El Niño events than usual, putting the Hawaii Islands within range. During the 1982–1982 El Niño, French Polynesia experienced six tropical cyclones – the first for 75 years – and Hawaii was affected by Hurricane Iwa, the first for 23 years (Couper-Johnston, 2000).

Fewer Pacific Islands are affected by tropical cyclones during La Niña periods, but because of the westward compression of the west Pacific warm pool, there is a 20% greater chance of tropical cyclones (typhoons) hitting islands in Japan and the Philippines at such times (Couper-Johnston, 2000). Generally however, Pacific-originating tropical-cyclone incidence is lower during La Niña events than at any other time. La Niña events are also associated with an increased tropical-cyclone frequency in the Atlantic (Bove et al., 1998). Just about the only occasions when tropical cyclones affect the Pacific Rim in central America is when La Niña conditions prevail in the Atlantic.

8.2. CLIMATE-LINKED ENVIRONMENTAL CHANGE DURING THE PERIOD OF RECENT WARMING IN THE PACIFIC BASIN

The range of forcing mechanisms involved in environmental change in the Pacific Basin during the period of Recent Warming became – in the course of the period – far greater than in earlier times. The reasons for this have to do with the increase in the numbers of humans in the region, the changes in the nature of human–environment interactions, and the effects of global (extra-region) forces. This section focuses on the environmental changes caused (largely) by climate change within the period of Recent Warming.

As in earlier times of the last millennium, one of the main environmental changes associated with temperature change was coastal change associated with sea-level change. As many data series record, the overall trend of sea level during the period of Recent Warming has been upward (Fig. 8.4), plausibly driven by the contemporaneous temperature rise (Warrick and Oerlemans, 1990).

Many long-term tide-gauge records in the Pacific Basin and elsewhere are located in places that are tectonically active, and doubts have therefore been cast as to whether these records are truly those of sea level alone (Douglas, 2001). For example, the relative sea-level falls recorded since the 1940s at Cebu and Jolo stations in the Philippines are contradicted by the sea-level rises recorded from other tide gauges in this archipelago (Siringan et al., 2000; see also Fig. 8.4B). The explanation is that the Cebu and Jolo sites are rising – something to be expected near convergent plate boundaries in the Pacific (Nunn, 1994). Other tide-gauge records, which show sea level to be rising faster than expected, might incorporate the effects of island subsidence. This is what has happened in the Hawaii Islands, where the long-term rate of relative sea-level rise at Hilo on the "Big Island" of Hawai'i, which is subsiding, is $\sim$4 mm/year which is significantly faster than that at Honolulu on the stable island O'ahu where sea-level rise is $\sim$1.6 mm/year (Wyrtki, 1990; see Fig. 8.4G).

The other important point is whether monitored sea-level time series are sufficiently long to allow the long-term (eustatic) signal to be separated from the shorter term influences on sea level (the noise). Most authorities would argue for a minimum of 30–40 years of record, preferably more (Pugh, 2004). This point is crucial to the proper interpretation of shorter term records. For example, it was once claimed that a 7-year record (1990–1997) of sea-level fall from Tuvalu (central Pacific) showed sea level in the Pacific to be falling, an erroneous conclusion that nevertheless made headlines worldwide. It is now realized that this time series – still too short to be sure – almost certainly exhibits the more common Pacific Basin trend of recent sea-level rise as shown in Fig. 8.4 (Hunter, 2002; Nunn, 2004).

Changes in precipitation have also occurred during the period of Recent Warming but, owing particularly to their variable spatial character, the environmental changes they produced are highly variable within the Pacific Basin – and therefore difficult to generalize about. Yet precipitation changes, especially changes in storminess, cannot be entirely sidelined since their environmental effects – amplified in many places by human impact on natural environments (particularly deforestation) – have been profound.

The effects of recent sea-level rise are being felt along most Pacific Basin coasts but only in some places is the issue at the forefront of scientific enquiry and public concern. Such places are usually those that are both vulnerable and densely populated (such as deltas) or vulnerable and circumscribed (such as low islands). This section looks first at climate-linked environmental changes along the Pacific Rim (Section 8.2.1) and in the Pacific Islands (Section 8.2.2) before going on to a discussion of the environmental effects of changed tropical-cyclone regime (Section 8.2.3).

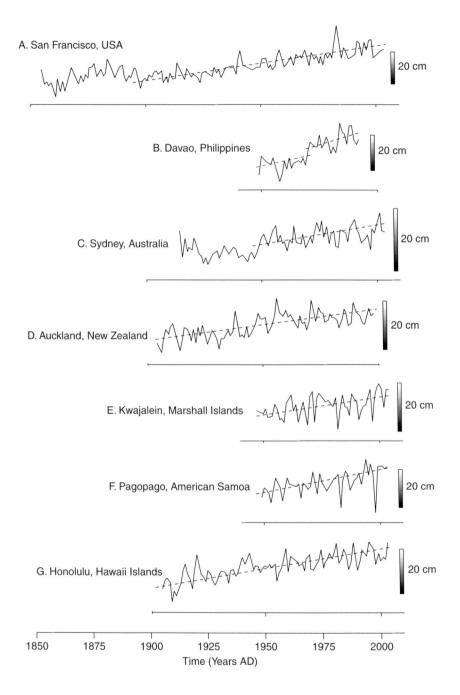

A. San Francisco, USA

20 cm

B. Davao, Philippines

20 cm

C. Sydney, Australia

20 cm

D. Auckland, New Zealand

20 cm

E. Kwajalein, Marshall Islands

20 cm

F. Pagopago, American Samoa

20 cm

G. Honolulu, Hawaii Islands

20 cm

1850   1875   1900   1925   1950   1975   2000

Time (Years AD)

8.2.1 Climate-linked environmental change along the Pacific Rim

Owing to their comparatively small areas of coastal land, coastal-environmental changes along many parts of the Pacific Rim have not attracted as much attention as elsewhere in the Pacific Basin.

Obviously the most vulnerable areas are the lowest ones. Of the comparatively few large deltas along the Pacific Rim, some of those in East Asia are experiencing problems associated with sea-level rise and associated shoreline erosion that are compounded by a reduction in terrestrial sediment inputs because of the effects of dams and the withdrawal of river water for agricultural, industrial and domestic purposes (Saito et al., 2001). The Huanghe (Yellow River) Delta in China is the best-studied example. Terrestrial (river-carried) sediment inputs here decreased from $\sim 1.3 \times 10^9$ tonnes/year in the 1950s to 0.287×10^9 tonnes/year in the 1990s, and large areas of wetlands have disappeared as a result of erosion and inundation in the past 40 years.

Sea-level rise in particular has been blamed for the increasing salinization of deltaic groundwater, something that is itself compounded by delta subsidence and deltaic-sediment compaction. On the nearby Laizhou Bay coastal plain, the net rate of saline intrusion landwards was 92 m/year in the early 1980s but rose to 404 m/year in the later part of that decade (Han et al., 1995).

Yet even comparatively high-elevation areas are proving vulnerable. The effects of sea-level rise on the coast of California are profound, with most sea cliffs retreating at an annual rate of 10–30 cm (Griggs and Patsch, 2004). Much depends on cliff bedrock. For example, those parts of the Monterey Peninsula that are composed of granite show few signs of erosion while those composed of sedimentary rocks are receding at ~ 30 cm/year, while dune coasts are retreating at 1–2 m/year.

FIGURE 8.4   Records of recent sea-level change in the Pacific Basin. All data from the Permanent Service for Mean Sea Level (www.pol.ac.uk/psmsl, accessed in November 2006) with minor discontinuities interpolated and trend lines added. All series plotted have been standardized by the PSMSL to revised local reference (RLR) form. (A) San Francisco, west coast United States, 1854–2003, is regarded as exhibiting a dominantly sea-level change signature (Smith, 1980). (B) Davao, Mindanao Island, Philippines, 1948–2004, is regarded as unaffected by significant tectonic activity or subsidence associated with groundwater extraction by Siringan et al. (2000). (C) Sydney (Fort Denison 2), east coast of Australia, 1914–2003, is regarded as a valid record of long-term sea-level change (Wyrtki, 1990). (D) Auckland (II), North Island, New Zealand, 1903–2000, is a record from which minor tectonic influences have been subtracted (Hannah, 2004). (E) Kwajalein, central Marshall Islands, NW Pacific Ocean, 1946–2004, is regarded as a valid record of long-term sea-level change (Wyrtki, 1990). (F) Pagopago, Tutuila Island, American Samoa, central Pacific Ocean, 1948–2003, is regarded as a valid record of long-term sea-level change (Wyrtki, 1990). (G) Honolulu, O'ahu Island, Hawaii Islands, NE Pacific Ocean, 1905–2003, is regarded as a valid record of long-term sea-level change (Wyrtki, 1990; Nunn, 1994).

8.2.2 Climate-linked environmental change in the Pacific Islands

The evidence for sea-level rise in the Pacific Islands region is manifest, ranging from directly monitored time series (see Fig. 8.4E–G) to coastal changes, the latter being more relevant to the inhabitants of this region as it often impacts directly on settlements or on food availability. The effects of net sea-level rise in many parts of the Pacific Islands have been varied, but most can be classified under land loss, shoreline erosion and vegetation change.

Land loss from sea-level rise has been reported from most parts of the Pacific Islands (Kaluwin and Smith, 1997). Within living memory, many rural coastal settlements extended farther seawards than they do today, and the fact that these examples occur in a variety of tectonic situations and across the spectrum of human–environment impacts (undeveloped areas to urban areas) suggests that a regional forcing mechanism (sea-level rise) underlies land loss.

It is widely accepted that sea-level rise along sandy shorelines results in their erosion (Nunn and Mimura, 2006). And shoreline erosion has been widespread in the Pacific Islands during the period of Recent Warming, probably largely as a result of sea-level rise. Some of the most widespread signs of shoreline erosion are the appearance of beachrock (Fig. 8.5A) and the undermining of the coconut-forest fringe along tropical island shores (Fig. 8.5B). On many atoll *motu* (enduring,

FIGURE 8.5   Recent coastal change in the Pacific Islands. (A) Beachrock exposed along the coast of Majuro Atoll, Marshall Islands [Photo courtesy of James Terry]. (B) Fallen and collapsing coconut palms along the south coast of Naigani Island, Fiji [Photo by Patrick Nunn]. (C) View along the lagoon shore of Luamotu, southwest Funafuti Atoll, Tuvalu, showing the composition of the atoll. The trees are growing in unconsolidated sediment overlying an emerged reef platform capped by conglomerate. Once sea level (at high tide) rises above the conglomerate surface, lateral erosion of such islands will accelerate (Dickinson, 1999) [Photo courtesy of William R. Dickinson].

partly armoured cays – Nunn, 1994) there is concern that once sea level rises above the limit of these islands' emerged reef cores, lateral erosion will accelerate and quickly lead to the loss of habitable land on many (Fig. 8.5C).

Vegetation change results both from land loss and flooding (inundation) and also from salinization of groundwater, all effects of which are widespread in the Pacific Islands. The recent conversion of many lowland forests in these islands from mixed forests to ones dominated by salt-tolerant species, particularly pandanus (*Pandanus caricosus* and others), manifests groundwater salinization of these areas.

Other environmental consequences of sea-level rise in the Pacific Islands have been generally more localized. They include beach erosion, particularly severe along certain managed coastlines such as those in Hawaii (Coyne et al., 1999). Smaller offshore islands, typically cays and *motu*, have frequently borne the brunt of the effects of sea-level rise leading in some cases to the disappearance of entire islands, such as Tebua in Kiribati (Moore, 2002).

It is also worth noting that the effects of net sea-level rise in the Pacific Islands region during the last few decades have induced responses from the islands' inhabitants that have sometimes exacerbated the environmental effects of sea-level rise. In particular, the construction of vertical impermeable seawalls along threatened shorelines has often changed nearshore water and sediment dynamics, resulting in erosion elsewhere along the coast and the loss of productive ecosystems. Seawalls of this kind – ubiquitous in the rural parts of the Pacific Islands – typically collapse a year or so after they are built because of undermining by water scour along the fronts and the effects of ponded water along their rears. Artificial structures that simulate the form and the permeability characteristics of a natural shoreline are much better, although the construction of any kind of artificial structure requires continuous human management thereafter. For tropical Pacific Islands, with their high coastline–land area ratios, replanting of mangrove forest as a form of natural shoreline protection is generally a better solution than seawall construction (Nunn et al., 2006).

8.2.3 Effects of changed tropical-cyclone regime

Of the climate-associated extreme events that affect the Pacific Basin, tropical cyclones (hurricanes) are those that have the greatest sustained impact on terrestrial environments. In recent decades, it is likely that the frequency, seasonality and geographical reach of tropical cyclones has increased owing to the rise in sea-surface temperatures and the consequent expansion of equatorial warm pools (delimited by the 27°C isotherm) in which most tropical cyclones develop.

Tropical cyclones have the potential to cause a massive amount of environmental change, particularly when affected landscapes have been rendered more vulnerable by other processes, such as human impact. Soils can be stripped off steep slopes, prodigious numbers of landslips may occur, river channels can become choked with sediment, and low islands (cays) can be alternately erased or created (Bayliss-Smith, 1988; Nunn, 1998; Terry et al., 2002).

It has been suggested that Pacific Island landscapes have largely evolved under a particular tropical-cyclone regime but that this has changed in the last 10–30 years and a new regime has been established (Nunn, 1998). In a general sense, this means that Pacific landscapes accustomed to, say, two cyclones every decade may now be experiencing eight, while other landscapes where no cyclones were formerly experienced now have two per decade. In terms of environmental change, this means that slope failure comes to dominate slope processes, while lowland river channels (and valleys) become sediment-choked and therefore more prone to (higher) floods, and that nearshore sediment mobility increases. All these amount to a changed process regime that has im-plications not only for the understanding of landscape evolution but also for environmental management in its broadest sense.

8.3. CLIMATE-LINKED SOCIETAL CHANGE DURING THE PERIOD OF RECENT WARMING IN THE PACIFIC BASIN

In this section only those societal changes that are directly and uncontroversially attributable to environmental changes driven by climate changes are considered. Many of these societal changes have been amplified or rendered quite different than they might otherwise have been because of the independent changes in Pacific Basin societies that occurred as a result of population increase, and increased societal complexity and interaction. In general, the more densely populated areas of the Pacific Basin have experienced greater societal change than the less densely populated areas which are therefore those where the effects of climate and environmental changes are most clearly seen. So, in the sections below, there is a focus on less densely populated areas of the Pacific Basin.

Among the major societal effects of climate change are those associated with climate extremes, such as tropical cyclones or severe droughts, that may force human abandonment of the worst affected areas. Similar societal effects arise from the impact of environmental changes associated with extreme events, such as storm surges. In contrast, it is almost impossible to separate the effects of slower, more long-term climate and environmental changes on Pacific Basin societies during the period of Recent Warming from those that were driven by other causes. One exception may be the emergence of new diseases and the redistribution and spread of old ones, particularly those like malaria and dengue that are vectored by mosquitoes (Epstein et al., 1998).

It is difficult to define increased climate variability meaningfully in isolation from those living things that have to endure the variability. During the past 30 years or so, humans have been massively impacted by El Niño, partly because the period 1920–1960 was a time of low El Niño incidence (Anderson, 1992; Juillet-Leclerc et al., 2006), partly because particular human societies (in the Pa-cific Basin and elsewhere) have today become far more vulnerable to climate variability than they were a few hundred years earlier. At that time, increased climate variability may have been much less of a problem to continental societies at lower population densities that were more mobile and had simpler, more sustainable systems of food production than today. Yet, on islands affected by El Niño, their circumscription may have reduced subsistence adaptation options and forced their societies into crisis, something that rarely happens today when global support mechanisms help island countries deal with climate crises (Pelling and Uitto, 2001).

Sections 8.3.1 and 8.3.2 describe societal changes within the Recent Warming period from the Pacific Rim and Pacific Islands respectively.

8.3.1 Climate-linked societal changes along the Pacific Rim

Throughout much of the Pacific Rim in East Asia, especially Japan, the cold period ~1890 brought about "the first economic panic" (Sakaguchi, 1983, p. 24) associated with rice shortages, and led to conflict and starvation. The first written accounts of colonial life in Australia began to appear in the second half of the nineteenth century, and some of the difficulties they record are the crop failures linked to the successive impacts of El Niño and La Niña. In April 1879 at the end of a great El Niño, there was the "most severe drought within the memory of the Colony's inhabitants" (Jenkins, 1975, p. 121) yet within a few months – with the arrival of a La Niña – the land was awash, and the colonists faced a third successive year of crop failure (Couper-Johnston, 2000).

Such effects do not apply only to agricultural societies but also to urban ones. The societal effects of the average 10–30 cm/year retreat of the California coast noted above can be gauged by understanding that 80% of California's 36 million people live less than 50 km from the shoreline, and that some 4 million people live less than 5 km away (Griggs and Patsch, 2004).

> "Coastal communities ... have lost entire oceanfront streets, utility lines and homes through cliff erosion over the last century. New developments on eroding or unstable bluff tops continue to be proposed, and small and older weekend beach cottages are still being extensively remodeled or torn down and replaced by larger new homes".
> (p. 39)

8.3.2 Climate-linked societal changes on Pacific Islands

During the period of Recent Warming, most Pacific Island people occupied island coasts, routinely acquiring food from these areas, particularly their offshore parts. While the spread of the cash economy and human-linked deterioration of these areas has lessened the dependence of people on coastal ecosystems, so too have the effects of climate change. These include the effects of increased sediment inputs to coastal areas, shoreline erosion, and coral-reef stress manifested by bleaching.

Both increased precipitation from more frequent tropical cyclones and sea-level rise have contributed to increased lowland sedimentation. For coastal and lowland communities in the Pacific Islands, this has meant more frequent flooding – from sea and land – and its many consequences for human occupation of such areas. Increased sediment inputs to coastal zones also affect areas offshore, particularly reef-bounded coasts where most such sediment ends up on lagoon floors or reef surfaces, with negative consequences for these ecosystems. Examples were given by Carpenter and Maragos (1989).

Shoreline erosion poses huge challenges for many coastal communities on Pacific Islands but, while claimed as widespread (Kaluwin and Smith, 1997), there are surprisingly few supporting data. Examples come from Fiji (Nunn, 1994) and Hawaii (Coyne et al., 1999).

Coral reefs have important protective functions for many Pacific Island coastal communities as well as being the lynchpin of offshore ecosystems that continue to sustain many of their inhabitants. Long-term deterioration of Pacific Island reefs has contributed to problems for their dependent human communities but it is the shorter term and more rapid falls in reef health over recent decades associated with El Niño events that provide a better illustration of climate-driven reef impacts on humans.

Intolerably high stress levels in coral will result in the ejection of symbiotic algae (zooxanthellae) and the bleaching of corals, something that often results in widespread reef death. In the 1998 mass bleaching event, some 16% of the world's reef-building corals died (Wilkinson, 2000). Most such stresses appear to be temperature linked. As seen in Fig. 8.6, the increasing incidence of (temperature-linked) coral bleaching over the past 40 years or so has been due to the superimposition of long-term temperature rise on short-term temperature

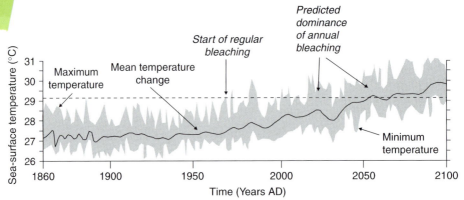

FIGURE 8.6   Sea-surface temperatures A.D. 1860–2100 for Tahiti incorporating ENSO effects compared to the thermal tolerance of corals (dashed line) (after Hoegh-Guldberg, 1999a). Note how for the first 100 years of record, sea-surface temperatures in Tahiti never rose above the coral tolerance limit, so there was no bleaching. Yet beginning ~1970 when mean sea-surface temperature had begun rising, temperature maxima during El Niño Events reached above the coral tolerance limit and resulted in bleaching. By ~2050, it is expected that mean sea-surface temperature around Tahiti will have reached close to the coral tolerance limit, so that bleaching episodes will become annual occurrences; "most evidence suggests that coral reefs will not be able to sustain this stress and a phase shift to algal dominated benthic communities will result" (Hoegh-Guldberg, 1999a, p. 15). A more recent assessment is that bleaching will become an annual or biannual event in the next 30–50 years, and that coral reefs in Micronesia and western Polynesia will be especially vulnerable (Donner et al., 2005).

maxima achieved during El Niño events, a trend that is predicted to lead to a steadily increasing incidence of bleaching episodes in the next 20–30 years. Human communities that subsist partly from coral-associated ecosystems have felt the impacts of bleaching episodes in the past, and must clearly expect more in the future (Hoegh-Guldberg et al., 2000).

While several studies in the Pacific have highlighted the importance of other environmental stressors that lead to coral bleaching, there is little doubt that it is sea-surface temperature that is the most widespread cause of most bleaching episodes. For example, the first large-scale coral bleaching in Hawaiian coastal waters in 1996 and the second in 2002 were triggered by high temperatures that were further elevated by 1–2°C in inshore areas by low winds and concentrated solar radiation. Other environmental factors recognized as influencing bleaching in Hawaii include those affecting irradiance and water movement (Jokiel and Brown, 2004). The bleaching outbreak in (Western) Samoa in 2006 is believed to have been triggered by heavy rainfall that decreased salinity of inshore reefal waters (Ed Lovell, personal communication, 2006).

El Niño-linked droughts have become a significant problem for Pacific Island societies recently, highlighting the imperative for improved water management practices. A good example of the effects of an El Niño-associated drought and societal collapse during the period of Recent Warming comes from New Caledonia where, during the late nineteenth century, the spread of ranching at the expense of subsistence cropping areas on La Grande Terre (island) sowed the seeds for conflict between French colonists and the indigenous Kanak population (Lyons, 1986). The 1877–1878 El Niño brought prolonged drought that raised societal stress, specifically when thirsty, hungry cattle encroached onto Kanak food gardens. Rebellion ensued, a clear example of how climate change led to conflict, as envisaged in the model of the A.D. 1300 Event (see Fig. 7.1).

Coming at the start of sustained European colonization of many Pacific Island groups, there is more information about the effects of the 1877–1878 El Niño than for many others during the nineteenth century. It seems to have been a comparatively severe event, with a warm pool being established in the central Pacific from which a number of tropical cyclones affected French Polynesia with devastating consequences (Kiladis and Diaz, 1986). The Tuuhora church on Anaa atoll in the Tuamotu Islands was destroyed, rebuilt subsequently, then destroyed again during the 1906 El Niño and then again during that in 1983.

8.4. FUTURE CLIMATE CHANGE AND ITS EFFECTS ON PACIFIC BASIN ENVIRONMENTS AND SOCIETIES

There is no shortage of sceptics on the subject of future climate change and its effects on Pacific environments and societies. Much of the scepticism derives from data series that can be portrayed as equivocal, particularly as regards to long-term trends in parts of the region where such data are comparatively sparse. Indeed, the masking effect of phenomena such as El Niño variations and the PDO/IPO make it exceedingly difficult to isolate long-term temperature and sea-level trends in the Pacific. Yet accepting – as most scientists working with these data in the Pacific do – that temperature and sea level have both been rising overall during the period of Recent Warming, then this seems a reasonable basis for projecting these trends into the future. More than this, it is important to understand what effects future changes in these variables will have on environments and societies.

Much of the guiding work in these fields has been carried out by the Intergovernmental Panel on Climate Change (IPCC), whose statement in 2001 projected that between A.D. 1990 and A.D. 2100, Earth-surface temperatures are likely to rise 1.4–5.8°C while average sea level will rise 9–88 cm. The large ranges reflect uncertainty, more recent work suggesting that the higher ends of these projections for A.D. 2100 are more probable (Stainforth et al., 2005).

Within the past 100 years or so, temperatures in the Pacific have risen by ~0.6°C and sea level by ~15 cm (see Figs. 8.1 and 8.4), so this means that within the next 100 years or so in the Pacific, there will likely be an acceleration of the rate of temperature rise by 2.3–9.7 times and an acceleration in the rate of sea-level rise by as much as 5.9 times (Fig. 8.7). It is in this context that we need to ponder future changes to environments and societies in the Pacific Basin.

It has been suggested that ENSO-driven interannual climate variability will increase over the next 100 years or so (Hoegh-Guldberg, 1999b), particularly in response to increasing temperatures associated with increased concentrations of greenhouse gases in the lower atmosphere (Collins, 2000).

Many studies have been undertaken to try and determine what will happen regarding tropical cyclones in the warmer world that is coming. One study of future tropical-cyclone frequency suggests that this will be reduced in the Pacific by ~30% in a warmer world (Oouchi et al., 2006). But the study of tropical-cyclone

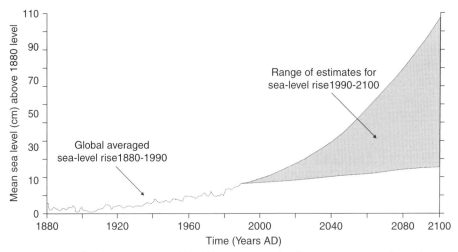

FIGURE 8.7   Sea-level changes, A.D. 1880–2100. Upper limit of estimates comes from the maximum sea-level rise under the A1F1 scenario of the IPCC. Lower limit comes from minimum sea-level rise under the B1 scenario.

intensity suggests, from both theoretical and modelling standpoints, that this may increase (Emanuel, 1987; Knutson et al., 1998; Oouchi et al., 2006). In summary, the present consensus is that there will be fewer but stronger tropical cyclones if temperatures continue to rise during the rest of the twenty-first century.

This section looks first at the effects of future climate change on Pacific Basin environments (Section 8.4.1), and then the effects of this and climate-driven environmental change on Pacific Basin societies (Section 8.4.2). Section 8.4.3 looks at how changes in Pacific Basin societies are themselves likely to feed back into changes of climate and environment during the foreseeable future in this region.

8.4.1 Effects of future climate change on Pacific environments

Since this account is not attempting to be comprehensive, only those environmental changes that are likely to occur in the Pacific Basin as a direct consequence of climate change are considered in this section. As in previous chapters, these will be those coastal-environmental changes associated with sea-level change.

Sea level is expected to continue rising during the twenty-first century reaching levels as high as 88 cm (above the 1990 level) by the year 2100 (IPCC, 2001). Yet there are signs that the projected rate of sea-level rise is faster than expected and may continue in this way as a result of hitherto unforeseen contributors to sea-level rise (Stainforth et al., 2005). On the other side of the equation, an increased rate of Antarctic ice formation by 32 mm water equivalent per year has been modelled, which could involve a sea-level fall of 1.2 mm/year by 2100 (Krinner et al., 2006).

The principal and most common effect of future sea-level rise will be inundation (or drowning) of low-lying coastal areas around the Pacific, changing not only the configuration of land and sea but also the nature of the processes affecting those environments. Inundation will result in the movement of the shoreline inland (unless prevented by human interference) which will mean that the "coast" as a landscape entity will be redefined. This will mean that tsunamis or storm surges will penetrate farther inland (by today's definition) in the future.

Other environmental effects of future sea-level rise on Pacific environments involve their vegetation and, by extension, the effects of vegetation change on ecosystems. Sea-level rise along low coastal areas will result in an increased inland penetration of seawater in the groundwater, resulting in the replacement of natural vegetation by more salt-tolerant species. For those coasts that are fringed by mangroves, there has been some discussion of how these will respond to sea-level rise (Gilman et al., 2006), some studies suggesting – perhaps somewhat optimistically – that mangrove fringes will migrate inland as this happens.

Coral reefs play important roles along many tropical Pacific coasts, particularly in protecting them from large-wave erosion and supplying them with calcareous sediment. Much debate has focused on how coral reefs will respond to higher sea-surface temperatures and a range of other stressors, one of the most compelling syntheses suggesting that there will be widespread coral death associated with increasingly frequent bleaching episodes within the present century (see Fig. 8.6). Yet countering this to some extent is the realization that, while an effectively dead coral reef may not be able to respond to future sea-level rise, it may become both a drawcard for calcareous algae and a source of increased (rather than decreased) amounts of sediment supply to adjoining coasts (Hoegh-Guldberg et al., 2000).

Many of the precise coastal-environmental changes associated with a future accelerated sea-level rise will depend on the nature of a particular coast (Woodroffe, 2002). Along rocky coasts, the effects are more straightforward to predict than along soft-sediment coasts where sea-level rise may result in more sediment circulating in nearshore areas that may actually reduce the erosional effects of sea-level rise. The case of reef-bounded coasts is the one that has received special attention because reefs prevent most such sediment from being permanently lost from adjoining coasts. Interest has also focused on the (non-human) biotic responses to future sea-level rise, particularly whether or not coral reefs could grow upwards and thereby offset some of the effects of sea-level rise on lagoons and adjacent coasts. In many Pacific Islands, the costs of building artificial structures to protect the long coastlines are prohibitive, so "natural" solutions, perhaps involving planting of a buffer vegetation zone, are more practical alternatives.

8.4.2 Effects of future climate and environmental change on Pacific societies

There have been numerous studies of the impacts of future climate change on Pacific Basin societies (e.g. Mote et al., 1999; Nunn et al., 2006). Global Climate Models (GCMs) have been used to simulate the most probable changes that the Pacific may experience, although recommendations are devalued in the eyes of some decision-makers by the unavoidable (and undisguised) scientific uncertainty. Much evidently hinges on the types of mitigation and adaptation carried out by Pacific people in response to threatened changes.

Unless countered by their inhabitants, inundation from sea-level rise is likely to displace huge numbers of people from low-lying areas of the Pacific that are currently densely populated. For example, the delta of the Changjiang in eastern China includes the largest economic centre in the country – Shanghai (Wang et al., 1995). The Shanghai municipality covers an area of some 6,185 km^2, is home to more than 15 million people and is located entirely on the Changjiang alluvial plain with an average elevation of <4 m above low-tide level. The effects of sea-level rise are exacerbated by land subsidence, the mean rate for the Changjiang Delta being 0.3 mm/year rising to 3–4 mm/year in Shanghai City as a result of groundwater extraction. Assuming no additional shoreline protection, a sea-level rise of 50 cm would result in land loss of 855 km^2, nearly 14% of the municipality; a sea-level rise of 2 m would see almost the entire area (96.1%) inundated. The issue of how to confront this threat – whether by substantial relocation or by creating an entirely artificial coastline – is one that is mirrored in many other similar situations in eastern China and elsewhere along the Pacific Rim.

Comparable situations are found in the Pacific Islands, although the numbers of people involved are far less. Figure 8.8A shows the most densely inhabited island (Tongatapu) in the Kingdom of Tonga with the modern shoreline and main population concentrations. Tongatapu is an emerged coral-reef island and much of its northern coast is low lying and/or made from unconsolidated sediments that might become mobile as sea level rises. Tonga is insufficiently wealthy to protect the island's present shoreline, so the inevitable consequence of projected twenty-first-century sea-level rise will be shoreline inundation (Mimura and Pelesikoti, 1997). One possible scenario is shown in Fig. 8.8B.

The disappearance or shrinkage of many ice caps, including those in the Pacific tropics (in New Guinea and the Andes), will not only alter mountain climates but will also have an effect on water supplies critical for drinking, irrigation, and even the generation of hydroelectric power. For example, the recession of the Qori Kalis glacier – part of the Andean Quelccaya ice cap – was occurring at a rate of 4 m annually in 1995 (Epstein et al., 1998) and may vanish entirely within the next few decades, depriving down-valley communities of regular water supply. Similar effects are being experienced in parts of western North America, where continued global warming is likely to cause earlier snow-melt and corresponding problems associated with summer water supply to the region's growing number of consumers (Knowles et al., 2006).

Many vector-borne diseases (mainly those involving insects as carriers) will spread as warming continues (McMichael et al., 1996; Epstein et al., 1998; Hopp and Foley, 2001). The spread will be especially noticeable with altitude in tropical areas. Many highland communities in Papua New Guinea, for example, currently live above the maximum altitude at which the malarial mosquito can survive, but this altitude is rising, with malaria now being reported as high as 2,100 m here.

8.4.3 Effects of future changes in Pacific societies on climate and environment

The externally driven changes to Pacific societies described above will be superimposed on societies that would be changing anyway as a result of unrelated internal changes. Paramount among these internal changes is population growth. Many Pacific populations are growing at rates that will strain food-production systems which will inevitably lead, particularly during times of climatic stress (such as during droughts or following storms), to crises of various kinds. How Pacific societies weather these crises will depend on their degree of preparedness (nationally and internationally) and their internal resilience. Such internally driven societal changes will also themselves impact on climate and coastal environments, in ways that are outlined below.

Increasing populations are likely to result – at a regional rather than a national scale – in increases in the emission of greenhouse gases through the combustion of fossil fuels, particularly in the industrial and transportation sectors. Increasing population is also likely to result in continued deforestation. Both these processes will increase the amount of greenhouse gases in the lower atmosphere and continue to enhance the greenhouse effect, thereby continuing to force temperature and sea-level rise into the twenty-second century. Present and proposed

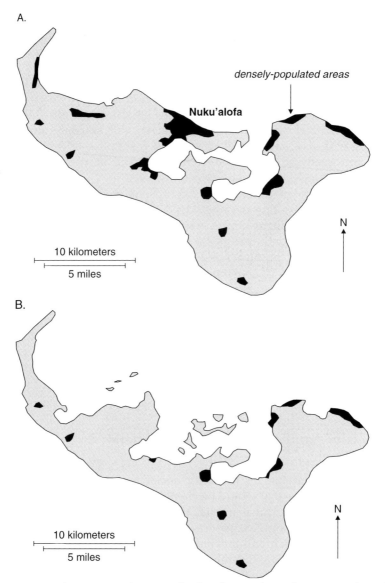

FIGURE 8.8 Maps of Tongatapu, the main island in the Kingdom of Tonga, South Pacific. Map A shows the geography of the modern island with densely populated areas (including the capital Nuku'alofa) shaded black. Map B shows the geography of the island were the sea level to rise an effective 4 m above its present level. Most of the presently densely populated areas would be submerged and the Nuku'alofa peninsula converted to a series of smaller islands.

international agreements to reduce greenhouse-gas emissions are unlikely to have a significant effect on this scenario.

Increasing population density is likely to put an increasing strain on environments, leading perhaps to the continued modification of many (Fig. 8.9). In

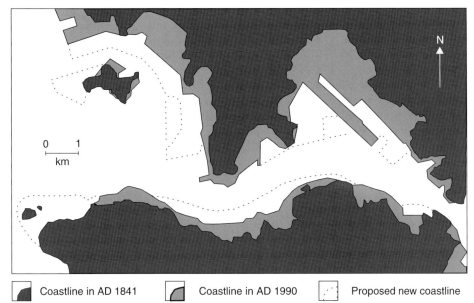

| | Coastline in AD 1841 | | Coastline in AD 1990 | | Proposed new coastline |

FIGURE 8.9   Map of Victoria Harbour, Hong Kong, showing the position of the coastline in A.D. 1841, the area which was reclaimed by A.D. 1990, and the area which it is proposed to reclaim for port and airport development (after Yim, 1995).

particular, it is likely that the combination of increased demand for coastal lowland for human use and accelerated sea-level rise will lead to an increase in coastal reclamation. Under conditions of accelerated sea-level rise, as suggested by the IPCC (2001) projections, reclaimed lands will become highly vulnerable and will need to be effectively managed. In some parts of the Pacific, this situation will be exacerbated by the effects of ground subsidence resulting from groundwater extraction.

Increased demand for foods and other materials from vulnerable natural resource bases (such as coral reefs and mangrove forests) will make their conservation – essential in some places for minimizing the effects of sea-level rise – even more challenging.

KEY POINTS

1. Since about A.D. 1800 in most of the Pacific Basin, there have been net rises in temperature and sea level.
2. Within this period of Recent Warming, sea-level rise has changed many Pacific Basin coasts as a result of inundation, groundwater salinization and the erosion of their shorelines.
3. Increased numbers of people in the Pacific Basin during the period of Recent Warming have exacerbated the effects of climate-driven change. In particular, human societies have become more vulnerable to interannual

climate variability (especially El Niño events) and have been impacted in various ways by temperature and sea-level rise.

4. It seems likely that the rates of both temperature and sea-level rise will accelerate in the twenty-first century posing even greater challenges for the inhabitants of the Pacific Basin.

CHAPTER 9

Teleconnections and Global Climate Trends

"Climate variability leading to harvest failures is just one cause of stress, like war or disease, but we delude ourselves if we do not assume that it is among the most important – especially in a society like that of preindustrial Europe, that devoted four-fifths of its labor just to keeping itself fed."
Fagan, *The Little Ice Age* (2000, p. 103)

Long-term climate changes commonly affect the entire Earth, not just part of it. So, if the climate changes described in the preceding chapters were driven by global forcing mechanisms, then it would be expected that their effects would have been experienced worldwide not just in the Pacific Basin. Yet, even if some of these climate changes were produced by regional (not global) forcing mechanisms, then they need not have been confined to the Pacific Basin. For while the Pacific Basin can be justifiably regarded as a discrete region – a vast island-peppered ocean bounded by a mountainous continental rim – and while many of the climatic processes that drive environmental change within it are generated within it, all are not. The Pacific Rim is a filter, not an impermeable barrier. Either way, it might be expected that some places outside the Pacific Basin experienced similar histories to those of last-millennium climate, environmental and societal changes described in Chapters 4–8.

The degree to which there are teleconnections between last-millennium climate histories within and beyond the Pacific Basin is key to understanding their cause(s). For if it can be demonstrated that there are very few similarities between last-millennium climate histories inside and outside the Pacific Basin, then clearly the dominant cause(s) of the former are not global and must be sought within this region. Alternatively, should ample evidence of teleconnections be found beyond the Pacific Basin, then global forcing mechanisms are likely to have dominated.

Of course, it is not really so black and white. Global forcing is likely in any part of the world to have been modulated by regional and/or local forcing mechanisms, and vice versa. The identification of teleconnections should help in the teasing apart of global from regional/local forcing mechanisms. For instance, it is quite possible that, while the broad picture of last-millennium climate change in the Pacific Basin described above may be shared by other parts of the world, its essential character may be unique to the Pacific.

Such considerations are important when the influences of climate change during the last millennium are extended to environmental and societal changes.

225

Pacific Basin environments, modulated by local climate, by geology and geography, may not have responded in the same way to similar forcing elsewhere in the world. Pacific Basin societies may also have responded differently to similar external forcing mechanisms for various reasons including their subsistence systems, their histories, interactions and structure. Yet with those caveats in mind, it would also be expected that subsistence-based societies would respond in broadly similar ways to varying degrees of external forcing, irrespective of their location.

Finally, by way of introduction, it should be noted that the essence of this book has been presented in Chapters 4–8. This book is primarily an empirical study, the validity and relevance of which does not hinge on the presence or absence of teleconnections (this chapter) or the identification of likely causes of observed climate changes (Chapter 10). Such considerations may even be premature in the sense that definitive data may not yet have been gathered.

It should also be appreciated that this chapter does not attempt a comprehensive account of last-millennium climate, environmental and societal changes outside the Pacific Basin. Rather it focuses on the key changes, highlighting parallels with the Pacific Basin, particularly those related to the A.D. 1300 Event described in Chapter 5. More comprehensive global surveys of last-millennium climate and environmental changes and their effects on human societies were given by Lamb (1977, 1982), Grove (1988) and Fagan (2000).

An account of last-millennium climate change outside the Pacific Basin is given in Section 9.1. Last-millennium environmental changes are discussed in Section 9.2, with an emphasis (as elsewhere in this book) on sea-level changes and their environmental effects. There follows a discussion of those last-millennium societal changes outside the Pacific Basin that can be plausibly attributed to climate and environmental forcings (Section 9.3).

It remains important to meld discussions of Pacific Basin climate change and those from regions beyond into global syntheses, not least to be able to measure the degree of concordance between various parts of the world. Unfortunately the uneven distribution of relevant data series is unable to provide truly global pictures that satisfy everyone, and there are significant differences in the interpretation of global climate change during the last millennium, something that has influenced the understanding of its causes. Global compilations of last-millennium climate change are discussed in Section 9.4.

9.1. CLIMATE CHANGE OUTSIDE THE PACIFIC BASIN DURING THE LAST MILLENNIUM

While there have been various statements doubting the reality of the Medieval Warm Period and Little Ice Age as discrete, globally occurring periods, most scientists would today see things otherwise, while perhaps allowing for more variability within these periods than was originally envisaged (Grove, 1988; Fagan, 2000; Broecker, 2001), a position taken at the beginning of this book (Section 1.3). It is worth reviewing briefly the chronology of the Medieval Warm Period and Little Ice Age within and beyond the Pacific Basin as a key to understanding their cause(s) in both the Pacific Basin and globally.

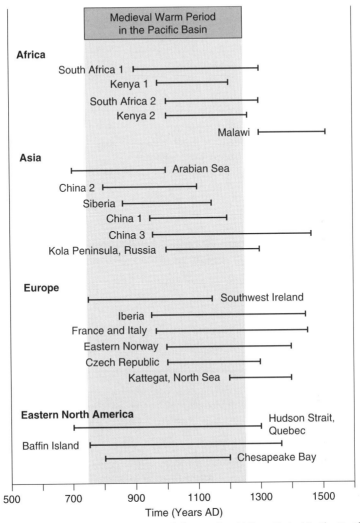

FIGURE 9.1 Comparison between the timing of the Medieval Warm Period in the Pacific Basin and at representative sites elsewhere. Data sources as follows: South Africa 1 (Huffman, 1996), Kenya 1 (Lamb et al., 2003), South Africa 2 (Holmgren et al., 2003), Kenya 2 (Verschuren et al., 2000), Malawi (Johnson et al., 2001), Arabian Sea (Doose-Rolinski et al., 2001), China 2 (Yang et al., 2002), Siberia (Naurzbaev and Vaganov, 2000), China 1 (Qian and Zhu, 2002), China 3 (Paulsen et al., 2003), Kola Peninsula (Hiller et al., 2001), Southwest Ireland (McDermott et al., 2001), Iberia (Desprat et al., 2003), France and Italy (Serre-Bachet, 1994), Eastern Norway (Nesje et al., 2001), Czech Republic (Bodri and Cermak, 1999), Kattegat (Fjellsa and Nordberg, 1996), Hudson Strait (Kasper and Allard, 2001), Baffin Island (Moore et al., 2001) and Chesapeake Bay (Willard et al., 2003).

Figure 9.1 shows the timing of the Medieval Warm Period in the Pacific Basin compared to selected records from outside. Most records from Asia and North America (east coast) coincide with the Medieval Warm Period within the Pacific Basin, probably because the causes were similar across these landmasses which

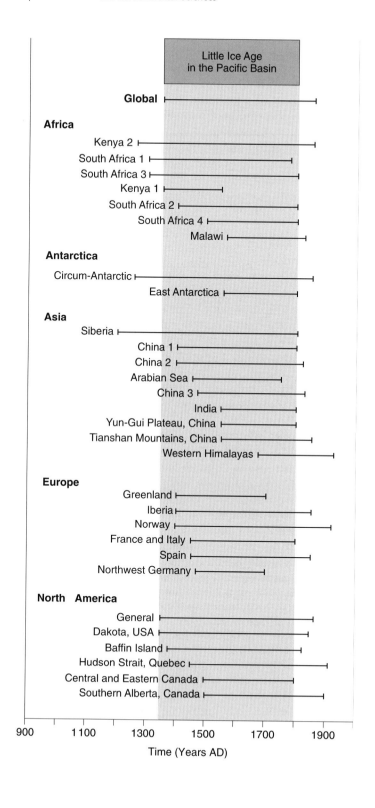

also extend into the Pacific Rim. Yet the records from Africa and Europe – the sites farthest from the Pacific Basin – seem to have modes that are ~150 years later than within the Pacific Basin suggesting that global forcing may have been significantly modulated by regional conditions.

The same is not so readily apparent for the Little Ice Age. Figure 9.2 shows the timing of the Little Ice Age within the Pacific Basin compared to continents beyond, and the coincidence is striking. This implies that, although the Medieval Warm Period may not have been globally synchronous and was therefore the product of global forcing significantly modulated by regional conditions, the Little Ice Age was globally synchronous and the product of global forcing mechanisms that largely overrode regional conditions.

For the purposes of discussing last-millennium climate change in the Pacific Basin, this book accepts the view that the Medieval Warm Period and Little Ice Age are distinct periods of climate as – by extension – are the A.D. 1300 Event and the period of Recent Warming. The present section highlights some of the principal environmental proxy datasets for last-millennium climate change from beyond the Pacific Basin (Section 9.1.1) with a brief review of directly monitored time series (Section 9.1.2). There follows a review of societal proxy data for climate change (Section 9.1.3) and, finally, an account of evidence for the A.D. 1300 Event beyond the Pacific Basin (Section 9.1.4).

9.1.1 Environmental proxies for last-millennium climate change

As in the Pacific Basin, ice coring has provided high-resolution records of climate change elsewhere in the world. Particularly influential have been the Greenland cores GRIP and Dye 3 (Dahl-Jensen et al., 1998) in which the Medieval Warm Period, Little Ice Age and Recent Warming periods are readily distinguished (see Fig. 1.5B). The Kilimanjaro (tropical Africa) ice-core record (Fig. 9.3) shows rapid change during the transition between the Medieval Warm Period and the Little Ice Age (A.D. 1300 Event) that was likely to have been associated with rapid cooling (Thompson et al., 2002).

Various palaeoclimate datasets have been acquired from isotope analyses of speleothems (cave dripstones). For example, evidence for both the Medieval

---

FIGURE 9.2 Comparison between the timing of the Little Ice Age in the Pacific Basin and at representative sites elsewhere. Data sources as follows: Kenya 2 (Verschuren et al., 2000), South Africa 1 (Huffman, 1996), South Africa 3 (Tyson et al., 2000), Kenya 1 (Karlén et al., 1999), South Africa 2 (Repinski et al., 1999), South Africa 4 (Holmgren et al., 2003), Malawi (Johnson et al., 2001), Circum-Antarctic (Domack et al., 2001), East Antarctica (Stenni et al., 2002), Siberia (Naurzbaev and Vaganov, 2000), China 1 (Qian and Zhu, 2002), China 2 (Yang et al., 2002), Arabian Sea (Doose-Rolinski et al., 2001), China 3 (Paulsen et al., 2003), India (Denniston et al., 1999), Yun-Gui Plateau (Chen et al., 2000), Tianshan Mountains (Chen, 1987), Western Himalayas (Yadav and Singh, 2002), Greenland (Fricke et al., 1995), Iberia (Desprat et al., 2003), Norway (Berstad et al., 2003), France and Italy (Serre-Bachet, 1994), Spain (Gutierrez-Elorza and Pena-Monne, 1998), Northwest Germany (Niggemann et al., 2003), North America general (from records of glacier expansion – Pielou, 1991), Dakota (Laird, 1996), Baffin Island (Moore et al., 2001), Hudson Strait (Kasper and Allard, 2001), Central and Eastern Canada (Beltrami and Mareschal, 1992) and Southern Alberta (Campbell, 2002).

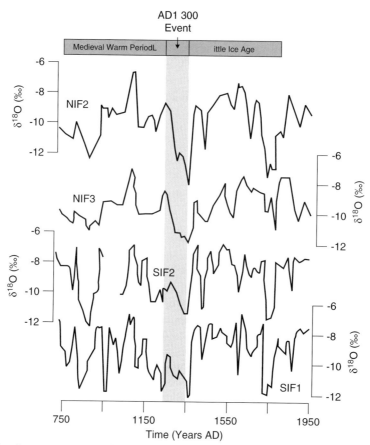

FIGURE 9.3 Oxygen-isotope analysis of four cores through the ice cap of Kilimanjaro in equatorial Africa shows clearly a rapid transition (equivalent to the A.D. 1300 Event) between the Medieval Warm Period and the Little Ice Age (adapted from Thompson et al., 2002).

Warm Period and the Little Ice Age (approximately A.D. 1500–1800) was found in speleothem records from caves in the Makapansgat Valley in South Africa (Holmgren et al., 2003). Influential Europe-based studies are those of McDermott et al. (2001) in southwest Ireland and Niggemann et al. (2003) in northwest Germany. The temperature record from the former study shows both the Medieval Warm Period (A.D. 750–1150) and a two-stage Little Ice Age. Similar results were obtained in the latter study, $\delta^{18}O$ minima A.D. 1470–1700 (480–250 cal BP) corresponding to the Little Ice Age.

Lake-level studies have been widely used to calculate palaeoprecipitation, particularly drought incidence. The Medieval Warm Period (approximately A.D. 1000–1270) and Little Ice Age (approximately A.D. 1270–1850) show up well in the climatically sensitive Lake Naivasha in Kenya (Verschuren et al., 2000). Studies of varved sediments on lake beds have also yielded information about last-millennium climates. For example, the Little Ice Age shows up as an arid period A.D. 1570–1850 in the Lake Malawi area (Johnson et al., 2001). Studies of a core from Lake Neuchatel, Switzerland, show that temperature here fell ~1.5°C during the transition from the Medieval Warm Period to the Little Ice Age and that this has not been completely reversed by temperature rise during the period of Recent Warming at this location (Filippi et al., 1999).

Among the many other environmental proxies used to reconstruct last-millennium climates, investigations of aeolian deposits (particularly sand dunes) have proved successful. A good example comes from Argentina where there is evidence for a southward tropical pulse – evidence for increased aridity – during the Medieval Warm Period and its contraction during the Little Ice Age (Iriondo and Garcia, 1993).

Biological proxies of last-millennium climate change – distinct from pheno-logical records in which data are filtered by subjective human observation – are widely used in palaeoclimate reconstruction. An example from Scotland used larval heads of Chironomidae (midges) preserved in lake sediments to identify both the Medieval Warm Period and the Little Ice Age (Brooks and Birks, 2001). Another chironomid sequence from Norway allowed the conclusion that summer temperatures during the Medieval Warm Period were ~0.4°C warmer than today. Perhaps the most widespread and compelling biological proxies of climate change used to reconstruct last-millennium climates are tree rings. A northern-hemisphere (mostly outside the Pacific) compilation of tree-ring-derived palaeotemperatures over the last 300–600 years found that the highest sustained temperatures were A.D. 1930–1945 (Briffa et al., 2004). A tree-ring record from the Asian subarctic showed a period of severe cooling A.D. 1750–1820 followed by warming A.D. 1820–1950 and, like other circum-Arctic records, subsequent cooling (Vaganov et al., 2000).

In recent decades, as the importance of oceanic forcing of global climate has become well appreciated, much effort has gone into obtaining high-resolution records of past sea-surface temperature (Bond et al., 1997; Delworth and Mann, 2000). Beyond the Pacific, these efforts have been focused in those parts of the oceans that are considered sensitive to global climate forcing such as the North Atlantic (Griffies and Bryan, 1997; Keigwin and Boyle, 2000). Other influential studies have been made in the Sargasso Sea (Keigwin, 1996), where sea-surface temperature fell ~2°C between A.D. 1100 and A.D. 1270 (850–680 cal BP), and off West Africa, where sea-surface temperatures during the Little Ice Age (A.D. 1300–1850) were 3–4°C lower than today (deMenocal et al., 2000). A selection of sea-surface temperature proxies is shown in Fig. 9.4.

Coral coring has been used to reconstruct sea-surface temperature fluctua-tions outside the Pacific. Many such records extend only into the later part of the Little Ice Age and provide confirmation of the subsequent warming trend. Ex-amples come from the Kenya coast of East Africa (Cole et al., 2000) and Puerto Rico in the Caribbean (Winter et al., 2000), where sea-surface temperatures dur-ing the Little Ice Age were 2–3°C cooler than today.

9.1.2 Direct measurements of recent climate change

There have been numerous analyses of directly monitored time series of climate data from outside the Pacific Basin, many of which have featured in regional and

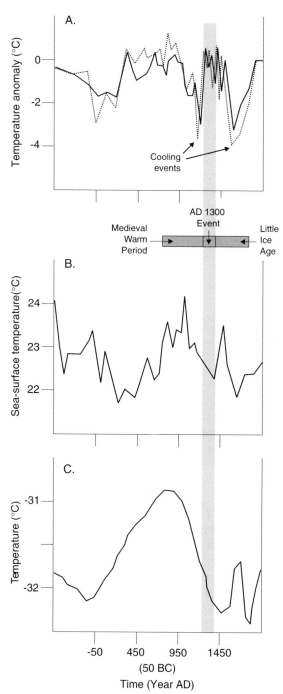

FIGURE 9.4 Palaeotemperature records for various North Atlantic locations since 2,500 BP showing rapid cooling equivalent within dating uncertainties to the A.D. 1300 event (adapted from deMenocal et al., 2000). A. Atlantic Ocean sea-surface temperature off West Africa (Hole 658C); note the two cooling events around the A.D. 1300 Event. B. Sea-surface temperatures above the Bermuda Rise obtained from oxygen-isotope analysis of planktonic foraminifera (Keigwin, 1996). C. The GRIP ice-core record from Greenland (see also Fig. 1.5B for comparison).

global syntheses (e.g. IPCC, 2001; Brohan et al., 2006). Numerous datasets are available on the internet through the World Data Center for Paleoclimatology (www.ncdc.noaa.gov/paleo), the Publishing Network for Geoscientific and Environmental Data (www.pangaea.de) and the Intergovernmental Panel on Climate Change (IPCC, www.ipcc.ch).

While most meteorological data series extending more than 40–50 years across the late twentieth century exhibit a warming trend, some clearly do not. For example, similar to proxy records in the circum-Arctic regions, the temperature records from three coastal stations in Greenland show no sustained warming trend (Chylek et al., 2004). The highest temperatures at these stations for more than 100 years of record occurred 1930–1940.

9.1.3 Societal proxies of last-millennium climate change

Some of the earliest compilations of last-millennium climate changes – that understandably proved influential – were for northwest Europe by Lamb (1965, 1977). Since there was no significant development of more rigorous scientific proxies, most reconstructions by Lamb and those who emulated his approach depended on human records of various phenomena, such as coastal sea ice extent, upper tree limits on mountains, agricultural records (particularly the geographical limits of indicator crops such as grapes), phenological data and more general historical records of flood and famine. In this way, the Medieval Warm Period and Little Ice Age were recognized for central England, for example, the transition between the two being marked by a temperature fall of 1.4°C between A.D. 1290 and A.D. 1430 (Lamb, 1965).

A similar approach proved helpful for understanding last-millennium climate variations in other parts of the world, particularly temperate locations where seasonal records were regularly kept; China and Japan are examples from within the Pacific Basin. Good examples from outside the Pacific Basin come from the records of diarists in the European Alps (Casty et al., 2005) and along the eastern seaboard of the United States, where, for example, observations of the decreasing numbers of days when ice affected river flow are consonant with the warming record (Hodgkins et al., 2005).

Within the last two decades, there has been an increasing emphasis on understanding the evolution of African climates, with numerous societal proxies being used in their reconstruction. Various studies show an enhanced variability of precipitation in the nineteenth century (Nicholson and Yin, 2001) and stress the role of drought in profound societal change (Holmgren and Öberg, 2006). For Uganda, the major climatic watershed within the last millennium was at A.D. 1328, which led to widespread societal changes (Taylor et al., 2000). In equatorial Africa, for example, the Little Ice Age shows up as a time punctuated by exceptional droughts that forced migration of people and initiated warfare (Nicholson and Yin, 2001).

9.1.4 Climate change during the A.D. 1300 Event beyond the Pacific Basin

There are numerous records of profound climate shifts around A.D. 1300 marking – as for the Pacific Basin – the abrupt transition (the A.D. 1300 Event) between the

Medieval Warm Period and the Little Ice Age. A good example is the reduction of drought severity and frequency marking the A.D. 1300 Event in the central United States (Laird et al., 1996b). The transition from the Medieval Warm Period to the Little Ice Age in Chesapeake Bay (eastern United States) was dated to A.D. 1300–1450 and represented a rapid cooling (Cronin et al., 2003). Farther south at a key indicator site, a change about A.D. 1320 in *Globigerina bulloides* (foraminifera) abundance in the Cariaco Basin of Venezuela manifests an abrupt shift in the variability of North Atlantic sea-surface temperatures (Black et al., 1999).

In Europe, the A.D. 1300 Event is marked by various records including extensive flood activity in Norway (Nesje et al., 2001), cooling of ~1.5°C in Scotland (Brooks and Birks, 2001) and Switzerland (Filippi et al., 1999), and an influx of cold-water taxa into the Baltic Sea (Andren et al., 2000). Winters throughout western Europe became abruptly more severe during the A.D. 1300 Event (Pfister et al., 1998).

For tropical Africa, there is evidence of an abrupt climate shift about A.D. 1270 at Lake Naivasha in Kenya (Verschuren et al., 2000). Similar evidence for rapid cooling is found in ice cores from Mt. Kilimanjaro (see Fig. 9.3). Around A.D. 1300, sea-surface temperatures off Mauritania (West Africa) fell to 3–4°C below today's levels (deMenocal et al., 2000). Cooling of perhaps 3–4°C also occurred in South Africa just after A.D. 1300 (Tyson et al., 2000).

At Taimyr on Russia's Arctic coast, temperature fell almost 1°C between about A.D. 1200 and A.D. 1300 (Naurzbaev et al., 2002). Along the Kola Peninsula (northwest Russia), treeline elevation dropped rapidly around A.D. 1300 as temperatures fell (Hiller et al., 2001).

Evidence for a period of rapid cooling around A.D. 1300 also comes from parts of the ocean basins (see Fig. 9.4), from cores through ice caps (see Figs. 1.5B and 4.1) and from deep boreholes (see Fig. 1.5A).

9.2. ENVIRONMENTAL CHANGE OUTSIDE THE PACIFIC BASIN DURING THE LAST MILLENNIUM

Climate-driven environmental changes during the last millennium are so diverse and generally so localized that no attempt is made to synthesize them here. Rather, as in earlier chapters, this section focuses on coastal-environmental changes associated with sea-level change, itself regarded as a likely outcome of climate forcing on this timescale. Section 9.2.1 looks at the evidence for sea-level change during the last millennium outside the Pacific Basin, focusing particularly on the A.D. 1300 Event. Section 9.2.2 reviews the evidence for net sea-level rise during the period of Recent Warming outside the Pacific Basin.

9.2.1 Sea-level changes A.D. 750–1800

Most scientists specialized in reconstructing past sea-level changes have shied away from the last millennium owing to the challenge of finding sufficiently high-resolution stratigraphic repositories in areas where tectonic (land-level) changes have either been negligible or comparatively easy to subtract from observations (detrend). The other issue has been the challenge in using radiocarbon dating to calibrate events (such as sea-level change) within the last millennium because of the existence of temporal and spatial variations in atmospheric ^{14}C that give rise to uncertainties, the significance of which is amplified on such comparatively short (millennial) timescales.

One of the first compilations of global sea-level changes for the Holocene showed a number of sea-level oscillations in its later part (Fairbridge, 1961). These included the Paria Emergence (a period of sea-level fall approximately A.D. 1200–1250) for which evidence was found in many parts of the world. Later global compilations generally discounted last-millennium sea-level changes, averring that the available data could not be uncontroversially interpreted at the high resolution necessary, preferring smoother curves that fitted the available evidence to the satisfaction of most scientists. This situation has changed in the past two decades as indicators of smaller sea-level changes within the last millennium (and at earlier times) have been found and dated with a higher degree of precision than was once possible.

Outside the Pacific Basin, detailed reconstructions of last-millennium sea-level fluctuations have been revealed by faunal and chemical indicators in salt-marsh sediments from Long Island Sound on the eastern seaboard of the United States. The first investigations found that sea level had fallen by as much as 1.1 m around A.D. 1200 (750 cal year BP: Colquhoun and Brooks, 1986) although this conclusion has been superseded. At Hammock River marsh (Clinton)

"real sea level ... was 25 ± 25 cm higher ca. A.D. 1050 than ca. A.D. 1650". (van de Plassche et al., 1998, p. 319)

At Farm River marsh, sea level fell ~40 cm around A.D. 1300 (650 cal year BP) before rising the same amount around A.D. 1350 (600 cal year BP) and then falling an estimated 80 cm in the period A.D. 1400–1525 (550–425 cal year BP: Varekamp et al., 1999). On the Maine coast, sea level at two sites was stable A.D. 800–1300 (1,150–650 cal year BP) but had fallen to a minimum 30–40 cm below present sea level by A.D. 1800 (150 cal year BP: Gehrels et al., 2002).

Farther south in the Gulf of Mexico, successions of beach ridges indicate a sea-level fall of ~1.05 m between about A.D. 1050 and A.D. 1300 (900–650 cal year BP: Tanner, 1991). Analysis of the beach-ridge data from the St. Vincent site showed that

"sea level was rising about 1000 years ago ... at about A.D. 1200 it began to fall, and somewhere near A.D. 1400 reached a low position". (Tanner, 1992, p. 302)

Similar work was carried out using granulometry on a series of 154 beach ridges at Jerup in northern Denmark dating back to ~5800 B.C. Rapid sea-level fall of at least 1 m is recorded at the time of the A.D. 1300 Event, specifically

"a sea-level drop starting around A.D. 1200–1300, bottoming about A.D. 1500". (Tanner, 1993, p. 226)

Other possible evidence comes from the coast of Israel where studies of former water levels in wells dug within the past 2,000 years show that during Fatimede time (A.D. 960–1001 or 990–949 cal year BP) sea level was ~51 cm above its present mean level and that during Crusader time (A.D. 1001–1265 or 949–685 cal year BP) sea level had fallen to a level 56 cm below present (Sivan et al., 2004).

A summary of these changes is shown in Fig. 9.5. Other examples, which cannot be plotted because of a lack of sufficient precise data and are therefore generally less compelling, are discussed briefly below.

Other studies from the eastern seaboard of the United States reach broadly the same conclusions as the more precise ones described above. These include a slowing of relative sea-level rise (interpreted as an absolute sea-level fall) from Guilford at the start of the Little Ice Age (Nydick et al., 1995) and the expansion of salt marshes in southeast Delaware as a result of sea-level regression at the same time (Schwimmer et al., 1997). A sea-level fall of at least 50 cm around this time is implicit in the data of Froede (2002) from Key Biscayne, Florida, and is also likely to have occurred in Chesapeake Bay (Kearney, 1996).

In northeast Spain, studies of the depositional history of alluvial valleys show that downcutting consistent with base-level (sea-level) fall occurred just prior to the start of the Little Ice Age, about A.D. 1450 (500 cal year BP) here (Gutierrez-Elorza and Pena-Monne, 1998). Coastal-environmental change along the Brittany (northern France) coast about A.D. 1370–1500 (580–450 BP) was interpreted as evidence for a sea-level fall of 0.5–1.0 m at the start of the Little Ice Age (Ters, 1987). There is considerable empirical evidence that sea level along the North Sea coasts of Europe rose in the two centuries after A.D. 1000 (950 cal year BP) to "as much as 40–50 cm above today's height" before falling subsequently as temperatures fell at the start of the Little Ice Age (Fagan, 2000, p. 63).

Carbon-isotope dating of tree stumps on the floor of the Knysna Lagoon (estuary) on the Cape coast of South Africa showed that they lived A.D. 1264–1297 (686–653 cal year BP) and represented a time when the sea level here was low enough to allow a freshwater swamp forest to develop (Marker, 1997). This is another possible indication of sea-level fall around the start of the Little Ice Age.

Given the various controls on sea-level change at any one place on the Earth's surface, it would be rash to link all the above changes to those that occurred in many parts of the Pacific at the same time. Yet, as argued in Chapter 7, it is plausible to suppose that many coastal-environmental and coastal-societal changes around the fourteenth century in the Pacific Basin were driven primarily by near-synchronous sea-level fall. The occurrence of a sea-level fall of comparable magnitude at several places outside the Pacific Basin suggests that similar environmental changes and societal changes may have occurred there. Figure 9.5 shows the highest resolution series compared to the Pacific Basin synthesis. The two vertical shaded bars represent the periods of sea-level fall within the Pacific Basin, and it can be seen that they coincide with (periods of) sea-level fall at most recording stations outside the Pacific Basin. On the face of it, this could be considered as a clue that sea level fell along many of the world's coasts around the transition between the Medieval Warm Period and the Little Ice Age, but it may be rash to read too much into this observation when it is based on so few records.

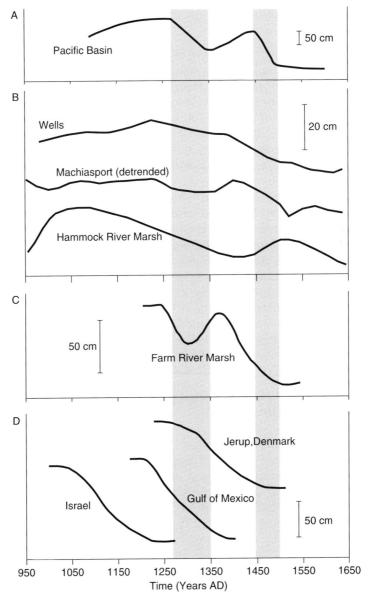

FIGURE 9.5   High-resolution sea-level records for the period A.D. 950–1650 from outside the Pacific Basin compared with the compilation from within (from Fig. 5.8). Graph A shows the Pacific Basin compilation, B and C show Eastern United States stations, and D shows various others. Data sources as follows: Wells and Machiasport (Gehrels et al., 2002), Hammock River Marsh (van de Plassche et al., 1998), Farm River Marsh (Varekamp et al., 1999), Jerup (Tanner, 1993), Gulf of Mexico (Tanner, 1992) and Israel (Sivan et al., 2004).

9.2.2 Sea-level rise during the period of Recent Warming (A.D. 1800 to present)

Global compilations of recent sea-level rise have been dominated by time series from outside the Pacific and in the northern hemisphere – something that has led to the justifiable questioning of their global applicability. Most data series involve sustained sea-level rise beginning in the second half of the nineteenth century and continuing at a rate of ~1.8 mm/year until the mid-twentieth century when rapid acceleration begins (Douglas, 2001). A spliced proxy and monitored data series for the Atlantic coast of Canada shows that sea level rose at a rate of 1.6 mm/year during the period A.D. 1800–1900 accelerating thereafter to 3.2 mm/year (Gehrels et al., 2005).

Environmental changes associated with sea-level rise during the period of Recent Warming have been many, varied, and almost everywhere enmeshed with the effects of intensified human occupation of coastal regions of the world. The most common environmental effects of recent sea-level rise have been inland movements of shorelines and coastal erosion. For example, one-fifth of the European Union's coastline is severely affected by erosion, shorelines receding at rates of 0.5–2.0 m/year on average; Poland has the highest proportion (55%) of eroding coasts (EU, 2004). As a result of sea-level rise, the delta lowlands of Bangladesh have experienced storm surges reaching progressively inland over the past few decades (Huq et al., 1995). Most soft-sediment (beach) coasts of the Indian Ocean are eroding; 73% of those of Peninsular Malaysia are classified as eroding (Midun and Lee, 1995).

9.3. SOCIETAL CHANGE OUTSIDE THE PACIFIC BASIN DURING THE LAST MILLENNIUM

Human societies outside the Pacific Basin changed in response to environmental and climate forcing during the last millennium, but the pace of change has everywhere accelerated during the period of Recent Warming. Within this period there have been profound, uncontroversially global, changes in human societies that have accompanied the emergence of the "global village" – a place where interaction between almost every society on Earth becomes routine, and where societies are in danger of losing those cultural characteristics that once made them readily distinguishable from one another.

This section focuses only on societal changes outside the Pacific Basin forced by climatic and environmental changes. Such societal changes have probably been underestimated in most parts of the world during at least the period of Recent Warming because of the dominance of human-associated changes, but some examples are still worth discussing.

This section proceeds on a case study basis, largely because there is too much material of only marginal relevance to synthesize in a book dealing with the Pacific Basin. Section 9.3.1 discusses the middle last-millennium changes in subsistence agriculture in the central United States, Section 9.3.2 the "calamitous fourteenth century" in Europe and its aftermath, and Section 9.3.3 the history of Norse exploration and settlement of the northernmost Atlantic.

9.3.1 Middle last-millennium changes in subsistence agriculture, central United States

During the first half of the last millennium in the southwest United States, a whole series of cultures evidently collapsed – expressed most clearly by permanent settlement abandonment. Climate is regarded as the principal driver of this societal collapse, specifically the lack of sufficient or reliable precipitation to sustain the agricultural basis of these societies, particularly through a reduction in winter precipitation that sharply reduced the area available for agriculture (Benson et al., 2002).

A good example is provided by the coincidence between a severe, prolonged drought A.D. 1276–1299 and the abandonment of the Four Corners area of the Colorado Plateau. Like many Pacific Islands, the agricultural basis of subsistence living during the Medieval Warm Period in this circumscribed area led to high population levels that could not be sustained by a return to hunting and gathering once agriculture failed (Jones et al., 1999). And, as in the Pacific Basin, societal collapse that followed crop failure was attended by intertribal warfare and reduced inter-regional interactions. In a final parallel with the Pacific situation it is clear that, even when conditions became more stable in the southwest United States around A.D. 1500, the societies that had existed during the Medieval Warm Period did not regroup, largely because of the environmental and societal outfall from their recent collapse.

On the face of it, this example does seem to fit the Diamond (2005) model in which internal societal changes are seen as the ultimate causes of cultural collapse while climate change is seen as a proximate cause. But we must be wary of representing data that are poor quality or even non-existent as exemplary, and therefore supportive of a supposedly self-evident model. There are no records of population during the Medieval Warm Period for the southwest United States, and ideas about unsustainably high levels of population pressure are inferential ones – outwardly appearing reasonable perhaps – but inferential nonetheless. It is possible that climate alone brought about the collapse of these societies (see Fig. 7.2), the tendency to involve humans being part of the desire to bolster the Diamond model and other similar models, which – indisputably – have a profound and important message for twenty-first-century humanity.

9.3.2 The "calamitous fourteenth century" in Europe and its aftermath

The well-documented nature and effects of climate change during the transition from the Medieval Warm Period to the Little Ice Age in Europe have led to the characterization of this time as the "calamitous fourteenth century" (Tuchman, 1978). Other book-length treatments to consider this conspicuous period are those of Lamb (1977), Jordan (1996) and Fagan (1999, 2000).

Tuchman's account attributes the societal disruption she describes – the Hundred Years' War, the Black Death, the papal schism and various popular uprisings – ultimately to climate changes across the transition from the Medieval Warm Period to the Little Ice Age. This transition involved a drop in temperature and increased climate variability under which the subsistence and associated systems that had developed during the Medieval Warm Period could no longer sustain Europe's population. Facing starvation, people sought to overthrow the existing order, and millions perished. Tuchman focuses on the Black Death, which in A.D. 1348–1350 killed one-third of all the people living between Iceland and India, most of whom had been made more vulnerable to disease by years of climate-associated famine.

The problems in Europe began during the last part of the Medieval Warm Period, when some 30 years of warm dry, frost-free years had seen an agricultural boom (Berglund, 2003). Then an exceptionally cold winter in A.D. 1309/1310 was followed by the onset of wet conditions in A.D. 1315 that were so intense and so sustained that they destroyed the standing crops, and effectively stopped significant interaction within Europe, bringing even military campaigns to a halt. Continuation of the rains in early A.D. 1316 led to widespread famine in Europe by the end of the year, a situation that prevailed until about A.D. 1322. In Fagan's (2000) words,

> "The settled climate of earlier years gave way to unpredictable, often wild weather, marked by warm and very dry summers in the late 1320s and 1330s and by a notable increase in storminess and wind strengths". (p. 44)

As the Little Ice Age took hold of Europe, the climatic stability that had characterized most of the Medieval Warm Period and had seen many societies burgeon in consequence returned only for short periods. More than 80% of people living in Europe subsisted entirely from the land, so were profoundly affected by the variability of Little Ice Age climate, effects that extended into the cities, and brought about massive economic, social and political disruption.

Among the long-term effects of Little Ice Age climate variability in Europe was the migration of people out of the worst affected areas in search of a better life. Towards the end of the period, the possibility of migration out of Europe altogether presented itself to many people, who moved to North America and the Pacific Basin, among other destinations. Potato blight broke out in North America around A.D. 1843 and found its way back across the Atlantic to Europe, where it wreaked havoc. This became particularly acute in Ireland, where potatoes – the principal staple – failed almost completely in the middle nineteenth century bringing a great famine (*An Ghorta Mór*). But the end of the Little Ice Age in Europe followed and with it the start of less variable climate conditions.

9.3.3 The history of Norse exploration and settlement of the northernmost Atlantic

A good example of societal response to warmer conditions during the Medieval Warm Period is provided by the Norse (Viking) voyages of the period that led settlers from continental Europe to colonize the hitherto uninhabited Faeroe Islands, Iceland and Greenland, and even to reach North America. These voyages began in the middle ninth century, when Vikings in Scotland and Norway set out for the islands to the west (Amorosi et al., 1997). According to oral traditions (sagas) and various palaeoenvironmental indicators, Iceland was reached about A.D. 874 and Greenland was colonized by Eric the Red about A.D. 985. Two settlements were eventually established in Greenland, a western and an eastern one. Leif Eriksson sailed west from Greenland in A.D. 1000 and landed in Vinland, a journey confirmed by the discovery of Viking artefacts and house foundations at L'Anse aux Meadows in Newfoundland (Ingstad and Ingstad, 2000). Journeys between these various outposts of Norse society, particularly between the Greenland settlements and the homelands in the east, continued throughout the Medieval Warm Period but came to an abrupt end by the mid-fifteenth century.

Direct evidence for warm conditions in the Greenland Viking settlements during the Medieval Warm Period comes from many sources, not least the remains of the settlements themselves and evidence of their regular interactions with Europe (Orlove, 2005). Palaeoclimate data have been obtained from coring of sediments in Nansen Fjord that reflect changes in the Greenland Current (Jennings and Weiner, 1996). During the period A.D. 1030–1220, conditions were as warm as they were during the warmest times of the twentieth century. More data confirming the existence of milder conditions in Greenland during the Medieval Warm Period (A.D. 900–1350 here) have been obtained from ice-core analysis, particularly the GISP2 record (Stuiver et al., 1995). The fact of the Vikings making successful long-distance voyages of the kind needed to cross the North Atlantic is commonly ascribed in part to the climate of the Medieval Warm Period, which – much as it did in the Pacific Islands (Section 4.3.7) – entailed clear skies, a low incidence of storms and constant winds (Jennings and Weiner, 1996). Winter storminess in the North Atlantic was at a 2,000-year minimum during the Medieval Warm Period after which time it increased to its present high level (Dawson et al., 2004a, 2004b).

Climate change is also implicated in the progressive abandonment of the Norse settlements in Greenland early in the Little Ice Age. A Greenland ice-sheet palaeotemperature record shows that temperature fell here from a maximum about A.D. 1000 to a minimum about A.D. 1500 (see Fig. 1.5B). Following the end of the Medieval Warm Period in the North Atlantic (about A.D. 1350 according to the GISP2 ice-core record), sailing conditions became more difficult because of the expansion of Arctic sea ice and the increased variability of winds and currents (Jennings and Weiner, 1996; Ogilvie and McGovern, 2000). There is evidence that the Greenland settlements lost contact with those in the east during the transition from the Medieval Warm Period to the Little Ice Age and finally ceased to function around A.D. 1430 (Arneborg et al., 1999).

Historical data (about dates of sea ice in harbours, for example) show that the Viking settlements in Greenland experienced clusters of exceptionally cold winters in A.D. 1308–1318, A.D. 1324–1329, A.D. 1343–1362 and A.D. 1380–1384 (Grove, 2001) which probably led to the deliberate abandonment of the Greenland settlements (Orlove, 2005). Confirmation of very cold winters in A.D. 1352 and A.D. 1355 comes from analysis of the GISP2 ice core (Stuiver et al., 1995). Changes in the nature of the Greenland Current, inferred from sediment analysis in Nansen Fjord, show that a cold period began about A.D. 1270 reaching a temperature minimum ~100 years later (Jennings and Weiner, 1996). Although it is less well documented, it appears that the onset of intensely cold conditions around the start of the Little Ice Age also caused the southward migration of Inuit (Eskimo) people from subarctic areas, resulting in their sustained presence around the Northern Settlement in Greenland at this time.

9.4. GLOBAL CLIMATE TRENDS DURING THE LAST MILLENNIUM

Largely because of the interest in understanding the causes and future of the recent warming trend across the Earth, much attention has been focused on reconstructing past climates, particularly during the last millennium. There have been numerous reviews of global (especially non-Pacific) last-millennium climate change, some of the most comprehensive being Bradley (2000) and various chapters in Alverson et al. (2003).

There is a fundamental disagreement among climate scientists and others regarding the validity and the nature of divisions of last-millennium climate. Many scientists would accept that the Medieval Warm Period and the Little Ice Age were global (or near-global) phenomena – from which it follows that the transition between the two (the A.D. 1300 Event) was also global – but an influential group regards these divisions as invalid and pictures last-millennium temperature change as essentially unvarying (actually slightly falling) prior to a rapid rise around A.D. 1900 (the so-called "hockey-stick" model). The implications of these contrasting interpretations for understanding the cause(s) and therefore future climate changes are important and discussed further in Chapter 10 but for now it is enough to report what these different reconstructions show.

The view of last-millennium climate as marked by several divisions (including the Medieval Warm Period and Little Ice Age) is long standing, and based on a host of inferential data – such as those described in this book – and a considerable number of more formal palaeoclimate records. In fact, up until 2001, this was the consensus view of most climate-interested scientists, as represented by its inclusion in the First Assessment Report of the IPCC (Houghton et al., 1990). This model has been broadly supported by other compilations of last-millennium temperatures. In that by Esper et al. (2002a), for example, both the Medieval Warm Period and the Little Ice Age are recognizable, temperature falling on average by 1.2°C between the two.

The hockey-stick model shows an overall fall in northern-hemisphere temperature of ~0.2°C from A.D. 1000 to the early 1900s (Mann et al., 1999). A contrasting reconstruction, that removed the biological growth function problem from tree-ring data series used in the hockey-stick model, is that of Briffa et al. (2001) in which the Little Ice Age shows up more clearly. Corrections to the dataset of Mann et al. (1999) by McIntyre and McKitrick (2003) allowed a temperature reconstruction from A.D. 1400 in which the Little Ice Age was even more readily recognizable. The implication is that the hockey-stick interpretation of last-millennium climate change is flawed because it does not capture short-term variability (Von Storch et al., 2004).

Yet the hockey-stick reconstruction has received broad endorsement, first from the IPCC when it became a centrepiece of their Third Assessment Report (IPCC, 2001), and second in 2006 from a specially constituted committee of the US National Academy of Sciences that agreed with its essence (Brumfiel, 2006).

There still remain legitimate questions about its global applicability and whether the proxies used are indeed able to identify smaller scale climate shifts such as those that are argued to have caused such profound and enduring environmental and societal changes in the Pacific Basin and elsewhere.

The most promising explanation for the divergence between models of global last-millennium climate that represent temperature changes as a hockey stick and those that in contrast show the Medieval Warm Period and Little Ice Age may be the way in which the former spliced long-term proxy records with short-term instrumental (directly monitored) records (Esper et al., 2005). Once the effects of splicing are removed from the hockey-stick reconstruction, most last-millennium temperature variations exhibit generally increased amplitudes and have roughly the same shape – and indeed show the Medieval Warm Period and Little Ice Age. This approach is more likely to reflect the correct situation for it acknowledges the influence of natural factors in pre-industrial temperature changes. In contrast, an approach that splices proxy and instrumental records reduces amplitudes for the pre-industrial part of the record by involving anthropogenic forcing in it. This is the best argument currently available for favouring a model in which the Medieval Warm Period and Little Ice Age are clearly shown rather than the hockey-stick model.

Finally, it is important to acknowledge that much of the debate about whether or not the Medieval Warm Period and the Little Ice Age were actually global phenomena, or even whether they existed at all, may derive from

- misinterpretations of climate signals from various proxy data sources, perhaps resulting from non-linearities "of biological, chemical and physical transfer functions necessary for temperature reconstruction" (Soon et al., 2003, p. 97);
- the mixing of results from proxy data sources in which the nature of response (particularly lag effects) to climate forcing is not fully understood (such as oxygen isotopes from ocean-surface waters and ice, vegetation change and glacier recession) (Barnett et al., 1999; Evans et al., 2002); and
- the inability of certain types of palaeoclimate indicator to register the effects of the comparatively small temperature change (perhaps 1–2°C averaged across a hemisphere) as may have occurred between the Medieval Warm Period and the Little Ice Age. A similar situation explains the controversy that initially surrounded the identification of the 8,200-BP Event (Morrill et al., 2003), now widely accepted.

With regard to the latter point, only two quantitative palaeoclimate indicators were considered by Broecker (2001) to be sufficiently precise to register centennial-scale temperature changes to within 0.5°C: mountain snowline measurements and borehole thermometry. While this view is regarded as unduly conservative by some (Esper et al., 2002b), it serves to emphasize that sometimes what we are looking for cannot be revealed because of the crudeness of the tools we are using. In such instances, inferential data, such as can be obtained from the study of environmental and societal changes across the transition between the Medieval Warm Period and the Little Ice Age, are valuable and should not be excluded from reconstructions of last-millennium climates simply on account of their largely qualitative nature.

KEY POINTS

1. Outside the Pacific Basin there are many palaeoclimate studies employing a range of approaches that show the existence of the Medieval Warm Period and the Little Ice Age. Rapid cooling occurred in many places during the transition between the Medieval Warm Period and the Little Ice Age (the A.D. 1300 Event).

2. Datasets from several parts of the world outside the Pacific Basin show that sea level was higher than present during the Medieval Warm Period and lower than present during the Little Ice Age, with some showing that the intervening sea-level fall (during the A.D. 1300 Event) was rapid.

3. Key areas of the world outside the Pacific Basin exhibit evidence for societal collapse (or profound change) around the time of the A.D. 1300 Event in the Pacific, suggesting that cooling, sea-level fall and the onset of increased climate variability posed similar challenges for societies worldwide at this time.

CHAPTER 10

Causes of Last-Millennium Climate Change

The fundamental argument of this book is that for the Pacific Basin within the last millennium or so, most of the most profound changes in human societies were ultimately driven by climate change, often through proximate climatic, environmental or societal filters. It is therefore of interest – not only in terms of the past but also the future – to understand what caused the climate changes in question. Whatever the answer to this, it is important that the correctness of the arguments concerning the climate–society relationships given elsewhere in this book should not be judged by them. Questions of ultimate and prox'imate causation are important for quite different reasons from those associated with empirical studies. In addition, the causes of last-millennium climate change are unable to be validated; all scientists can do in this regard is to evaluate probabilities based on a limited series of observations while lacking a complete understanding of the Earth's climate system.

The degree to which external forcing controls Earth climate and the ways in which this happens are subjects of some debate. In reference to the last millennium, much depends on the position that an individual takes regarding the shape and the amplitude of temperature fluctuations (discussed in Section 9.4). Clearly, if you believe – as is implicit in the hockey-stick model – that temperatures fell slowly ~0.2°C in the period A.D. 1000–1900 and that the amplitudes of variations around the mean for this period were small compared to those for the period A.D. 1900 to present during which temperatures have risen sharply, then it seems reasonable to suppose that global external natural forcing mechanisms have been subordinate both to regional/local external natural forcing mechanisms and to anthropogenic forcing mechanisms.

Yet if one disengages instrumental from proxy records, and accepts – as is explicit in this book – that amplitudes of temperature variation during the last millennium were considerable and that the Medieval Warm Period and Little Ice Age were real and global, then there is considerable reason to look to variations in external climatic variables – both global (ultimate) and regional/local (proximate) – to understand how forcing may have occurred. This chapter adopts this approach.

That said, it may come as something of an anticlimax to learn that there is actually widespread agreement among those climate scientists who accept that the Medieval Warm Period and Little Ice Age were global as to the principal cause of last-millennium climate change. While recognizing – more or less – the influence of other contributor causes, numerous studies have concluded that solar forcing is the principal cause of last-millennium climate change (Chambers et al., 1999; Perry and Hsu, 2000; Shindell et al., 2003; Weber et al., 2004). In this context, solar forcing means changes in the amount of solar irradiance (solar radiation) received at the Earth's surface through time, changes that have the potential to "force" climate change.

This chapter first discusses the ultimate causes of last-millennium climate change (Section 10.1), evaluating the likely contribution of various factors both globally and within the Pacific Basin. In Section 10.2, there is a discussion of ultimate causes of sea-level change during the last millennium. In Section 10.3, there is a discussion of the main proximate causes of last-millennium climate and sea-level changes within the Pacific Basin. Section 10.4 looks specifically at the issue of abrupt climate change during the late Quaternary, arguing that the A.D. 1300 Event has numerous earlier analogues. The final Section 10.5 is a last word on this book and how it might be read.

10.1. ULTIMATE CAUSES OF LAST-MILLENNIUM CLIMATE CHANGE

It is widely accepted by climate scientists that long-term climate change is essentially cyclic at a number of levels (Rampino et al., 1987). Long-term climate-change cycles (20,000–100,000 years in duration) are caused by orbital variations but this explanation is inapplicable to shorter term cycles, such as those that appear to characterize the last 2,000 years or so (see Fig. 1.3A, for example). At such timescales, however simple the record may appear, complexities must be expected because of the effects of multiple forcings and the overlaying of cycles with differing wavelengths. It must also be expected that amplitudes of particular cycles will vary through time and that the effects of these variations will show up on shorter timescales. Two ultimate (rather than proximate) causes of last-millennium temperature change are considered here, namely solar forcing and volcanic forcing.

Solar forcing results from variability of solar irradiance which is linked to variations in the outputs of solar energy from the Sun. These variations can be measured by either sunspot observations or cosmogenic isotopes – the commonest methods of reconstructing long-term variability – or by direct measurement from satellites – something that has been carried out for $\sim$25 years. Solar forcing appears to be the only sustained and tenable explanation for observed global temperature variations within the last millennium.

At a millennial timescale, the long-term cycle that is visible in most Holocene records has an $\sim$1,300/1,500-year wavelength, meaning that events like the Little Ice Age recur every 1,300–1,500 years (Stuiver et al., 1995; Bond et al., 1997; Campbell et al., 1998). This cycle has been compellingly linked to the rises and falls of human civilizations (Perry and Hsu, 2000). It has also been suggested to account for the change within the last millennium from the Medieval Warm Period (during which most societies flourished) and the Little Ice Age (during which most societies deteriorated). In this scenario, the Little Ice Age is seen as the last in the line of innumerable cool excursions during the late Quaternary and undoubtedly earlier (Gauthier, 1999). Shorter period interannual cycles are discussed with particular reference to the Pacific Basin in Section 10.3.1.

Yet it is also clear that global temperatures at various times – typically for periods of 2–6 years – have been altered in the aftermath of large volcanic eruptions. This kind of volcanic forcing may be more important regionally or locally but has on occasions undoubtedly also had global effects.

There is speculation that various other processes have had significant effects on last-millennium climate change. These include changes in amounts of atmospheric carbon dioxide, but this is generally regarded as more important on longer timescales. Then there are questions about the long-term variability of the oceans (Broecker et al., 1999) but these have been discussed largely in reference to the Atlantic, not the Pacific, so are not recounted in detail in the present section. The possibility of an oceanic tidal cycle having been responsible for last-millennium climate fluctuations is also recognized (Keeling and Whorf, 2000) although there appears to be little evidence at present to suggest that it was more important than solar forcing.

Then there is the wild card of the El Niño-Southern Oscillation (ENSO) phenomenon, which appears to operate independently of any potential forcing mechanism at the millennium-century scale (Pierrehumbert, 2000; Cobb et al., 2003). In those parts of the world where ENSO effects are especially pronounced – such as the tropical Pacific Ocean – local climate forcing may be almost wholly due to this, accounting for a misfit between climate history in such places and those in other parts of the world.

This section looks first at solar irradiance variations during the last millennium, making the case for these as having been the principal ultimate cause of observed global temperature variations (Section 10.1.1). It then examines volcanic forcing, demonstrating with selected examples how global climate has been temporarily altered by this during the last millennium (Section 10.1.2). Finally, in recent decades, anthropogenic forcing of Earth climate has come to dominate natural forcing mechanisms; this is considered in Section 10.1.3.

10.1.1 Solar irradiance variations

Solar irradiance variation is generally accepted to be the principal cause of last-millennium climate change. Examples of empirical data series agreeing with variations in solar forcing abound, both within the Pacific Basin (Davis, 1994; Hong et al., 2000; Wang et al., 2005; Polissar et al., 2006) and beyond (Black et al., 1999; Yu and Ito, 1999; Verschuren et al., 2000).

Figure 10.1 shows the variations in sunspot number (as a proxy for solar irradiance) during the last millennium, and is labelled with the principal maxima and minima, dates for which are given in Table 10.1. The Medieval Warm Period coincides with a time of comparatively high sunspot activity while the Little Ice Age – except for its last part – was generally a time of comparatively low activity. The A.D. 1300 Event represents the start of the low-activity period marking the Little Ice Age but it is likely that the response of temperature and other variables

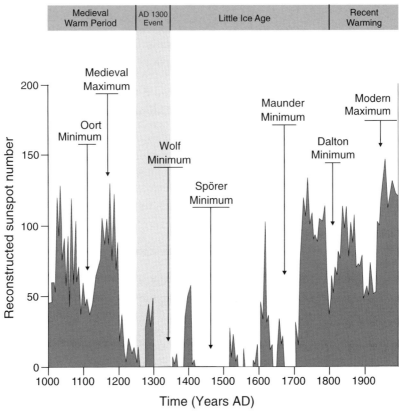

FIGURE 10.1 Reconstructed sunspot numbers for the last millennium showing the dates of the main maxima and minima (after Rigozo et al., 2001). For dates, see Table 10.1.

TABLE 10.1 Periods of distinct sunspot activity within the last millennium defined by reconstructed sunspot numbers (after Rigozo et al., 2001).

Epoch	Time	Duration (years)	Reconstructed sunspot number
Oort	A.D. 1090–1140	50	24.00 ± 20.00
Medieval	A.D. 1140–1200	60	53.00 ± 38.30
Wolf	A.D. 1300–1386	86	0.46 ± 1.87
Spörer	A.D. 1410–1515	105	0.06 ± 0.59
Maunder	A.D. 1641–1715	74	3.56 ± 8.72
Dalton	A.D. 1790–1825	35	26.10 ± 23.40
Modern	A.D. 1900–1999	99	57.54 ± 36.45

(such as sea level) to solar-irradiance change was subject to a lag. A lag of ~100 years sees the A.D. 1300 Event coincide with the rapid fall in sunspot activity – unprecedented within the last millennium – marking the end of the Medieval Maximum.

Within the Little Ice Age there were two notable sunspot minima – the Spörer Minimum (A.D. 1400–1510) and Maunder Minimum (A.D. 1675–1715) – that have been suggested as influencing Earth-surface climate in the Pacific Basin and thereby forcing changes to the societies of the region (Shindell et al., 2001; Schimmelmann et al., 2003). This was certainly the case elsewhere in the world (Bradley and Jones, 1992) and seems a well-established fact for western North America. An example is shown in Fig. 10.2A of temperatures in the Canadian Rockies where the similarities between the shapes of the two curves points to the considerable influence of solar-irradiance changes on temperature change there, particularly during the Spörer and Maunder Minima. Another dataset from the Pacific Rim in Canada that supports a connection between sunspot activity and palaeoclimate comes from Dog Lake, British Columbia, where lake levels increased and fire activity decreased notably during the Maunder Minimum (Hallett et al., 2003). Comparable studies of high-resolution palaeoclimate series come from the tropical Andes (Polissar et al., 2006). Ice-core records from north and south polar regions show that the Little Ice Age was a period of variable climate compared to the Medieval Warm Period and that the coldest time was approximately A.D. 1680–1730 (Kreutz et al., 1997) which corresponds to the Maunder Minimum (A.D. 1675–1715).

Along the western Pacific Rim, the few palaeoclimate series that extend across the last millennium and are of sufficiently high resolution show less clear agreements with sunspot activity. Both palaeotemperature series from Buddha Cave (Fig. 10.2B) have been shifted left (earlier) by 98 years and show a degree of agreement but significantly less than for the eastern Rim. The palaeoprecipitation series from Shihua Cave (Fig. 10.2C) has been shifted left (earlier) by 60 years to improve agreement with solar-irradiance changes.

There is considerable evidence that solar irradiance is an important control on tropical precipitation (Haug et al., 2001), something that shows up with the time series for the western Pacific Rim in Fig. 10.2B and C. Both these palaeoprecipitation series have been shifted left to match as many wiggles as possible. It is likely that these series have also been significantly influenced by monsoonal variations attributable to shorter term fluctuations of Pacific Basin climate such as the Pacific Decadal Oscillation and Interdecadal Pacific Oscillation (PDO/IPO).

10.1.2 Contribution of volcanic forcing to climate change

There are a few comparatively well-documented instances from the last millennium of how large volcanic eruptions have input such large quantities of gas and aerosols into the atmosphere that the amount of solar radiation reaching the entire Earth's surface has been reduced, sometimes for several years.

A good example from a Pacific volcano is the A.D. 1453 eruption of Kuwae Volcano in Vanuatu (Fig. 10.3A) which ejected so much material into the atmosphere that climate was affected for several years in antipodean Europe (Pang, 1993). We are still learning about the effects of this enormous eruption, which expelled $32–39\,km^3$ rock equivalent (Monzier et al., 1994), and was the largest volcanic sulphate eruption within the past 700 years (Gao et al., 2006). A similar situation followed the A.D. 1815 eruption of Mt. Tambora in Indonesia which caused the

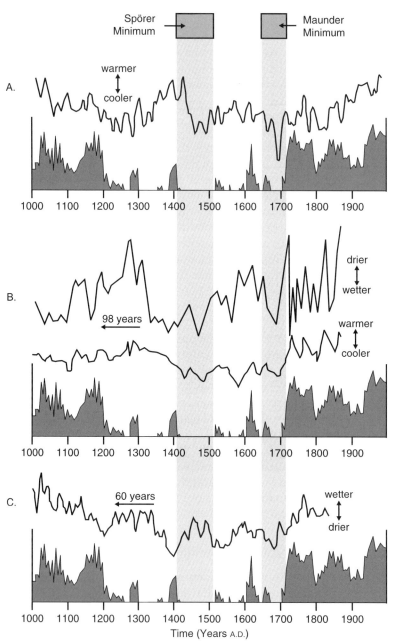

FIGURE 10.2 Possible correlations between sunspot numbers (from Fig. 10.1) with selected Pacific Basin palaeoclimate series (all in this book, details not repeated). Dates for the Spörer and Maunder minima from Table 10.1. (A) Summer temperature reconstruction (RCS2004) for the Canadian Rockies (as shown in Fig. 4.2B). In a general sense there is a good agreement between the timing of sunspot maxima and high temperatures and vice versa. Note that this agreement is not improved by shifting the data series to the right as in (B) and (C).

FIGURE 10.3 Large volcanoes of the Pacific Basin. (A) The shallow-water submarine eruption of Karua Volcano, central Vanuatu, on 22 February 1971, with the southeast tip of Epi Island in the background. Karua rises from the submerged rim of the Kuwae Volcano which, when it erupted in A.D. 1453, proved to be the largest volcanic sulphate event of the past 700 years [photo by Don Mallick, used with permission of Vanuatu Cultural Centre]. (B) Eruption of Mt. Pinatubo, Philippines, in June 1991 [photo by R.S. Culbreth, U.S. Air Force, courtesy of United States Geological Survey].

FIGURE 10.2 *Continued.*

(B) Precipitation (upper) and temperature (lower) from Buddha Cave, eastern China (from Fig. 4.4), both series minus 98 years. Wiggle matching improved the agreement between the timing of sunspot maxima and those of precipitation and temperature. Positive wiggles matched throughout. (C) Precipitation from Shihua Cave (near Beijing), China (from Fig. 5.3B), minus 60 years. Wiggle matching improved the agreement between the timing of sunspot maxima and that of precipitation. Positive sunspot wiggles match with negative precipitation wiggles.

"Year without a Summer", the effects of which were also felt worldwide. These included poor harvests in China in 1816 and 1817, and the same throughout much of Europe and North America (C. Oppenheimer, 2003). The A.D. 1835 eruption of Coseguina Volcano in Nicaragua led to widespread famine in Japan in 1836 after crops failed, and much of North America where 1837 was the coldest year within the past 250 years (Bradley and Jones, 1992). Similar effects were experienced when the Huanyaputina Volcano (Peru) erupted in A.D. 1600 (De Silva and Zielinski, 1998).

Volcanic forcing naturally occurs more frequently at a regional scale and, owing to the concentration of active volcanoes along the Pacific Rim, this is an important factor in short-term (decadal-century) climate change in the Pacific Basin. This was demonstrated by the April 1991 eruption of Mt. Pinatubo (Philippines), shown in Fig. 10.3B. In the first 9 h of the eruption, more than $5 \, km^3$ of ash and other material were ejected into the atmosphere by way of an eruption column 35 km tall. Taking almost 1 year to spread around the entire planet, the presence of Pinatubo ash is believed to have been responsible for a global 18-month temperature fall of $\sim 0.5°C$, something that in the Pacific Basin was less than it might have been otherwise because of the warming influence of the 1992 El Niño event (Soden et al., 2002).

It is also worth noting – in the context of the intimate relationship proposed between temperature and sea level throughout this book – that cooling following large volcanic eruptions can also lead to a temporary sea-level fall. Following the 1991 Pinatubo eruption, global sea level fell by $\sim 5 \, mm$ as a result of the thermal contraction of the upper ocean (Church et al., 2005). The subsequent rate of sea-level rise was faster than it would otherwise have been because of the recovery from Pinatubo cooling.

10.1.3 Contribution of anthropogenic forcing to recent climate change

There is a conspicuous correlation between carbon-dioxide concentrations in the lower atmosphere and global warming over the past 50 years or so. This underpins the broad consensus that accelerated warming since the 1970s in many parts of the world is linked to increased amounts of greenhouse gases, principally carbon dioxide, present in the lower atmosphere (IPCC, 2001). It therefore may come as some surprise to find that carbon-dioxide variations do not seem to so readily explain decadal-century temperature fluctuations earlier during the last millennium and before (Perry and Hsu, 2000; Shindell et al., 2003).

The key issue is whether or not atmospheric carbon dioxide has fluctuated independently of solar irradiance and therefore has independently driven climate change. Most datasets suggest it has not but there are exceptions in various parts of the world. These include Antarctica (Cuffey and Vimeux, 2001) and the tropical Pacific (Lea, 2004). The latter study showed that atmospheric carbon-dioxide abundance best explains tropical climate variations on the orbital (100,000-year) timescale but this throws little light on what is happening here at the millennial timescale.

Unarguably, the increased amounts of greenhouse gases that have been emitted into the Earth's atmosphere since the mid-nineteenth century, primarily as consequences of fossil-fuel combustion and deforestation, have affected temperatures in the Pacific Basin as much as most other places, even though comparatively few of the sharply-increased emissions during the earliest part of the period of Recent Warming originated in the Pacific Basin.

The increase in El Niño frequency in the tropical Pacific beginning ~1976, shown in many coral cores, has been interpreted as an indication of the commencement of anthropogenic forcing on climate variability (Hughes et al., 2003). Yet work on Maiana Atoll in Kiribati showed that this degree of climate variability was not unprecedented within the 155-year palaeoclimate record obtained (shown in Fig. 8.3), although the warming and salinity decrease since 1976 were unprecedented (Urban et al., 2000).

In contrast, there is nothing equivocal about the records of upper-ocean (upper 3,000 m) warming over the past 60 years or so (Levitus et al., 2000). Studies have shown that this warming cannot be satisfactorily explained by internal climate variability, solar or volcanic forcing, and is almost certainly a result of anthropogenic warming (Barnett et al., 2005).

Recent work using time series analysis hints that both solar irradiance and atmospheric carbon dioxide may in fact be forcing global warming along the lines that various studies have shown (Kaufmann et al., 2006). Regression results showed that there is a statistically significant relationship between solar forcing and temperature rise during the period of Recent Warming. Yet these results also suggest that temperature rise has altered the process network by which carbon dioxide moves in and out of the atmosphere in ways that have increased the atmospheric concentration of carbon dioxide. This conclusion could well account for a three-way correlation between increasing solar irradiance, warming and rising atmospheric carbon-dioxide levels.

10.2. ULTIMATE CAUSES OF LAST-MILLENNIUM SEA-LEVEL CHANGE

It seems beyond reasonable doubt that the pattern of last-millennium sea-level changes proposed for the Pacific Basin is similar to regional compilations and well-constrained local-area studies from most parts of the world and therefore requires a global explanation. This is not to discount the effects of local influences, be they tectonic, atmospheric or oceanographic, but only to infer that these are subordinate to a global overprint. It seems likeliest that this global overprint was temperature change, given the manifest similarities between warm conditions and higher sea level during the Medieval Warm Period, cool conditions and lower sea level during the Little Ice Age, and warming conditions and sea-level rise during the period of Recent Warming.

Many writers agree that temperature from solar irradiance (solar forcing) is the principal control of sea-level change at a variety of timescales. For example, there is a broad correlation in the shape of the curves of solar irradiance and sea-level change during the entire Holocene (Fig. 10.4A) that shows the effect of the 1,500/1,300-year solar cycle on sea level, something that also extends to the rises and falls of human civilizations within this period (Perry and Hsu, 2000).

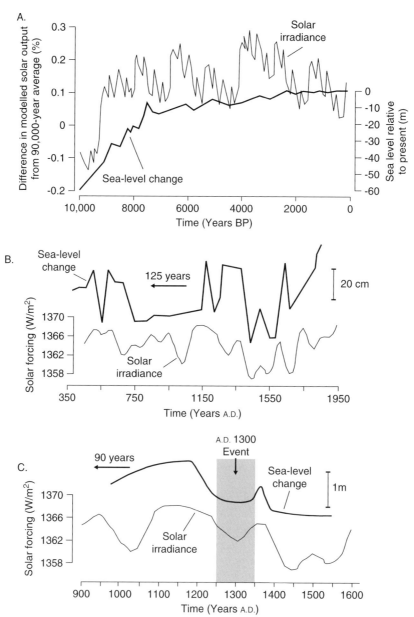

FIGURE 10.4   Links between solar irradiance and sea-level change. (A) Solar output within the past 10,000 years compared to sea-level changes along the French Atlantic coast (after Ters, 1987; Perry and Hsu, 2000). There is broad agreement in the shape of the two curves, suggesting that solar forcing is the principal control of sea-level change at this site over this time period. (B) Solar-irradiance reconstruction based on the Δ^{14}C record (Stuiver et al., 1998) compared to sea-level change at Farm River Marsh, east-coast United States (after Varekamp et al., 1999) for the period A.D. 350–1950. Figure based on that in van de Plassche et al. (1998). Note that the two curves have been wiggle matched, with the sea-level curve being moved

For shorter time periods, an important breakthrough in understanding the links between solar forcing and sea-level change came with the work by van de Plassche et al. (2003) in which wiggle matching was used to optimize the agreement between the shapes of the solar-irradiance curve and high-resolution sea-level curves (Fig. 10.4B). It was found that by shifting the latter 125 years back in time an almost perfect match between the two curves was obtained. This implies that at these particular coastal sites on the eastern seaboard of the United States, sea level took ~125 years to respond to solar forcing.

Although comparable high-resolution series do not exist for the Pacific Basin, a similar exercise was undertaken using the generalized sea-level curve for the middle part of the last millennium (Fig. 10.4C). Wiggle matching (by eye) shows that the best agreement between the shapes of the two curves comes when the sea-level curve is shifted 90 years back in time, suggesting that this is the average response time of Pacific sea level to solar forcing. This is the first time that matching of solar irradiance and Pacific sea level has been attempted, and not too much should be read into it. That said, the agreement in shape around the A.D. 1300 Event is very good.

Sea-level rise during the period of Recent Warming is generally accepted to have been a result of two main processes (Munk, 2002; Casenave and Nerem, 2004). The first is thermal expansion of the upper ocean resulting from global warming. This is referred as a steric change, and is predicted accurately by coupled ocean–atmosphere models. The second is the variation in ocean mass brought about largely by the melting of land-grounded ice. This is referred as a eustatic change. Throughout the period of Recent Warming, sea-level rise has risen as a result of both factors with steric changes coming to dominate ~1950. Yet there is clearly much not well understood about the proximate causes of recent and future sea-level rise for there is a considerable misfit between observations and calculated contributions to sea-level rise.

10.3. PROXIMATE CAUSES OF LAST-MILLENNIUM CLIMATE AND SEA-LEVEL CHANGES IN THE PACIFIC BASIN

Within the Pacific Basin, there are many possible ways in which global climate forcing mechanisms have been modified to produce differences between what happened here at particular times and what happened elsewhere in the world. In general, global temperature change during the last millennium would have altered thermal gradients between the equator and poles in the Pacific Basin. Warmer temperatures would have reduced thermal gradients, lowering the rate

---

FIGURE 10.4 *Continued.*

left (back in time) by 125 years to maximize agreement between them. (C) Solar-irradiance reconstruction based on the $\Delta^{14}C$ record (Stuiver et al., 1998) compared to Pacific Basin composite sea level for the period A.D. 900–1600 (from Fig. 5.8). Note that the two curves have been wiggle matched, with the sea-level curve being moved left (back in time) by 90 years to maximize agreement between them. Positive wiggles are matched with positive.

of latitudinal heat exchange, and shifting climate zones generally polewards; circum-polar vortices would have contracted. Conversely, during cooler times, latitudinal thermal gradients would have steepened as circum-polar vortices expanded, and heat transfer and wind speed generally increased.

All these effects would in turn have been influenced by the morphology of the Pacific Basin, movements of ocean-surface water, and deep ocean heat exchange about which much is still unknown. As a result, while it is possible to identify probable regional climate modifiers of global forcing in the Pacific Basin, it is not yet possible to make a comprehensive list of all these and evaluate their relative importance to last-millennium climate change.

This section provides a brief review of the key proximate causes of last-millennium climate change in the Pacific Basin, beginning with climate modifiers (Section 10.3.1) and then oceanic modifiers (Section 10.3.2). Proximate causes of sea-level change are considered in Section 10.3.3.

10.3.1 Regional climate modifiers of global climate forcing

During the Medieval Warm Period, it is likely that a contracted circum-polar vortex over both the North and South Poles led to a poleward extension of the warmer climate zones. This explains why western Canada and Alaska became unusually wet and California unusually dry during the Medieval Warm Period (Stine, 1998).

The most likely explanation for the increased incidence of droughts during the later Medieval Warm Period along the eastern Pacific Rim is an intensification of the Pacific High. This could have been because of reduced equator–pole temperature gradients associated with warming, an explanation which is also consistent with the suggestion that a reduced frequency of storms entering the northernmost Pacific during the Medieval Warm Period contributed to the aridity of this area at the time (Mann et al., 2002). Along the Pacific coast of South America, the contracted South Polar vortex produced strong westerly air flow over southern Patagonia, causing the Pacific side of the Andes Mountains farther north to be in a "strong" rainshadow and consequently unusually dry (Stine, 1998).

In the Pacific Basin, the climate changes that mark the A.D. 1300 Event may have been driven by changes in atmospheric circulation. Data from ice cores in the southern polar region (Siple Dome, Antarctica) and northern polar region (central Greenland) show that meridional atmospheric circulation intensity increased in the surrounding oceans during the A.D. 1300 Event and start of the Little Ice Age about A.D. 1400 (Kreutz et al., 1997).

The circum-polar vortices that had contracted during the Medieval Warm Period are likely to have begun expanding towards its end, across the A.D. 1300 Event, reaching a new equilibrium position early in the Little Ice Age. This A.D. 1300 Event represents a time of atmospheric reorganization, one manifestation of which is the "aberrant" winter wetness during the A.D. 1300 Event in Alaska and western Canada which may be explained by

"a strong ridge setting up and persisting over the eastern Pacific … [which] … would have deflected Pacific storms to the north of California and Oregon and into Alaska and British Columbia". (Stine, 1998, p. 62)

Along the Pacific coast of South America, a weakening of the "strong" rain-shadow effect is indicated by the rise in lake levels from around A.D. 1130, a result of increased moist winds from the eastern side of the Andes, itself a result of an expanding South Polar vortex (Stine, 1998) consistent with cooling during the A.D. 1300 Event, and the steepening of equator–pole temperature gradients.

In their study of abrupt changes in the Asian summer monsoon during the Holocene, Morrill et al. (2003) found that the precipitation changes associated with the A.D. 1300 Event were similar to those being experienced today as a result of increased winter snowcover in Eurasia. They concluded that it is possible that this factor may also have been important in the change in monsoon strength observed around A.D. 1300 (see also Fig. 5.5).

The Inter-Tropical Convergence Zone (ITCZ) is an important controller of Pacific Basin climate and there is evidence that its position changed during the last millennium. It has been suggested that orbital changes during the late Holocene led to a general southward shift in the ITCZ that may in turn have led to increased ENSO frequency (Haug et al., 2001). This may provide the gross context for last-millennium climate change in the Pacific although there has been modulation by other factors.

Yet the position of the ITCZ, particularly in the eastern Pacific Basin (including the eastern Pacific Rim), is also influenced by ENSO. During El Niño events, the ITCZ fails to move as far north as it does normally, meaning that areas in the northeast Pacific usually experience droughts at such time. In the western Pacific, the northeast shift of the South Pacific Convergence Zone (SPCZ) renders the central part of the western Pacific Rim drier than normal.

The influences of short-term (decadal-century) interannual cycles of climate in modifying the effects of global forcing in the Pacific Basin during the last millennium are also worth considering briefly. In a 1,270-year long palaeoclimate record from China, Paulsen et al. (2003) found a 33-year cycle, which has been recognized in other such datasets from the same region (Bradley, 1999), and which is best explained by large pressure anomalies across East Asia. The PDO/IPO is a comparable phenomenon, similar in manifestation to ENSO, but with positive and negative phases lasting 20–30 years (see Section 8.1.4).

10.3.2 Oceanic influences on climate change

Some of the most important proximate causes of millennial-scale climate change in the Atlantic Ocean have to do with cessations or slowdowns of deepwater formation that have affected surface-water movements and therefore altered the amount of heat being moved around the Earth, especially from the tropics to high latitudes (Broecker, 1997; Broecker et al., 1999). Specifically, these authors suggest that the Little Ice Age was a result of reduced deepwater formation in the North Atlantic (compared to the Southern Ocean) with opposing conditions prevailing during the Medieval Warm Period. Such shifts may have occurred through time, perhaps following the 1,300/1,500-year cycle (see Section 10.4). If this is correct, then empirical support comes from the record of periods of ice rafting in the North Atlantic (Bond et al., 1997).

Deepwater does not form in the North Pacific although during extremely cold periods there has been localized ventilation of the Pacific thermocline in places such as the Santa Barbara Basin (Pisias, 1978; Lehodey et al., 2006). So there is little reason for supposing that a similar oscillation to that which periodically affects the Atlantic Ocean also occurs in the Pacific Ocean.

10.3.3 Proximate causes of last-millennium sea-level change

Within the last decade, satellite altimetric measurements have demonstrated considerable spatial variation of sea-level change (Casenave and Nerem, 2004). For while global mean sea level has been rising at 2.8 ± 0.4 mm/year over the past decade, sea level in the western Pacific has been rising at almost 10 times this rate while sea level in the eastern Pacific exhibits a net fall over this time period. The best solution to this dilemma is that, until we have several decades of satellite altimetry data, we cannot be certain that we have identified a long-term trend, as Casenave and Nerem (2004) concede. It is likely that such spatial variations are due to cyclical causes of variation in sea level, such as those associated with ENSO and the PDO/IPO, and that they balance each other over longer time periods.

Tectonic influences on last-millennium sea-level change in the Pacific Basin are mostly negligible, except locally. A few parts of the Pacific Rim, especially those close to convergent plate boundaries, are rising rapidly, a process typically accomplished through bursts of co-seismic uplift that accompany large-magnitude earthquakes. Such areas include much of the northeast Pacific Rim, from Alaska to California, as well as parts of the Peru–Chile coast. Along the western Rim, parts of the Philippines and Papua New Guinea are rising rapidly, as is some of the east coast of North Island in New Zealand (Nunn, 1994). Some of the most tectonically active island groups are in the southwest Pacific; Solomon Islands, Tonga and Vanuatu are good examples.

10.4. ABRUPT CLIMATE CHANGE

There is evidence from many parts of the world including the Pacific Basin of an enduring 1,300/1,500-year climate cycle (see Section 10.1). It has been suggested that this cycle may be "the pacemaker of rapid climate change" (Bond et al., 1997, p. 1264) inasmuch as the ends/starts of such cycles are sometimes comparatively swift. Such climate changes have been discovered only quite recently for, in earlier times, there was simply not the resolution in palaeoclimate data series to be able to identify these, although their existence had been inferred from catastrophic phenomena such as mass extinctions (Donovan, 1989) and abrupt societal collapse across large regions (Weiss, 2000; Weiss and Bradley, 2001).

Today abrupt climate changes have been recognized throughout the geological record, especially from those times for which most high-resolution data series exist. Most reviews of the causes and consequences of abrupt climate change therefore target the Holocene (Berger and Labeyrie, 1987; National Research Council, 2002; Alley et al., 2003; Thompson et al., 2006). Yet surprisingly few of these acknowledge, even suggest, that there may have been periods of abrupt climate change within the last millennium. Some compilations regard the Little Ice Age as a period of abrupt climate change (Adams et al., 1999) but this book takes the view that it was not the Little Ice Age itself but the transition from the Medieval Warm Period that preceded it which is the abrupt climate change – the A.D. 1300 Event – in question. This section looks first at abrupt climate-change events during the late Quaternary (Section 10.4.1) and compares their nature and effects to those of the A.D. 1300 Event. Section 10.4.2 looks at whether the A.D. 1300 Event signal was amplified in the Pacific Basin and how this might have happened.

10.4.1 Abrupt climate-change events during the late Quaternary

One of the best-studied abrupt climate changes during the late Quaternary was the Younger Dryas, which began with an abrupt cooling ~12,900 BP (Edwards et al., 1993). The Younger Dryas lasted ~1,300 years and saw a change from warm to cold in many parts of the world, a change that brought about profound societal changes in many places. The Younger Dryas is best marked along the periphery of the North Atlantic Ocean, where there was a shutdown in deepwater formation and a corresponding reduction in heat transport between low and high latitudes here. There seems irrefutable evidence that the Younger Dryas cold reversion was experienced in most other parts of the world, including the Pacific Basin (Nunn, 1999). The cold reversions that occurred in the North Atlantic – and possibly throughout the world – during the late Quaternary are termed Dansgaard–Oeschger events, and they represent the cool phase of the 1,300/1,500-year duration climate oscillations described above (Section 10.1).

The 8,200-BP cooling event also appears to have been global in extent, although this was not always certain (Alley et al., 1997; Baldini et al., 2002). Many subsequent cool excursions have had profound effects on human societies, albeit sometimes not globally, but are still regarded as having been global in extent (Perry and Hsu, 2000). The 5,200-year event was widespread through the tropics (Thompson et al., 2006).

The A.D. 536 Event, for which global evidence was assembled by Gunn (2000), is worthy of renewed attention in order to help evaluate the global range and outfall of the A.D. 1300 Event. Although apparently part of a longer term climate deterioration, marking the transition from the Roman Warm Period to the Dark Ages Cold Period, the A.D. 536 Event was a time of rapid cooling that heralded – much as the A.D. 1300 Event did – a period of comparatively variable climate conditions. The societal outfall of the A.D. 536 Event varied but was marked in most parts of the world where its influence has been documented by profound societal change. Examples come from central America where the Maya civilization experienced a widespread cultural hiatus (Robichaux, 2000) and throughout the eastern hemisphere (Africa, China and Europe) where the event was followed by "substantial socio-cultural and economic instability" (Mathis, 2000, p. 111).

The A.D. 1300 Event falls neatly into the 1,500/1,300-year cycle that explains the timing of earlier warm–cool alternations, and its abruptness has been commented on by many authors (see Section 5.1). For some, the Little Ice Age had the most abrupt onset of any of the Holocene rapid climate-change events (O'Brien et al., 1995), a view bolstered by recognition that the A.D. 1300 Event (as defined in this book) marked "the most dramatic change in atmospheric circulation and surface temperature conditions in the last 4000 years" (Kreutz et al., 1997, p. 1294). Records of sea-salt increase in polar ice cores show that the A.D. 1300 Event (called the modern millennial [MM] Event) was "an example of a rapid climate change under near modern boundary conditions" (Meyerson et al., 2003, p. 1). It therefore seems valid to group the A.D. 1300 Event together with earlier periods of analogous rapid cooling, particularly in order to discover its cause(s). The most likely ultimate cause of such oscillations is solar irradiance (see Section 10.1.1) but there may be contributory causes that influenced the abruptness of the cooling at the end of warm events. And by comparison with others, the A.D. 1300 Event does appear uncommonly rapid, so it is legitimate to ask what made it so.

There are many precedents for the transitions between warm and cool climate oscillations having a 1,300/1,500-year recurrence varying in their inferred amplitudes. For example, the 8,200-year event was comparatively rapid and of large amplitude, the latter reflecting "a mechanism that in some way amplified the climate signal at that time" (Bond et al., 1997, p. 1264). Time series like those for temperature and sea level in Fig. 1.3A show that the amplitudes of oscillations in the northwest Pacific varied considerably, while data covering the last two oscillatory cycles like those in Fig. 9.4A show that the amount of cooling around A.D. 1300 was greater than that ~100 B.C. off West Africa.

Certain aspects of the ocean–atmosphere system are known to be more readily tipped from one mode to another than others. Research in the North Atlantic suggests that its thermohaline circulation is very sensitive, and can be shifted to another mode (typically involving a shutdown of deepwater production) by only slight external forcing (such as temperature rise or salinity fall arising from sustained excessive precipitation). Yet it seems unlikely that this could evoke a response throughout the Pacific, which is a net exporter of heat to the Atlantic, unless the effects are registered in deep ocean water transport, about which comparatively little is known. It has been speculated that ENSO may be a driver rather than a consequence of long-term climate oscillations in the Pacific (Pierrehumbert, 2000). Perhaps it is changes in ENSO frequency and intensity that have the ability to force the tropical Pacific into an alternative state where temperature gradients between low and high latitudes are increased compared to today.

Of course, the answer as to why some warm–cool transitions were more abrupt than others may lie simply in the nature of their ultimate forcing – solar-irradiance changes. Using residual ^{14}C as a proxy for solar forcing, Renssen et al. (2000) suggested that the abrupt reduction in ^{14}C at the start of the Younger Dryas signals a prominent role for solar forcing in causing this rapid cooling. There is a similar dip in ^{14}C during the A.D. 1300 Event (see Fig. 10.4B and C); a reduction in sunspot number at roughly the same time can also be seen in Fig. 10.1. These are

independent pieces of evidence for abrupt (not gradual) reductions in solar irradiance across the A.D. 1300 Event – compelling evidence that this was the dominant
reason for its abrupt/rapid nature.

10.4.2 Amplification of the A.D. 1300 Event signal in the Pacific?

It is possible that the A.D. 1300 Event – the period of rapid climate change marking
the transition between the Medieval Warm Period and the Little Ice Age – may
have been amplified in both time and space in the Pacific Basin, making it more
readily identifiable here than elsewhere. The best evidence comes from studies of
sea-level change during the A.D. 1300 Event. In the eastern United States, data such
as those illustrated in Fig. 9.5 from the eastern seaboard show that sea-level fall
here was about half of what it appears to have been in the Pacific (also shown
in Fig. 9.5). So perhaps there was a mechanism operating in the Pacific by which
sea-level fall around A.D. 1300 was amplified relative to the eastern seaboard of the
United States.

In searching for a cause of such amplification, a seemingly analogous situation
from northwest Europe during the late Holocene is worth examining (van Geel
and Renssen, 1998). Here ~800 B.C. (2,650 BP), sea-level fall and water-table
fall exposed large areas of what are now subtidal lowlands in the northern
Netherlands, permitting their initial human settlement. The authors show that this
abrupt sea-level fall was ultimately driven by temperature fall resulting from
reduced solar irradiance. An important factor in this analysis was the role of a
thermohaline circulation in the North Atlantic that was weakened as a result
of cooling and the accompanying change in storm tracks in this sensitive region.
The weakened thermohaline circulation amplified the magnitudes of both the
cooling and sea-level fall.

Possible analogues during the A.D. 1300 Event in the Pacific may have been
associated with El Niño frequency which increased sharply at this time. Just as we
extrapolate temperature extremes associated with El Niño onto future sea-surface
temperature rise (as in Fig. 8.6), it is possible that increased El Niño frequency
during the A.D. 1300 Event was superimposed on the falling sea-level trend,
resulting in prolonged periods of low sea level (especially in the western tropical
Pacific) that foreshadowed for short periods what was to become enduring within
a few decades and made their effects appear more rapid.

10.5. A LAST WORD

This book makes a number of assertions that some readers will inevitably regard
sceptically and critically. Foremost among these assertions is that a period of
rapid cooling and sea-level fall affected the entire Pacific Basin during the A.D.
1300 Event. The evidence presented in favour of this scenario is compelling, and
a scatter of corroborative data from outside the Pacific Basin suggests that it may
have been a global event.

The second – and probably more generally unpalatable – assertion is that these climate and environmental (largely sea-level) changes led to fundamental changes in many of the human societies in the Pacific Basin. The suggestion that external forcing could bring about societal change offends some by its very nature: the implication that modern humankind has not always controlled its own destiny. Many such critics will have been inculcated (as many of their students are still being inculcated) with the orthodoxy that pre-modern societal change came only from within, and that the environment is a mere passive backdrop to the human drama. This view is currently proving untenable in many parts of the world besides the Pacific Basin, but it is unlikely to die with just a whimper. As Fagan (2000) notes, "the ghosts of environmental determinism ... still haunt scholarly opinion" (p. 102).

The societal changes that took place in the Pacific Basin during the A.D. 1300 Event cry out for a regional explanation, as is given here, rather than the local explanations, typically involving unverifiable scenarios of population growth, that are favoured by most archaeologists and other human-focused scientists. Perhaps they should ponder the words of Carl Sagan, one of the greatest scientists of recent times, who wrote in his 1996 book *The Demon-Haunted World*:

> "In those cultures lacking unfamiliar challenges, external or internal, where fundamental change is unneeded, novel ideas need not be encouraged ... But under varied and changing environmental or biological or political circumstances, simply copying the old ways no longer works. Then, a premium awaits those who, instead of blandly following tradition, or trying to foist their preferences onto the physical or social Universe, are open to what the Universe teaches. Each society must decide where in the continuum between openness and rigidity safety lies". (Sagan, 1996, p. 311)

Sagan's words explain why societies responded in the ways that they did to external forcing around A.D. 1300. Fortuitously, his words also explain the imperative for modern scientists to think beyond the confines of cultural determinism in their search for explanations of last-millennium societal change.

KEY POINTS

1. Whether or not the causes of last-millennium climate change are correctly identified does not affect the validity of the empirical-based arguments about climate forcing of societal change in the Pacific Basin elsewhere in this book.
2. Most scientists agree that solar forcing is the ultimate cause of last-millennium climate changes with possible shorter term contributions from volcanic forcing. Solar and volcanic forcings are likewise ultimately responsible for last-millennium sea-level changes.
3. Within the Pacific Basin, global climate forcing was modified by various regional climate filters, particularly the effects of changed equator–pole temperature gradients.
4. The A.D. 1300 Event is one of a number of abrupt climate changes marking the end/start of 1,300/1,500-year duration solar-forced climate cycles.

Abbott, I. (1991). Polynesian uses of seaweed, in Cox, P. and Banack, S. (eds), *Islands, Plants, and Polynesians: An Introduction to Polynesian Ethnobotany*, Dioscorides Press, Portland, pp. 135–145.

Ackerman, R. E. (1998). Early maritime traditions in the Bering, Chukchi and East Siberian Seas. *Arctic Anthropology*, 35, 247–262.

Adams, J. (1995). *Risk*, UCL Press, London.

Adams, J. M., Maslin, M. and Thomas, E. (1999). Sudden climate transitions during the Quaternary. *Progress in Physical Geography*, 23, 1–36.

Akazawa, T., Aoki, K. and Kimura, T. (eds). (1992). *The Evolution and Dispersal of Modern Humans in Asia*. Hokusen-sha, Japan.

Allen, J. (1997). Pre-contact landscape transformation and cultural change in windward O'ahu, in Kirch, P. V. and Hunt, T. L. (eds), *Historical Ecology in the Pacific Islands*, Yale University Press, New Haven, pp. 230–247.

Allen, J. and White, J. P. (1989). The Lapita homeland: Some new data and an interpretation. *Journal of the Polynesian Society*, 98, 129–146.

Allen, M. S. (1992). Temporal variation in Polynesian fishing strategies: the southern Cook Islands in regional perspective. *Asian Perspectives*, 31, 183–204.

Allen, M. S. (1994). The chronology of coastal morphogenesis and human settlement on Aitutaki, Southern Cook Islands, Polynesia. *Radiocarbon*, 36, 59–71.

Allen, M. S. (1997). Coastal morphogenesis, climatic trends, and Cook Islands prehistory, in Kirch, P. V. and Hunt, T. L. (eds), *Historical Ecology in the Pacific Islands*, Yale University Press, New Haven, pp. 124–146.

Allen, M. S. (2002). Resolving long-term change in Polynesian marine fisheries. *Asian Perspectives*, 41, 195–212.

Allen, M. S. and Johnson, K. T. M. (1997). Tracking ancient patterns of interaction: recent geochemical studies in the southern Cook Islands, in Weisler, M. I. (ed), *Prehistoric Long-Distance Interaction in Oceania: an interdisciplinary approach*, New Zealand Archaeological Association, Auckland, pp. 111–133, Monograph 21.

Alley, R. B., Marotzke, J., Nordhaus, W. D., Overpeck, J. T., Peteet, D. M., Pielke, R. A., Pierrehumbert, R. T., Rhines, P. B., Stocker, T. F., Talley, L. D. and Wallace, J. M. (2003). Abrupt climate change. *Science*, 299, 2005–2010.

Alley, R. B., Mayewski, P. A., Sowers, T., Stuiver, M., Taylor, K. C. and Clark, P. U. (1997). Holocene climate instability: A prominent widespread event 8200 yr. ago. *Geology*, 25, 483–486.

Alverson, K. D., Bradley, R. S. and Pedersen, T. F. (eds). (2003). *Paleoclimate, Global Change, and the Future*. Springer, Berlin.

Ambrose, S. H. (1998). Late Pleistocene human population bottlenecks. *Journal of Human Evolution*, 34, 623–651.

Ambrosiani, B. (1984). Settlement expansion–settlement contraction: A question of war, plague or climate?, in Mörner, N.-A. and Karlén, W. (eds), *Climatic Changes on a Yearly to Millennial Basis*, Reidel, Dordrecht, pp. 241–247.

Ames, K. M. (1995). Chiefly power and household production on the northwest coast, in Price, T. D. and Feinman, G. M. (eds), *Foundations of Social Inequality*, Plenum Press, New York, pp. 155–187.

Amesbury, J. R. (1999). Changes in species composition of archaeological marine shell assemblages in Guam. *Micronesica*, 31, 347–366.

Amherst, L., and Thomson, B. (1901). *The Discovery of the Solomon Islands by Alvaro de Mendana in 1568.* Hakluyt Society, London (2 volumes).

Amorosi, T., Buckland, P., Dugmore, A., Ingimundarson, J. H. and McGovern, T. H. (1997). Raiding the landscape: Human impact in the Scandinavian North Atlantic. *Human Ecology*, 25, 491–518.

An, Z. (1991). Radiocarbon dating and the prehistoric archaeology of China. *World Archaeology*, 23, 193–200.

Anderson, A. (1989). *Prodigious Birds*, Cambridge University Press, Cambridge.

Anderson, A. (1991). The chronology of colonisation in New Zealand. *Antiquity*, 65, 767–795.

Anderson, A. (2002). Faunal collapse, landscape change and settlement history in Remote Oceania. *World Archaeology*, 33, 375–390.

Anderson, A. (2003). Initial human dispersal in Remote Oceania: Pattern and explanation, in: C. Sand (Ed), *Pacific Archaeology: Assessments and Prospects (Proceedings of the International Conference for the 50th Anniversary of the First Lapita Excavation, Kone-Nouméa, 2002)*. Nouméa: Services des Musées et du Patrimoine, pp. 71–84.

Andersen, S. T. and Berglund, B. E. (1994). Maps for terrestrial non-tree pollen (NAP) percentages in north and central Europe 1800 and 1450 yr BP. *Paläoklimaforschung*, 12, 119–134.

Anderson, A., Chappell, J., Gagan, M. and Grove, R. (2006). Prehistoric maritime migration in the Pacific islands: An hypothesis of ENSO forcing. *The Holocene*, 16, 1–6.

Anderson, A., Conte, E., Kirch, P. V. and Weisler, M. (2003). Cultural chronology in Mangareva (Gambier Islands), French Polynesia: Evidence from recent radiocarbon dating. *Journal of the Polynesian Society*, 112, 119–140.

Anderson, A. and McFadgen, B. (1990). Prehistoric two-way voyaging between New Zealand and East Polynesia: Mayor Island obsidian on Raoul Island, and possible Raoul Island obsidian in New Zealand. *Archaeology in Oceania*, 25, 24–37.

Anderson, A. and White, J. P. (2001). Approaching the prehistory of Norfolk Island, in Anderson, A. J. and White, J. P. (eds), *The Prehistoric Archaeology of Norfolk Island, Southwest Pacific*, Australian National Museum, Canberra, pp. 1–10.

Anderson, P. M., Bartlein, P. J. and Brubaker, L. B. (1994). Late Quaternary history of tundra vegetation in northwestern Alaska. *Quaternary Research*, 41, 306–315.

Anderson, R. Y. (1992). Long-term changes in the frequency of occurrence of El Niño events, in Diaz, H. F. and Markgraf, V. (eds), *El Niño: Historical and Paleoclimate Aspects of the Southern Oscillation*, Cambridge University Press, New York, pp. 193–200.

Andren, E., Andren, T. and Sohlenius, G. (2000). The Holocene history of the southwestern Baltic Sea as reflected in a sediment core from the Bornholm Basin. *Boreas*, 29, 233–250.

Andres, M. S., Bernasconi, S. M., McKenzie, J. A. and Röhl, U. (2003). Southern Ocean deglacial record supports global Younger Dryas. *Earth and Planetary Science Letters*, 216, 515–524.

Aristarain, A. J., Jouzel, J. and Lorius, C. (1990). A 400 years isotope record of the Antarctic Peninsula climate. *Geophysical Research Letters*, 17, 2369–2372.

Arneborg, J., Heinemeier, J., Lynnerup, N., Nielsen, H. K., Rud, N. and Sveinbjörnsdóttir, A. E. (1999). Change of diet of the Greenland Vikings determined from stable carbon isotope analysis and ^{14}C dating of their bones. *Radiocarbon*, 41, 157–168.

Arnold, J. E. (1992). Complex hunter-gatherer-fishers of prehistoric California: Chiefs, specialists and maritime adaptations of the Channel Islands. *American Antiquity*, 57, 60–84.

Athens, J. S. (1995). *Landscape Archaeology: Prehistoric Settlement, Subsistence, and Environment of Kosrae, Eastern Caroline Islands, Micronesia*, International Archaeological Research Institute, Honolulu.

Athens, J. S. (1997). Hawaiian native lowland vegetation in prehistory, in Kirch, P. V. and Hunt, T. L. (eds), *Historical Ecology in the Pacific Islands*, Yale University Press, New Haven, pp. 248–270.

Athens, J. S. (2001). *Identification of Fishpond Sediments, Loko Pa'au'au, Pearl City Peninsula, O'ahu, Hawai'i*, International Archaeological Research Institute, Honolulu.

Athens, J. S., Tuggle, H. D., Ward, J. V. and Welch, D. J. (2002). Avifaunal extinctions, vegetation change, and Polynesian impacts in prehistoric Hawai'i. *Archaeology in Oceania*, 37, 57–78.

Athens, J. S. and Ward, J. V. (1997). The Maunawili core: Prehistoric inland expansion of settlement and agriculture, O'ahu, Hawai'i. *Hawaiian Archaeology*, 6, 37–51.

Athens, J. S. and Ward, J. V. (1998). *Paleoenvironment and Prehistoric Landscape Change: A Sediment Core Record from Lake Hagoi, Tinian, CNMI*, International Archaeological Research Institute, Honolulu.

Bahn, P. G. and Flenley, J. (1992). *Easter Island, Earth Island*, Thames and Hudson, London.

Baker, B. H. and Wohlenberg, J. (1971). Structure and evolution of the Kenya Rift Valley. *Nature*, 229, 538–542.

Baldini, J. U. L., McDermott, F. and Fairchild, I. J. (2002). Structure of the 8200-year cold event revealed by a speleothem trace element record. *Science*, 296, 2203–2206.

Bamforth, D. B. (1994). Indigenous people, indigenous violence: Pre-contact warfare on the North American Great Plains. *Man*, 29, 95–115.

Bao, Y., Braeuning, A., Johnson, K. R., and Shi, Y. (2002). General characteristics of temperature variation in China during the last two millennia. *Geophysical Research Letters*, doi:10.1029/2001GL014485.

Bar-Yosef, O. (1998). The Natufian culture in the Levant, threshold to the origins of agriculture. *Evolutionary Anthropology*, 6, 159–177.

Barlow, L. K., Sadler, J. P., Ogilvie, A. E. J., Buckland, P. C., Amorosi, T., Ingimundarson, J. H., Skidmore, P., Dugmore, A. J. and McGovern, T. H. (1997). Interdisciplinary investigations of the end of the Norse western settlement in Greenland. *The Holocene*, 7, 489–499.

Barnett, J. and Adger, W. N. (2003). Climate dangers and atoll countries. *Climatic Change*, 61, 321–337.

Barnett, T., Hasselmann, K., Chelliah, M., Delworth, T., Hegerl, G., Jones, P., Rasmusson, E., Roeckner, E., Ropelewski, C., Santer, B. and Tett, S. (1999). Detection and attribution of recent climate change: A status report. *Bulletin of the American Meteorological Society*, 80, 2631–2659.

Barnett, T. P., Pierce, D. W., AchutaRao, K. M., Gleckler, P. J., Santer, B. D., Gregory, J. M. and Washington, W. M. (2005). Penetration of human-induced warming into the world's oceans. *Science*, 309, 284–287.

Baroni, C. and Orombelli, G. (1994a). Abandoned penguin rookeries as Holocene paleoclimatic indicators in Antarctica. *Geology*, 22, 23–26.

Baroni, C. and Orombelli, G. (1994b). Holocene glacier variations in the Terra Nova Bay area (Victoria Land, Antarctica). *Antarctic Science*, 6, 497–505.

Baumgartner, T. R., Soutar, A. and Ferreira-Bartrina, V. (1992). Reconstructions of the history of Pacific sardine and northern anchovy populations over the past two millennia from sediments of the Santa Barbara Basin, California. *California Cooperative Oceanic Fisheries Investigation Report*, 33, 24–40.

Bayliss-Smith, T. (1988). The role of hurricanes in the development of reef islands, Ontong Java atoll, Solomon Islands. *The Geographical Journal*, 154, 377–391.

Beaton, J. M. (1985). Evidence for a coastal occupation time-lag at Princess Charlotte Bay, north Queensland, and the implications for coastal colonisation and population growth theories for Aboriginal Australia. *Archaeology in Oceania*, 20, 1–20.

Bellwood, P. S. (1978). *Archaeological research in the Cook Islands*, BP Bishop Museum, Honolulu, (Pacific Anthropological Records 27).

Bellwood, P. (1979). *Man's Conquest of the Pacific: The Prehistory of Southeast Asia and Oceania*, Oxford University Press, New York.

Beltrami, H. and Mareschal, J.-C. (1992). Ground temperature histories for central and eastern Canada from geothermal measurements: Little Ice Age signature. *Geophysical Research Letters*, 19, 689–692.

Bennett, J. (1974). *Cross-cultural influences on village relocation on the Weather Coast of Guadalcanal, Solomon Islands, c. 1870–1953*. Unpublished MA thesis, University of Hawaii, Hawaii.

Benoist, J. P., Jouzel, J., Lorius, C., Merlivat, L. and Pourchet, M. (1982). Isotope climatic record over the last 2.5 ka from Dome C, Antarctica, ice cores. *Annals of Glaciology*, 3, 17–22.

Bensa, A. and Goromido, A. (1997). The political order and corporal coercion in Kanak societies of the past (New Caledonia). *Oceania*, 68, 84–106.

Benson, L., Kashgarian, M., Rye, R., Lund, S., Paillet, F., Smoot, J., Kester, C., Mensing, S., Meko, D. and Lindstrom, S. (2002). Holocene multidecadal and multicentennial droughts affecting northern California and Nevada. *Quaternary Science Reviews*, 21, 659–682.

Berg, M. L. (1992). Yapese politics, Yapese money, and the Sawei tribute network before World War I. *Journal of Pacific History*, 27, 150–164.

Berger, W. H. and Labeyrie, L. D. (eds). (1987). *Abrupt Climatic Change*. Reidel, Dordrecht.

Berglund, B. E. (2003). Human impact and climate changes – synchronous events and a causal link? *Quaternary International*, 105, 7–12.

Berland, J. C. and Rao, A. (eds). (2004). *Customary Strangers: New Perspectives on Peripatetic Peoples in the Middle East, Africa, and Asia*. Praeger, Westport, CT.

Berstad, I. M., Sejrup, H. P., Klitgaard-Kristensen, D. and Haflidason, H. (2003). Variability in temperature and geometry of the Norwegian Current over the past 600 yr; stable isotope and grain size evidence from the Norwegian margin. *Journal of Quaternary Science*, 18, 591–602.

Best, S. (1998). Tokelau archaeology: A preliminary report of an initial survey and excavations. *Bulletin of the Indo-Pacific Prehistory Association*, 8, 104–118.

Best, S. (2002). *Lapita: a view from the east*, New Zealand Archaeological Association Monograph 24.

Best, S.B. (1984). *Lakeba: the prehistory of a Fijian island*, Unpublished PhD dissertation, University of Auckland, New Zealand.

Best, S. B., Sheppard, P. J., Parker, R. J. and Green, R. C. (1992). Necromancing the stone: Archaeologists and adzes in Samoa. *Journal of the Polynesian Society*, 101, 45–84.

Binford, M. W., Kolata, A. L., Brenner, M., Janusek, J., Seddon, M., Abbott, M. and Curtis, J. (1997). Climate variation and the rise and fall of an Andean civilization. *Quaternary Research*, 47, 235–248.

Birdsell, J. B. (1977). The recalibration of a paradigm for the first peopling of Greater Australia, in Allen, J., Golson, J. and Jones, R. (eds), *Sunda and Sahul*, Academic Press, London, pp. 113–168.

Black, D. E., Peterson, L. C., Overpeck, J. T., Kaplan, A., Evans, M. N. and Kashgarian, M. (1999). Eight centuries of North Atlantic ocean atmosphere variability. *Science*, 286, 1709–1713.

Blanchon, P. and Shaw, J. (1995). Reef drowning during the last deglaciation: Evidence for catastrophic sea-level rise and ice-sheet collapse. *Geology*, 23, 4–8.

Blust, R. A. (2004). *Austronesian Languages*, Cambridge University Press, Cambridge.

Boaz, N. T. and Ciochon, R. L. (2004). *Dragon Bone Hill*, Oxford University Press, Oxford.

Bodri, L. and Cermak, V. (1999). Climate change of the last millennium inferred from borehole temperatures: Regional patterns of climatic changes in the Czech Republic – Part III. *Global and Planetary Change*, 21, 225–235.

Bond, G., Showers, W., Cheseby, M., Lotti, R., Almasi, P., deMenocal, P., Priore, P., Cullen, H., Hajdas, I. and Bonani, G. (1997). A pervasive millennial-scale cycle in North Atlantic Holocene and glacial climates. *Science*, 278, 1257–1266.

Bove, M. C., Elsner, J. B., Landsea, C. W., Niu, X. and O'Brien, J. (1998). Effect of El Niño on U.S. landfalling hurricanes, revisited. *Bulletin of the American Meteorological Society*, 79, 2477–2482.

Bowdler, S. (1976). Hook, line, and dilly bag: An interpretation of an Australian coastal shell midden. *Mankind*, 10, 248–258.

Bowdler, S. (1997). The Pleistocene Pacific, in Denoon, D. (ed), *The Cambridge History of the Pacific Islanders*, Cambridge University Press, Cambridge, pp. 41–50.

Boyd, B. and Torrence, R. (1996). Periodic erosion and human land-use on Garua Island, PNG: A progress report. *Tempus*, 6, 265–274.

Bradley, R. S. (1999). *Paleoclimatology: Reconstructing Climates of the Quaternary*, Academic Press, San Diego.

Bradley, R. S. (2000). 1000 years of climate change. *Science*, 288, 1353–1354.

Bradley, R. S. and Jones, P. D. (eds). (1992). *Climate since A.D. 1500*. Routledge, London.

Brewster, A. B. (1922). *The Hill Tribes of Fiji*, Seeley Service, Philadelphia.

Bridgman, H. A. (1983). Could climatic change have had an influence on the Polynesian migrations? *Palaeogeography, Palaeoclimatology, Palaeoecology*, 41, 193–206.

Briffa, K. R., Osborn, T. J. and Schweingruber, F. H. (2004). Large-scale temperature inferences from tree rings: A review. *Global and Planetary Change*, 40, 11–26.

Briffa, K. R., Osborn, T. J., Schweingruber, F. H., Harris, I. C., Jones, P. D., Shiyatov, S. G. and Vaganov, E. A. (2001). Low-frequency temperature variations from a northern tree ring density network. *Journal of Geophysical Research*, 106(D3), 2929–2941.

Brodie, J. W. (1957). Late Pleistocene beds, Wellington Peninsula. *New Zealand Journal of Science and Technology*, B38, 623–643.

Broecker, W. S. (1997). Thermohaline circulation, the Achilles heel of our climate system: Will man-made CO_2 upset the current balance? *Science*, 278, 1582–1588.

Broecker, W. S. (2001). Was the Medieval Warm Period global? *Science*, 291, 1497–1499.

Broecker, W. S., Sutherland, S. and Peng, T.-H. (1999). A possible 20th-century slowdown of Southern Ocean deep water formation. *Science*, 286, 1132–1135.

Brohan, P., Kennedy, J. J., Haris, I., Tett S. F. B., and Jones, P. D. (2006). Uncertainty estimates in regional and global observed temperature changes: a new dataset from 1850. *Journal of Geophysical Research*, 111, D12106, doi:10.1029/2005JD006548.

Brookfield, H. C. (1989). Frost and drought through time and space, Part III: What were conditions like when the high valleys were first settled? *Mountain Research and Development*, 9, 306–321.

Brooks, S. J. and Birks, H. J. B. (2001). Chironomid-inferred air temperatures from Lateglacial and Holocene sites in north-west Europe: progress and problems. *Quaternary Science Reviews*, 20, 1723–1741.

Brumfiel, G. (2006). Academy affirms hockey-stick graph. *Nature,* published online 28 June 2006, doi:10.1038/4411032a.

Buck, P. (Te Rangi Hiroa). (1949). *The Coming of the Maori.* Whitcombe and Tombes, Christchurch.

Bull, W. B. (1991). *Geomorphic Responses to Climatic Change*, Oxford University Press, New York.

Bunn, A. G., Graumlich, L. J. and Urban, D. L. (2005). Trends in twentieth-century tree growth at high elevations in the Sierra Nevada and White Mountains, USA. *The Holocene*, 15, 481–488.

Burley, D. V. (1998). Tongan archaeology and the Tongan past, 2850–150 BP. *Journal of World Prehistory*, 12, 337–392.

Burley, D. V. (2003). Dynamic landscapes and episodic occupations: archaeological interpretation and implications in the prehistory of the Sigatoka Sand Dunes, in: C. Sand (ed), *Pacific Archaeology: Assessments and Prospects (Proceedings of the International Conference for the 50th Anniversary of the First Lapita Excavation, Kone-Nouméa 2002)*. Nouméa: Services des Musées et du Patrimoine, pp. 307–315.

Burley, D. V., and Clark, J. T. (2003). The archaeology of Fiji/West Polynesia in the post-Lapita era, in: C. Sand (ed), *Pacific Archaeology: Assessments and Prospects (Proceedings of the International Conference for the 50th Anniversary of the First Lapita Excavation, Kone-Nouméa 2002)*. Nouméa: Services des Musées et du Patrimoine, pp. 235–254.

Burley, D. V., Dickinson, W. R., Barton, A. and Shutler, R. Jr. (2001). Lapita on the periphery: New data on old problems in the Kingdom of Tonga. *Archaeology in Oceania*, 36, 89–104.

Butler, V. L. (2001). Changing fish use on Mangaia, southern Cook Islands: Resource depression and the prey choice model. *International Journal of Osteoarchaeology*, 11, 88–100.

Byrami, M., Ogden, J., Horrocks, M., Deng, Y., Shane, P. and Palmer, J. (2002). A palynological study of Polynesian and European effects on vegetation in Coromandel, New Zealand, showing the variability between four records from a single swamp. *Journal of the Royal Society of New Zealand*, 32, 507–531.

Calkin, P. E., Wiles, G. C. and Barclay, D. J. (2001). Holocene coastal glaciation of Alaska. *Quaternary Science Reviews*, 20, 449–461.

Calvin, W. H. (2002). *A Brain for All Seasons: Human Evolution and Abrupt Climate Change*, University of Chicago Press, Chicago.

Campbell, A. R. T. (1985). Social relations in ancient Tongareva. *Pacific Anthropological Records*, 36.

Campbell, C. (2002). Late Holocene lake sedimentology and climate change in southern Alberta, Canada. *Quaternary Research*, 49, 96–101.

Campbell, I. D., Campbell, C., Apps, M. J., Rutter, N. W. and Bush, A. B. G. (1998). Late Holocene approximately 1500 yr climatic periodicities and their implications. *Geology*, 26, 471–473.

Carlson, R. C. (1994). Before Malaspina: The archaeology of northwest coast Indian cultures, in Baquero, M. P. and Acuaviva, N. O. (eds), *Malaspina'92*, Real Academia Hispano-Americana, Cadiz, pp. 34–42.

Carpenter, R. A. and Maragos, J. E. (1989). *How to Assess Environmental Impacts on Tropical Islands and Coastal Areas*, Environment and Policy Institute, East-West Center, Honolulu.

Carson, M. T. (2003). Integrating fragments of a settlement pattern and cultural sequence in Wainiha Valley, Kaua'i, Hawaiian Islands. *People and Culture in Oceania*, 19, 83–105.

Carson, M. T. (2004). Resolving the enigma of early coastal settlement in the Hawaiian Islands: The stratigraphic sequence of the Wainiha Beach site in Kaua'i. *Geoarchaeology*, 19, 98–118.

Carson, M. T. (2006). Chronology in Kaua'i: Colonization, land use, demongraphy. *Journal of the Polynesian Society*, 115, 173–185.

Casenave, A. and Nerem, R. S. (2004). Present-day sea level change: Observations and causes. *Review of Geophysics*, 32, RG3001.

Casty, C., Wanner, H., Luterbacher, J., Esper, J. and Böhm, R. (2005). Temperature and precipitation variability in the European Alps since 1500. *International Journal of Climatology*, 25, 1855–1880.

Catto, N. and Catto, G. (2004). Climate change, communities, and civilizations: driving force, supporting player, or background noise? *Quaternary International*, 123-125, 7–10.

Cawte, J. (1978). Gross stress in small islands: A study in macro-psychiatry, in Laughlin, C. and Brady, I. (eds), *Extinction and Survival in Human Populations*, Columbia University Press, New York, pp. 95–121.

Chagnon, N. (1990). Reproduction and somatic conflicts of interest in the genesis of violence and warfare among tribesmen, in Haas, J. (ed), *The Anthropology of War*, Cambridge University Press, Cambridge, pp. 77–104.

Chambers, F. M., Ogle, M. I. and Blackford, J. J. (1999). Palaeoenvironmental evidence for solar forcing of Holocene climate: Linkages to solar science. *Progress in Physical Geography*, 23, 181–204.

Chan, J. C. L. and Shi, J. E. (2000). Frequency of typhoon landfall over Guangdong province of China during the period 1490–1931. *International Journal of Climatology*, 20, 183–190.

Chappell, J. (1982). Sea levels and sediments: Some features of the context of coastal archaeological sites in the tropics. *Archaeology in Oceania*, 17, 69–78.

Chartkoff, J. L. and Chartkoff, K. K. (1984). *The Archaeology of California*, Stanford University Press, Stanford.

Chase, D. Z. and Chase, A. F. (2006). Framing the Maya Collapse: Continuity, discontinuity, method, and practice in the Classic to Postclassic Southern Maya Lowlands, in Schwartz, G. and Nichols, J. (eds), *After the Collapse: The Regeneration of Complex Societies*, University of Arizona Press, Tucson, pp. 168–187.

Chen, J. (1987). Preliminary research on the mid- and late-Holocene glacial fluctuations in Tianger Peak II regions, Tianshan Mountains. *Bingchuan Dongtu (Journal of Glaciology and Cryopedology)*, 9, 347–356.

Chen, J., Wan, G. and Tang, D. (2000). Recent climate changes recorded by sediment grain sizes and isotopes in Erhai Lake. *Progress in Natural Science*, 10, 54–61.

Chen, Z. Y. and Daniel, J. S. (1998). Sea-level rise on eastern China's Yangtze Delta. *Journal of Coastal Research*, 14, 360–366.

Chen Chung-Yu, J. (2002). Sea nomads in prehistory on the southeast coast of China. *Indo-Pacific Prehistory Association Bulletin*, 22, 51–54.

Chepstow-Lusty, A. and Winfield, M. (2000). Inca agroforestry: Lessons from the past. *Ambio*, 29, 322–328.

Chepstow-Lusty, A., Frogley, M. R., Bauer, B. S. and Bush, M. B. (2003). A late Holocene record of arid events from the Cuzco region, Peru. *Journal of Quaternary Science*, 18, 491–502.

Childe, V. G. (1951). *Man Makes Himself*, Watts, London.

Chu, P.-S. and Clark, J. D. (1999). Decadal variations of tropical cyclone activity over the central North Pacific. *Bulletin of the American Meteorological Society*, 80, 1875–1881.

Chu, G., Liu, J., Sun, Q., Lu, H., Gu, Z., Wang, W. and Liu, T. (2002). The 'Mediaeval Warm Period' drought recorded in Lake Huguangyan, tropical South China. *The Holocene*, 12, 511–516.

Church, J. A., White, N. J. and Arblaster, J. M. (2005). Significant decadal-scale impact of volcanic eruptions in sea level and ocean heat content. *Nature*, 438, 74–77.

Chylek, P., Box, J. E. and Lesins, G. (2004). Global warming and the Greenland ice sheet. *Climatic Change*, 63, 201–221.

Clague, J. J., Wohlfarth, B., Ayotte, J., Eriksson, M., Hutchinson, I., Mathewes, R. W., Walker, I. R. and Walker, L. (2004). Late Holocene environmental change at treeline in the northern Coast Mountains, British Columbia, Canada. *Quaternary Science Reviews*, 23, 2413–2431.

Clark, J. T., Cole, A. O. and Nunn, P. D. (1999). Environmental change and human prehistory on Totoya island, Fiji, in Galipaud, J.-C. and Lilley, I. (eds), *The Pacific from 5000 to 2000 BP: Colonizations and Transformations*, Editions de IRD (Institut de Recherche pour le Développement), Paris, pp. 227–240.

Clark, J. T., Wright, E. and Herdrich, D. J. (1997). Interactions within and beyond the Samoan archipelago: Evidence from basaltic rock geochemistry, in Weisler, M. I. (ed), *Prehistoric Long-Distance Interaction in Oceania: An Interdisciplinary Approach*, New Zealand Archaeological Association, Auckland, pp. 68–84.

Cleghorn, P. (1987). *Prehistoric Cultural Resources and Management Plan for Nihoa and Necker Islands, Hawai'i*, BP Bishop Museum, Honolulu.

Cobb, K. M., Charles, C. D., Cheng, H. and Edwards, R. L. (2003). El Niño/Southern Oscillation and tropical Pacific climate during the last millennium. *Nature*, 424, 271–276.

COC (Consular Outward Correspondence) (1864). *Untitled*. National Archives of Fiji, Suva.

Cochrane, E. E. (2004). Archaeological investigations on Waya Island: The 2001 University of Hawai'i Archaeological Field School. *Domodomo*, 17, 7–13.

Cochrane, E. E., Pietrusewsky, M. and Douglas, M. T. (2004). Culturally modified human remains recovered from an earth-oven interment on Waya Island, Fiji. *Archaeology in Oceania*, 39, 54–59.

Cole, J. E., Dunbar, R. B., McClanahan, T. R. and Muthiga, N. A. (2000). Tropical Pacific forcing of decadal SST variability in the western Indian Ocean over the past two centuries. *Science*, 287, 617–619.

Cole, J. E., Fairbanks, R. G. and Shen, G. T. (1993). Recent variability in the Southern Oscillation: Isotopic results from a Tarawa Atoll coral. *Science*, 260, 1790–1793.

Collins, M. (2000). The El Niño-southern oscillation in the second Hadley Centre coupled model and its response to greenhouse warming. *Journal of Climate*, 13, 1299–1313.

Colquhoun, D. J. and Brooks, M. J. (1986). New evidence from the southeastern US for eustatic components in the late Holocene sea levels. *Geoarchaeology*, 1, 275–291.

Comiso, J. C. (2000). Variability and trends in Antarctic surface temperatures from in situ and satellite infrared measurements. *Journal of Climate*, 13, 1674–1697.

Cook, E., Bird, T., Peterson, M., Barbetti, M., Buckley, B., D'Arrigo, R. and Francey, R. (1992). Climatic change over the last millennium in Tasmania reconstructed from tree-rings. *The Holocene*, 2, 205–217.

Cordy, R. (1996). The rise and fall of the O'ahu Kingdom: a brief overview of O'ahu's history, in Davidson, J. M., Irwin, G., Leach, B. F., Pawley, A. and Brown, D. (eds), *Oceanic Culture History: Essays in Honour of Roger Green*, New Zealand Journal of Archaeology Special Publication, pp. 591–613.

Corrège, T., Quinn, T., Delcroix, T., Le Cornec, F., Récy, J. and Cabioch, G. (2001). Little Ice Age sea surface temperature variability in the southwest tropical Pacific. *Geophysical Research Letters*, 28, 3477–3480.

Couper-Johnston, R. (2000). *El Niño: The Weather Phenomenon that Changed the World*, Hodder and Stoughton, London.

Coyne, M. A., Fletcher, C. H. and Richmond, B. M. (1999). Mapping coastal erosion hazard areas in Hawaii: Observations and errors. *Journal of Coastal Research*, 28, 171–184 Special Issue.

Craib, J.L. (1990). Results and conclusions. In: Craib, J.L. (ed). Archaeological Investigations at Mochong, Rota, Mariana Islands. *Unpublished Report for Historic Preservation Division, Commonwealth of the Northern Mariana Islands, Saipan*, pp. 126–150.

Crealock, W. B. (1955). *Towards Tahiti*, Peter Davies, London.

Cronin, T. M. (1983). Rapid sea level and climate change: Evidence from continental and island margins. *Quaternary Science Reviews*, 1, 177–214.

Cronin, T. M., Dwyer, G. S., Kamiya, T., Schwede, S. and Willard, D. A. (2003). Medieval Warm Period, Little Ice Age and 20th century temperature variability from Chesapeake Bay. *Global and Planetary Change*, 36, 17–29.

Cronin, T. M., Thunell, R., Dwyer, G. S., Saenger, C., Mann, M. E., Vann, C., and Seal II, R. R. (2005). Multiproxy evidence of Holocene climate variability from estuarine sediments, eastern North America. *Paleoceanography*, 20, PA4006, doi:10.1029/2005PA001145.

Crosby, A. (1988). *Beqa: archaeology, structure and history in Fiji*, Unpublished MA thesis, University of Auckland. New Zealand.

Crowley, T. J. (2000). Causes of climate change over the past 1000 years. *Science*, 289, 270–277.

Cuffey, K. M. and Vimeux, F. (2001). Covariation of carbon dioxide and temperature from the Vostok ice core after deuterium-excess correction. *Nature*, 412, 523–527.

Culliney, J. L. (1988). *Islands in a Far Sea: Nature and Man in Hawaii*, Sierra Club Books, San Francisco.

Currey, B. (1980). Famine in the Pacific: losing the chances for change. *GeoJournal*, 4, 447–466.

Curtis, J. H., Hodell, D. A. and Brenner, M. (1996). Climate variability on the Yucatan Peninsula (Mexico) during the past 3500 years, and implications for Maya cultural evolution. *Quaternary Research*, 46, 37–47.

D'Arrigo, R., Mashig, E., Frank, D., Wilson, R. and Jacoby, G. (2005). Temperature variability over the past millennium inferred from northwestern Alaska tree rings. *Climate Dynamics*, 24, 227–236.

D'Arrigo, R. D., Cook, E. R., Salinger, M. J., Palmer, J., Krusic, P. J., Buckley, B. M. and Villalba, R. (1998). Tree-ring records from New Zealand: Long-term context for recent warming trend. *Climate Dynamics*, 14, 191–199.

Dahl-Jensen, D., Mosegaard, K., Gundestrup, N., Clow, G. D., Johnsen, S. J., Hansen, A. W. and Balling, N. (1998). Past temperatures directly from the Greenland Ice Sheet. *Science*, 282, 268–271.

Dalfes, H. N., Kukla, G. and Weiss, H. (eds). (1997). *Third Millennium BC Climate Change and Old World Collapse*. Springer, Berlin.

Dalzell, P. (1998). The role of archaeological and cultural–historical records in long-range coastal fisheries resources management strategies and policies in the Pacific Islands. *Ocean and Coastal Management*, 40, 237–252.

Davidson, J. (1969). Settlement patterns in Samoa before 1840. *Journal of the Polynesian Society*, 78, 44–82.

Davidson, J. (1984). *The Prehistory of New Zealand*, Longman Paul, Auckland.

Davis, O. K. (1992). Rapid climatic change in coastal southern California inferred from pollen analysis of San Joaquin marsh. *Quaternary Research*, 37, 89–100.

Davis, O. K. (1994). The correlation of summer precipitation in the southwestern USA with isotopic records of solar activity during the Medieval Warm Period. *Climatic Change*, 26, 271–287.

Dawson, A., Elliott, L., Noone, S., Hickey, K., Holt, T., Wadhams, P. and Foster, I. (2004a). Historical storminess and climate 'see-saws' in the North Atlantic region. *Marine Geology*, 210, 247–259.

Dawson, S., Smith, D. E., Jordan, J. and Dawson, A. G. (2004b). Late Holocene coastal sand movements in the Outer Hebrides, N.W. Scotland. *Marine Geology*, 210, 281–306.

De Silva, S. L. and Zielinski, G. A. (1998). Global influence of the A.D. 1600 eruption of Huanyaputina, Peru. *Nature*, 393, 455–458.

Dean, J. S. (1994). The Medieval Warm Period on the southern Colorado Plateau. *Climatic Change*, 26, 225–241.

Dega, M. F. and Kirch, P. V. (2002). A modified cultural history of Anahulu Valley, O'ahu, Hawai'i, and its significance for Hawaiian prehistory. *Journal of the Polynesian Society*, 111, 107–126.

Delworth, T. L. and Mann, M. E. (2000). Observed and simulated multidecadal variability in the northern hemisphere. *Climate Dynamics*, 16, 661–676.

deMenocal, P., Ortiz, J., Guilderson, T. and Sarnthein, M. (2000). Coherent high- and low-latitude climate variability during the Holocene warm period. *Science*, 288, 2198–2202.

deMenocal, P. B. (2001). Cultural responses to climate change during the late Holocene. *Science*, 292, 667–673.

Demezhko, D. Yu. and Shchapov, V. A. (2001). 80,000 years ground surface temperature history inferred from the temperature-depth log measured in the superdeep hole SG-4 (the Urals, Russia). *Global and Planetary Change*, 29, 167–178.

Denniston, R. F., Gonzalez, L. A., Reagan, M. K. and Asmerom, Y. (1999). A high-resolution record of Indian summer monsoon variability preserved by speleothem carbonate mineralogy. *Geological Society of America Abstracts with Programs*, 3, 153.

Derrick, R. A. (1953) (for 1940–1944). Fijian warfare. In: *Transactions and Proceedings of the Fiji Society of Science and Industry*, 2, 137–146.

Desprat, S., Goñi, M. F. S. and Loutre, M.-F. (2003). Revealing climatic variability of the last three millennia in northwestern Iberia using pollen influx data. *Earth and Planetary Science Letters*, 213, 63–78.

Di Piazza, A. and Pearthree, E. (2001a). Voyaging and basalt exchange in the Phoenix and Line archipelagoes: The view point from three Mystery Islands. *Archaeology in Oceania*, 36, 146–152.

Di Piazza, A. and Pearthree, E. (2001b). An island for gardens, an island for birds and voyaging: A settlement pattern for Kiritimati and Tabuaeran, two "mystery islands" in the northern Lines, Republic of Kiribati. *Journal of the Polynesian Society*, 110, 149–170.

Di Piazza, A. and Pearthree, E. (2004). *Sailing Routes of Old Polynesia: The Prehistoric Discovery, Settlement and Abandonment of the Phoenix Islands*, B.P. Bishop Museum, Honolulu, (Bulletin in Anthropology II).

Diamond, J. (1999). *Guns, Germs, and Steel: The Fates of Human Societies*, Norton, New York.

Diamond, J. M. (2005). *Collapse: How Societies Choose to Fail or Succeed*, Viking, New York.

Dickinson, W. R. (1999). Holocene sea-level record on Funafuti and potential impact of global warming on central Pacific atolls. *Quaternary Research*, 51, 124–132.

Dickinson, W. R. (2001). Paleoshoreline record of relative Holocene sea levels on Pacific Islands. *Earth-Science Reviews*, 5, 191–234.

Dickinson, W. R. (2003). Impact of mid-Holocene hydro-isostatic highstand in regional sea level on habitability of islands in Pacific Oceania. *Journal of Coastal Research*, 19, 489–502.

Dickinson, W. R. and Green, R. C. (1998). Geoarchaeological context of Holocene subsidence at the Ferry Berth Lapita site, Mulifanua, Upolu, Samoa. *Geoarchaeology*, 13, 239–263.

Dillehay, T. (ed), (1989). *Monte Verde: A Late Pleistocene Settlement in Chile*. Smithsonian Institution, Washington.

Dillehay, T. D. (1997). *Monte Verde: A Late Pleistocene Settlement in Chile*. Volume 2, Smithsonian Institution, Washington.

Dillehay, T. D. (1999). The late Pleistocene cultures of South America. *Evolutionary Anthropology*, 7, 206–216.

Dillehay, T. D. and Kolata, A. L. (2004). Long-term human response to uncertain environmental conditions in the Andes. *Proceedings of the National Academy of Sciences*, 101, 4325–4330.

Dirks, R. (1978). Resource fluctuations and competitive transformations in West Indian slave societies, in Laughlin, C. and Brady, I. (eds), *Extinction and Survival in Human Populations*, Columbia University Press, New York, pp. 122–180.

Dodson, J. R. and Intoh, M. (1999). Prehistory and palaeoecology of Yap, Federated States of Micronesia. *Quaternary International*, 59, 17–26.

Dollar, S. J. and Tribble, G. W. (1993). Recurrent storm disturbance and recovery: A long-term study of coral communities in Hawaii. *Coral Reefs*, 12, 223–233.

Domack, E., Leventer, A., Dunbar, R., Taylor, F., Brachfeld, S. and Sjunneskog, C.ODP Leg 178 Scientific Party (2001). Chronology of the Palmer Deep site, Antarctic Peninsula: A Holocene palaeoenvironmental reference for the circum-Antarctic. *The Holocene*, 11, 1–9.

Domack, E. W., Leventer, A., Root, S., Ring, J., Williams, E., Carlson, D., Hirshorn, E., Wright, W., Gilbert, R. and Burr, G. (2003). Marine sedimentary record of natural environmental variability and recent warming in the Antarctic Peninsula. *Antarctic Research Series*, 79, 205–224.

Donner, S. D., Skirving, W. J., Little, C. M., Oppenheimer, M. and Hoegh-Guldberg, O. (2005). Global assessment of coral bleaching and required rates of adaptation under climate change. *Global Change Biology*, 11, 2251–2265.

Donovan, S. K. (ed), (1989). *Mass Extinctions: Processes and Evidence*. Belhaven, London.

Doose-Rolinski, H., Rogalla, U., Scheeder, G., Luckge, A. and von Rad, U. (2001). High-resolution temperature and evaporation changes during the late Holocene in the northeastern Arabian Sea. *Paleoceanography*, 16, 358–367.

Douglas, B. C. (2001). Sea level change in the era of the recording tide gauge, in Douglas, B. C., Kearney, M. S. and Leatherman, S. P. (eds), *Sea Level Rise: History and Consequences*, Academic Press, San Diego, pp. 37–64.

Druffel, E. R. M. and Griffin, S. (1993). Large variations of surface ocean radiocarbon: Evidence of circulation changes in the southwestern Pacific. *Journal of Geophysical Research*, 98, 20249–20259.

Dunbar, R. B., Wellington, G. M., Colgan, M. W. and Glynn, P. W. (1994). Eastern Pacific sea surface temperature since 1600 A.D.: The $\delta^{18}O$ record of climate variability in Galápagos corals. *Paleoceanography*, 9, 291–315.

Duncan, C. C. and Turcotte, D. L. (1994). On the breakup and coalescence of continents. *Geology*, 22, 103–106.

Durham, W. H. (1976). Resource competition and human aggression, part I. *Quarterly Review of Biology*, 51, 385–415.

Dye, T. and Cleghorn, P. L. (1990). Prehistoric use of the interior of southern Guam. *Micronesica*, 2(Suppl.), 261–274.

Dye, T. S. and Komori, E. (1992). A pre-censal population history of Hawaii. *New Zealand Journal of Archaeology*, 14, 113–128.

Earle, T. (1980). Prehistoric irrigation in the Hawaiian Islands: An evaluation of evolutionary significance. *Archaeology and Physical Anthropology in Oceania*, 15, 1–28.

Eden, D. N. and Page, M. J. (1998). Palaeoclimatic implications of a storm erosion record from late Holocene lake sediments, North Island, New Zealand. *Palaeogeography, Palaeoclimatology, Palaeoecology*, 139, 37–58.

Edwards, R. L., Beck, J. W., Burr, G. S., Donahue, D. J., Chappell, J. M. A., Bloom, A. L., Druffel, E. R. M. and Taylor, F. W. (1993). A large drop in atmospheric ^{14}C/^{12}C and reduced melting in the Younger Dryas, documented with ^{230}Th ages of corals. *Science*, 260, 962–968.

Elias, S. A., Short, S. K. and Birks, H. H. (1997). Late Wisconsin environments of the Bering land bridge. *Palaeogeography, Palaeoclimatology, Palaeoecology*, 136, 293–308.

Elston, R. G., Xu, C., Madsen, D. B., Zhong, K., Bettinger, R. L., Li, J., Brantingham, P. J., Wang, H. and Yu, J. (1997). New dates for the North China Mesolithic. *Antiquity*, 71, 985–993.

Emanuel, K. A. (1987). The dependence of hurricane intensity on climate. *Nature*, 326, 483–485.

Ember, C. R. and Ember, M. (1992). Resource unpredictability, mistrust, and war: A cross-cultural study. *Journal of Conflict Resolution*, 36, 246–262.

Emory, K. P. (1928). *Archaeology of Nihoa and Necker Islands*, BP Bishop Museum, Honolulu.

Empson, L., Flenley, J., and Sheppard, P. (2000). Mayor Island – a strategic site of late Maori occupation, in: M. Roche, M. McKenna, & P. Hesp (eds), *Proceedings of the Twentieth Conference of the New Zealand Geographical Society*, Palmerston North, New Zealand, July 1999. Auckland, New Zealand: Geographical Society, pp. 33–34.

Epstein, P. R., Diaz, H. F., Elias, S., Grabherr, G., Graham, N. E., Martens, W. J. M., Mosley-Thompson, E. and Susskind, J. (1998). Biological and physical signs of climate change: Focus on mosquito-borne diseases. *Bulletin of the American Meteorological Society*, 79, 409–418.

Erickson, C. (1999). Neo-environmental determinism and agrarian 'collapse' in Andean prehistory. *Antiquity*, 73, 634–642.

Esaka, T. (1943). Minami Kanto Shinsekki jidai kaizuka yori kantaru Chusekisei ni okeru kaishin kaitai (Holocene sea transgressions and regressions as seen in the Neolithic shellmounds of South Kanto). *Kodai Bunka*, 14(4) (in Japanese).

Esaka, T. (1954). Kagansen no shintai kara mita Nihon no Shinsekki jidai (The Japanese Neolithic as seen from the changing coastline). *Kagaku Asahi*, 14(3) (in Japanese).

Esaka, T. (1967). Seikatsu butai (Context of livelihood), in Kamaki, Y. (ed), *Nihon no Kokogaku, II, Jomon Jidai (Japanese Archaeology, II, The Jomon Period)*, Kawade Shobo, Tokyo, pp. 399–415 (in Japanese).

Esper, J., Schweingruber, F. H. and Fritiz, H. (2002a). Low-frequency signals in long tree-ring chronologies for reconstructing past temperature variability. *Science*, 295, 2250–2254.

Esper, J., Schweingruber, F. H. and Winiger, M. (2002b). 1300 years of climatic history for Western Central Asia inferred from tree-rings. *The Holocene*, 12, 267–277.

Esper, J., Wilson, R. J. S., Frank, D. C., Moberg, A., Wanner, H. and Luterbacher, J. (2005). Climate: Past ranges and future changes. *Quaternary Science Reviews*, 24, 2164–2166.

EU (European Union) (2004). *Living with Coastal Erosion in Europe: Sediment and Space for Sustainability*, EU, Brussels.

Evans, M. N., Kaplan, A., and Cane, M. A. (2002). Pacific sea surface temperature field reconstruction from coral δ^{18}O data using reduced space objective analysis. *Paleoceanography*, 17, 1007, doi:10.1029/2000PA000590.

Fagan, B. (1999). *Floods, Famines, and Emperors: El Niño and the Fate of Civilizations*, Basic Books, New York.

Fagan, B. (2000). *The Little Ice Age: How Climate Made History, 1300–1850*, Basic Books, New York.

Fagan, B. (2004). *The Long Summer: How Climate Changed Civilization*, Basic Books, New York.

Fairbridge, R. W. (1964). Eustatic changes in sea level, in Ahrens, L. H., Press, F., Rankama, K. and Runcorn, S. K. (eds), Physics and Chemistry of the Earth. Pergamon Press, New York, Vol. 4, pp. 99–185.

Fairbridge, R. W. (1992). Holocene marine coastal evolution of the United States, in Fletcher, C. H. and Wehmiller, J. F. (eds), *Quaternary Coasts of the United States: Marine and Lacustrine Systems*, Society for Sedimentary Geology, Tulsa, pp. 9–20.

Fang, J.-Q. (1992). Establishment of a data bank from records of climatic disasters and anomalies in ancient Chinese documents. *International Journal of Climatology*, 12, 499–519.

Fang, J.-Q. (1993). Lake evolution during the last 3000 years in China and its implications for environmental change. *Quaternary Research*, 39, 175–185.

Feng, Z., Thompson, L. G., Mosley-Thompson, E. and Yao, T. (1993). Temporal and spatial variations of climate in China during the last 10 000 years. *The Holocene*, 3, 174–180.

Ferguson, R. B. (1998). Violence and war in prehistory, in Martin, D. and Frayer, D. (eds), *Troubled Times: Violence and Warfare in the Past*, Gordon and Breach, Langhorne, pp. 321–355.

Field, J. S. (2004). Environmental and climatic considerations: A hypothesis for conflict and the emergence of social complexity in Fijian prehistory. *Journal of Anthropological Archaeology*, 23, 79–99.

Filippi, M. L., Lambert, P., Hunziker, J., Kubler, B. and Bernasconi, S. (1999). Climatic and anthropogenic influence on the stable isotope record from bulk carbonates and ostracodes in Lake Neuchatel, Switzerland, during the last two millennia. *Journal of Paleolimnology*, 21, 19–34.

Finney, B. (1976). *Pacific Navigation and Voyaging*, Polynesian Society, Wellington.

Finney, B. (1985). Anomalous westerlies, El Niño, and the colonization of Polynesia. *American Anthropologist*, 87, 9–26.

Finney, B. P., Gregory-Eaves, I., Douglas, M. S. V. and Smol, J. P. (2002). Fisheries production in the northeastern Pacific Ocean over the past 2,200 years. *Nature*, 416, 729–733.

Firth, R. (1959). *Social Change in Tikopia*, George, Allen and Unwin, London.

Fitzhugh, W. W. (1997). Biogeographical archaeology in the eastern North American Arctic. *Human Ecology*, 25, 385–418.

Fitzhugh, B. (2002). Residential and logistical strategies in the evolution of complex hunter-gatherers on the Kodiak Archipelago, in Fitzhugh, B. and Habu, J. (eds), *Beyond Foraging and Collecting: Evolutionary Change in Hunter-Gatherer Settlement Systems*, Kluwer, New York, pp. 157–304.

Fjellsa, A. and Nordberg, K. (1996). Toxic dinoflagellate "blooms" in the Kattegat, North Sea, during the Holocene. *Palaeogeography, Palaeoclimatology, Palaeoecology*, 124, 87–105.

Flannery, T. (1994). *The Future Eaters*, Reed Books, Port Melbourne.

Flenley, J. and Bahn, P. (2003). *The Enigmas of Easter Island: Island on the Edge*, Oxford University Press, New York.

Flenley, J., Parkes, A. and Teller, J. T. (1991). A 500-year climate change record from Tahiti, in Hay, J. E. (ed), *South Pacific Environments: Interactions with Weather and Climate*, Auckland University, Auckland, p. 31. Environmental Science Occasional Publication 6.

Folland, C. K., Renwick, J. A., Salinger, M. J., and Mullan, A. B. (2002). Relative influences of the Interdecadal Pacific Oscillation and ENSO on the South Pacific Convergence Zone. *Geophysical Research Letters*, 29, 1643, doi:10.1029/2001GL014201.

Fornander, A. (1969). *An Account of the Polynesian Race, Its Origins and Migrations*. Volume 2, Tuttle, Tokyo.

Frakes, L. A., Francis, J. E. and Syktus, J. I. (1992). *Climate Modes of the Phanerozoic*, Cambridge University Press, Cambridge.

Fricke, H. C., O'Neil, J. R. and Lynnerup, N. (1995). Oxygen isotope composition of human tooth enamel from medieval Greenland; linking climate and society. *Geology*, 23, 869–872.

Froede, C. R. Jr. (2002). Rhizolith evidence in support of a late Holocene sea-level highstand at least 0.5 m higher than present at Key Biscayne, Florida. *Geology*, 30, 203–206.

Frost, E. (1979). Fiji, in Jennings, J. (ed), *The Prehistory of Polynesia*, Harvard University Press, Cambridge, pp. 69–81.

Fukui, E. (1977). *The Climate of Japan*, Elsevier, Amsterdam.

Gao, C., Robock, A., Self, S., Witter, J., Steffenson, J. P., Clausen, H. B., Siggaard-Andersen, M.-L., Johnsen, S., Mayewski, P. A., and Ammann, C. (2006). The 1452 or 1453 A.D. Kuwae eruption signal derived from multiple ice core records: Greatest volcanic sulfate event of the past 700 years. *Journal of Geophysical Research*, doi:10.1029/2005JD006710.

Gauthier, J. H. (1999). Unified structure in Quaternary climate. *Geophysical Research Letters*, 26, 763–766.

Ge, Q., Zheng, J., Fang, X., Man, Z., Zhang, X., Zhang, P. and Wang, W.-C. (2003). Winter half-year temperature reconstruction for the middle and lower reaches of the Yellow River and Yangtze River, China, during the past 2000 years. *The Holocene*, 13, 933–940.

Gehrels, W. R., Belknap, D. F., Black, S. and Newnham, R. W. (2002). Rapid sea-level rise in the Gulf of Maine, USA, since A.D. 1800. *The Holocene*, 12, 383–389.

Gehrels, W. R., Kirby, J. R., Prokoph, A., Newnham, R. W., Acterberg, E. P., Evans, H., Black, S. and Scott, D. B. (2005). Onset of recent rapid sea-level rise in the western Atlantic Ocean. *Quaternary Science Reviews*, 24, 2083–2100.

Gibb, J. (1986). A New Zealand regional Holocene eustatic sea-level curve and its application to determination of vertical tectonic movements. *Royal Society of New Zealand, Bulletin*, 24, 377–395.

Gifford, E. W. (1951). *Archaeological Excavations in Fiji*. University of California Anthropological Records 13.

Gill, R. B. (2000). *The Great Maya Droughts: Water, Life and Death*, University of New Mexico Press, Albuquerque.

Gilman, E., Ellison, J. C., Jungblut, V., Van Lavieren, H., Wilson, L., Areki, F., Brighouse, G., Bungitak, J., Dus, E., Henry, M., Kilman, M., Matthews, E., Sauni, I. Jr., Teariki-Ruatu, N., Tukia, S. and Yuknavage, K. (2006). Adapting to Pacific Island mangrove responses to sea level rise and climate change. *Climate Research*, 32, 161–176.

Glasser, N. F., Hambrey, M. J. and Aniya, M. (2002). An advance of Soler Glacier, North Patagonian Icefield, at c. A.D. 1222–1342. *The Holocene*, 12, 113–120.

Goff, J. R. and McFadgen, B. G. (2003). Large earthquakes and the abandonment of prehistoric coastal settlements in 15th century New Zealand. *Geoarchaeology*, 18, 609–623.

Goland, C. (1991). The ecological context of hunter-gatherer storage: Environmental predictability and environmental risk, in Miracle, P. T., Fisher, L. E. and Brown, J. (eds), *Foragers in Context: Long-Term, Regional, and Historical Perspectives in Hunter-Gatherer Studies*, University of Michigan, Ann Arbor, pp. 107–125 (Michigan Discussions in Anthropology 10).

Golson, J. (1982). Kuk and the history of agriculture in the New Guinea highlands, in May, R. J. and Nelson, H. (eds), *Melanesia: Beyond Diversity*, Australian National University Press, Canberra, pp. 297–307.

Goodwin, I. D. and Grossman, E. E. (2003). Middle to late Holocene coastal evolution along the south coast of Upolu Island, Samoa. *Marine Geology*, 202, 1–16.

Goodwin, I. D., van Ommen, T. D., Curran, M. A. J. and Mayewski, P. A. (2004). Mid latitude winter climate variability in the South Indian and southwest Pacific regions since 1300 AD. *Climate Dynamics*, 22, 783–794.

Graf, K. (1981). Palynological investigation of two post-glacial peat bogs near the boundary of Bolivia and Peru. *Journal of Biogeography*, 8, 353–368.

Graham, N. E. (2004). Late-Holocene teleconnections between tropical Pacific climatic variability and precipitation in the western USA: Evidence from proxy records. *The Holocene*, 14, 436–447.

Grant, P. J. (1981). Major periods of erosion and sedimentation in the North Island, New Zealand, since the 13th century, in Davies, R. H. and Pearce, A. J. (eds), *Erosion and Sediment Transport in Pacific Rim Steeplands*, International Association of Hydrological Sciences, Christchurch, pp. 288–304 (Publication 132).

Grant, P. J. (1994). Late Holocene histories of climate, geomorphology and vegetation, and their effects on the first New Zealanders, in Sutton, D. (ed), *The Origins of the First New Zealanders*, Auckland University Press, Auckland, pp. 164–194.

Grant-Taylor, T. L. and Rafter, T. A. (1971). New Zealand radiocarbon measurements 6. *New Zealand Journal of Geology and Geophysics*, 14, 364–402.

Graumlich, L. J. (1993). A 1000-year record of temperature and precipitation in the Sierra Nevada. *Quaternary Research*, 39, 249–255.

Graves, M., Hunt, T. and Moore, D. (1990). Ceramic production in the Marianas Islands: Explaining change and diversity in prehistoric interaction and exchange. *Asian Perspectives*, 29, 211–233.

Green, R. C. (2002). A retrospective view of settlement pattern studies in Samoa, in Ladefoged, T. N. and Graves, M. W. (eds), *Pacific Landscapes: Archaeological Approaches*, Easter Island Books, Los Osos, pp. 125–152.

Green, R. C. (2003). The Lapita horizon and traditions – signature for one set of oceanic migrations, in Sand, C. (ed), *Pacific Archaeology: Assessments and Prospects (Proceedings of the International Conference for the 50th Anniversary of the First Lapita Excavation, Kone-Nouméa 2002)*, Services des Musées et du Patrimoine, Nouméa, New Caledonia, pp. 95–120.

Green, R. C. and Weisler, M. I. (2002). The Mangarevan sequence and dating of the geographic expansion into southeast Polynesia. *Asian Perspectives*, 41, 213–241.

Green, R. C. and Weisler, M. I. (2004). Prehistoric introduction and extinction of animals in Mangareva, southeast Polynesia. *Archaeology in Oceania*, 39, 34–41.

Griffies, S. M. and Bryan, K. (1997). Predictability of North Atlantic multidecadal climate variability. *Science*, 275, 181–184.

Griggs, G. B. and Patsch, K. (2004). Cliff erosion and bluff retreat along the California coast. *Sea Technology*, 45, 36–40.

Grossman, E., Fletcher, C. and Richmond, B. (1998). The Holocene sea-level highstand in the Equatorial Pacific: Analysis of the insular paleosea-level database. *Coral Reefs*, 17, 309–327.

Groube, R. C. (1971). The origin and development of earthwork fortifications in the Pacific. *Studies in Oceanic Culture History, Pacific Anthropological Association Records*, 1, 133–164.

Grove, J. M. (1988). *The Little Ice Age*, Methuen, London.

Grove, J. M. (2001). The initiation of the "Little Ice Age" in regions around the North Atlantic. *Climatic Change*, 48, 53–82.

Grove, J. M. and Switsur, R. (1994). Glacial geological evidence for the Medieval Warm Period. *Climatic Change*, 26, 143–169.

Guilcher, A. (1973). Lord Howe, l'île à récifs coralliens la plus méridionale du monde (Mer de Tasman, 31°30′S, 158°E). *Bulletin, Association des Géographes Français*, 405, 427–437.

Gunn, J. D. (2000). *The Years without Summer: Tracing AD 536 and its Aftermath*, Archaeopress, Oxford, (British Archaeological Series 872).

Gupta, A. K., Anderson, D. M. and Overpeck, J. T. (2003). Abrupt changes in the Asian Southwest Monsoon during the Holocene and their links to the North Atlantic Ocean. *Nature*, 421, 354–357.

Gutierrez-Elorza, M. and Pena-Monne, J. L. (1998). Geomorphology and late Holocene climatic change in northeastern Spain, in Lavee, L. H. and Yair, A. (eds), *Geomorphic Response of Mediterranean and Arid Areas to Climate Change*, Elsevier, Amsterdam, pp. 205–217.

Haberle, S. G. and Chepstow-Lusty, A. (2000). Can climate influence cultural development? A view through time. *Environment and History*, 6, 349–369.

Haberle, S. G. and David, B. (2003). Climates of change: Human dimensions of Holocene environmental change in low latitudes of the PEPII transect. *Quaternary International*, 118–119, 165–179.

Haberle, S. G. and Ledru, M.-P. (2001). Correlations among charcoal records of fires from the past 16,000 years in Indonesia, Papua New Guinea, and Central and South America. *Quaternary Research*, 55, 97–104.

Hallam, A. (1984). Pre-Quaternary sea-level changes. *Annual Review of Earth and Planetary Science*, 12, 205–243.

Hallett, D. J., Mathewes, R. W. and Walker, R. C. (2003). A 1000-year record of forest fire, drought and lake-level change in southeastern British Columbia, Canada. *The Holocene*, 13, 751–761.

Hallett, D. J. and Walker, R. C. (2000). Paleoecology and its application to fire and vegetation management in Kootenay National Park, British Columbia. *Journal of Paleolimnology*, 24, 401–414.

Hamilton, S. and Shennan, I. (2005). Late Holocene relative sea-level changes and the earthquake deformation cycle around upper Cook Inlet, Alaska. *Quaternary Science Reviews*, 24, 1479–1498.

Han, M., Hou, J. and Wu, L. (1995). Potential impacts of sea-level rise on China's coastal environment and cities: a national assessment. *Journal of Coastal Research*, 14, 79–95 Special Issue.

Handy, E. S. C. (1930). Marquesan Legends. *B.P. Bishop Museum, Honolulu, Bulletin* 69.

Hannah, J. (2004). An updated analysis of long-term sea level change in New Zealand. *Geophysical Research Letters*, 31, L03307, doi:10.1029/2003GL019166.

Hansen, B. C. S., Seltzer, G. O. and Wright, H. E. (1994). Late Quaternary vegetational change in the central Peruvian Andes. *Palaeogeography, Palaeoclimatology, Palaeoecology*, 109, 263–285.

Haq, B. U., Hardenbol, J. and Vail, P. R. (1987). Chronology of fluctuating sea levels since the Triassic. *Science*, 235, 1156–1167.

Harris, D. R. (ed), (1996). *The Origins and Spread of Agriculture and Pastoralism in Eurasia*. University College London Press, London.

Hashiguchi, N. (1994). The Izu islands: their role in the historical development of ancient Japan. *Asian Perspectives*, 33, 121–149.

Haug, G. H., Hughen, K. A., Sigman, D. M., Peterson, L. C. and Röhl, U. (2001). Southward migration of the intertropical convergence zone through the Holocene. *Science*, 293, 1304–1308.

Haug, G. H., Günther, D., Peterson, L. C., Sigman, D. M., Hughen, K. A. and Aeschlimann, B. (2003). Climate and the collapse of Maya civilization. *Science*, 299, 1731–1735.

Haynes, C. V. (1980). The Clovis culture. *Canadian Journal of Anthropology*, 1, 115–121.

Hendy, E. J., Gagan, M. K., Alibert, C. A., McCulloch, M. T., Lough, J. M. and Isdale, P. J. (2002). Abrupt decrease in tropical Pacific sea surface salinity at the end of Little Ice Age. *Science*, 295, 1511–1514.

Henry, D. O. (1995). *Prehistoric Cultural Ecology and Evolution: Insights from Southern Jordan*, Plenum Press, New York.

Henry, T. (1951). *Tahiti aux Temps Anciens*, Société des Océanistes, Paris.

Herdrich, D. J. and Clark, J. T. (1993). Samoan Tia 'Ave and social structure: methodological and theoretical considerations, in Graves, M. W. and Green, R. C. (eds), *The Evolution and Organization of Prehistoric Society in Polynesia*, New Zealand Archaeological Association (Monograph 19), Auckland, pp. 52–63.

Heyerdahl, T. and Ferndon, E. N. (eds). (1961). *Reports of the Norwegian Archaeological Expedition to Easter Island and the East Pacific. Volume 1. Archaeology of Easter Island*. Forum Publishing House, Stockholm.

Higham, C. F. W. and Lu, T. L.-D. (1998). The origins and dispersal of rice cultivation. *Antiquity*, 72, 867–877.

Hiller, A., Boettger, T. and Kremenetski, C. (2001). Medieval climatic warming recorded by radio-carbon dated alpine tree-line shift on the Kola Peninsula, Russia. *The Holocene*, 11, 491–497.

Hiscock, P. and Kershaw, A. P. (1992). Paleoenvironments and prehistory of Australia's tropical Top End, in Dodson, J. (ed), *The Naive Lands: Prehistory and Environmental Change in Australia and the South-West Pacific*, Longman Cheshire, Melbourne, pp. 43–75.

Hodell, D. A., Brenner, M. and Curtis, J. H. (2005). Terminal Classic droughts in the northern Maya lowlands inferred from multiple sediment cores in Lake Chichacanab (Mexico). *Quaternary Science Reviews*, 24, 1413–1427.

Hodgkins, G. A., Dudley, R. W. and Huntington, T. G. (2005). Changes in the number and timing of days of ice-affected flow on northern New England rivers, 1930–2000. *Climatic Change*, 71, 319–340.

Hoegh-Guldberg, O. (1999a). *Climate Change, Coral Bleaching, and the Future of the World's Coral Reefs*, Greenpeace, Amsterdam.

Hoegh-Guldberg, O. (1999b). Coral bleaching, climate change and the future of the world's coral reefs. *Review of Marine and Freshwater Research*, 50, 839–866.

Hoegh-Guldberg, O., Hoegh-Guldberg, H., Stout, D. K., Cesar, H. and Timmerman, A. (2000). *Pacific in Peril: Biological, Economic and Social Impacts of Climate Change on Pacific Coral Reefs*, Greenpeace, Amsterdam.

Hogg, A. G., Higham, T. F. G., Lowe, D. J., Palmer, J., Reimer, P. and Newnham, R. M. (2003). A wiggle-match date for Polynesian settlement of New Zealand. *Antiquity*, 77, 116–125.

Holdaway, S. J., Fanning, P. C., Jones, M., Siner, J., Witter, D. C. and Nicholls, G. (2002). Variability in the chronology of late Holocene aboriginal occupation on the arid margin of southeastern Australia. *Journal of Archaeological Science*, 29, 351–363.

Holdsworth, G., Krouse, H. R. and Nosal, M. (1992). Ice core climate signals from Mount Logan, Yukon A.D. 1700–1897, in Bradley, R. S. and Jones, P. D. (eds), *Climate since A.D. 1500*, Routledge, New York, pp. 483–504.

Holmgren, K., Lee-Thorp, J. A., Cooper, G. R. J., Lundblad, K., Partridge, T. C., Scott, L., Sithaldeen, R., Talma, A. S. and Tyson, P. D. (2003). Persistent millennial-scale climatic variability over the past 25,000 years in Southern Africa. *Quaternary Science Reviews*, 22, 2311–2326.

Holmgren, K. and Öberg, H. (2006). Climate change in southern and eastern Africa during the past millennium and its implications for societal development. *Environment, Development and Sustainability*, 8, 185–195.

Hommon, R. J. (1986). Social evolution in ancient Hawaii, in Kirch, P. V. (ed), *Island Societies: Archaeological Approaches to Evolution and Transformation*, Cambridge University Press, Cambridge, pp. 55–68.

Hong, Y. T., Jiang, H. B., Liu, T. S., Qin, X. G., Zhou, L. P., Beer, J., Li, H. D. and Leng, X. T. (2000). Response of climate to solar forcing recorded in a 6000-year $\delta^{18}O$ time-series of Chinese peat cellulose. *The Holocene*, 10, 1–7.

Hopkins, D. M. (1982). Aspects of the paleogeography of Beringia during the late Pleistocene, in Hopkins, D. M., Matthews, J. V., Schweger, C. E. and Young, S. B. (eds), *Paleoecology of Beringia*, Academic Press, New York, pp. 3–28.

Hopp, M. J. and Foley, J. A. (2001). Global-scale relationships between climate and the dengue fever vector, *Aedes aegypti*. *Climatic Change*, 48, 441–463.

Houghton, J. T., Jenkins, G. J. and Ephraums, J. J. (eds). (1990). *Climate Change, The IPCC Assessment*. Cambridge University Press, Cambridge.

Howe, K. R. (1984). *Where the Waves Fall: A New South Seas History from First Settlement to Colonial Rule*, Allen and Unwin, Sydney.

Hsu, K. J. (2000). *Climate and Peoples: A Theory of History*, Orell Fussli, Zurich.

Hu, F. S., Ito, E., Brown, T. A., Curry, B. B. and Engstrom, D. R. (2001). Pronounced climatic variations in Alaska during the last two millennia. *Proceedings of the National Academy of Sciences*, 98, 10552–10556.

Huang, S., Pollack, H. N. and Shen, P.-Y. (2000). Temperature trends over the past five centuries reconstructed from borehole records. *Nature*, 403, 756–758.

Hubbard, C. B. and Neall, V. E. (1980). A reconstruction of late Quaternary erosional events in the West Tamaki River catchment, Southern Ruahine Range, North Island, New Zealand. *New Zealand Journal of Geology and Geophysics*, 23, 587–593.

Huffman, T. N. (1996). Archaeological evidence for climatic change during the last 2000 years in southern Africa. *Quaternary International*, 33, 55–60.

Hughes, M. K., and Diaz, H. F. (1994). Was there a 'Medieval Warm Period', and if so, where and when? *Climatic Change*, 26, 109–142.

Hughes, M. K. and Graumlich, L. J. (1996). Multimillenial dendroclimatic studies from the western United States, in Jones, P. D., Bradley, R. S. and Jouzel, J. (eds), *Climatic Variations and Forcing Mechanisms of the Last 2000 Years*, Springer-Verlag, Berlin, pp. 109–124.

Hughes, P. J., and Djohadze, V. (1980). Radiocarbon dates from archaeological sites on the south coast of New South Wales and the use of age/depth curves. *Occasional Papers in Prehistory No. 1*. Department of Prehistory, Research School of Pacific Studies, The Australian National University, Canberra.

Hughes, P. J., Hope, G., Latham, M., and Brookfield, M. (1979). Prehistoric man-induced degradation of the Lakeba landscape: evidence from two inland swamps, in: *UNESCO/UNFPA Fiji Island Report*. Canberra: Australian National University, pp. 93–110.

Hughes, T. P., Baird, A. H., Bellwood, D. R., Card, M., Connolly, S. R., Folke, C., Grosberg, R., Hoegh-Guldberg, O., Jackson, J. B. C., Kleypas, J., Lough, J. M., Marshall, P., Nyström, M., Palumbi, S. R., Pandolfi, J. M., Rosen, B. and Roughgarden, J. (2003). Climate change, human impacts, and the resilience of coral reefs. *Science*, 301, 929–933.

Hunt, T. L., Aronson, K. F., Cochrane, E. E., Field, J. S., Humphrey, L. and Rieth, T. M. (1999). A preliminary report on archaeological research in the Yasawa Islands, Fiji. *Domodomo*, 12, 5–43.

Hunt, T. L. and Lipo, C. P. (2006). Late colonization of Easter Island. *Science*, 311, 1603–1606.

Hunter, J. R. (2002). *Note on relative sea level change, Funafuti, Tuvalu*. Unpublished paper. Antarctic Cooperative Research Centre, Hobart, p. 25.

Hunter-Anderson, R. L. (1998). Human vs. climatic impacts at Easter Island: did the people really cut down all those trees?, in Stevenson, C. M., Lee, G. and Morin, F. J. (eds), *Easter Island in Pacific Context, South Seas Symposium, Proceedings of the Fourth International Conference on Easter Island and East Polynesia, University of New Mexico*, The Easter Island Foundation, Los Osos, pp. 85–99.

Hunter-Anderson, R. L., and Butler, B. M. (1995). *An Overview of Northern Marianas Prehistory*. Micronesian Archaeological Survey Report 31, Saipan.

Huq, S., Ali, S. I. and Rahman, A. A. (1995). Sea-level rise and Bangladesh: a preliminary analysis. *Journal of Coastal Research, Special Issue*, 14, 44–53.

Ikawa-Smith, F. (2004). Humans along the Pacific margin of Northeast Asia before the Last Glacial Maximum, in Madsen, D. B. (ed), *Entering America: Northeast Asia and Beringia before the Last Glacial Maximum*, University of Utah Press, Salt Lake City, pp. 285–309.

Imbrie, J., Hays, J. D., Martinson, D. G., McIntyre, A., Mix, A. C., Morley, J. J., Pisias, G., Prell, W. L. and Shackleton, N. J. (1984). The orbital theory of Pleistocene climate: Support from a revised chronology of the marine δ^{18}O record, in Berger, A. (ed), *Milankovitch and Climate, Part 1*, Reidel, Dordrecht, pp. 269–305.

Ingstad, H. and Ingstad, A. S. (2000). *The Viking Discovery of America: The Excavation of a Norse Settlement in L'Anse aux Meadows*, Breakwater Books, Newfoundland.

IPCC (Working Group I) (2001). *Climate Change 2001, Scientific Basis*, Cambridge University Press, Cambridge.

Iriondo, M. H. and Garcia, N. O. (1993). Climatic variations in the Argentine plains during the last 18,000 years. *Palaeogeography, Palaeoclimatology, Palaeoecology*, 101, 209–220.

Irwin, G. (1992). *The Prehistoric Exploration and Colonisation of the Pacific*, Cambridge University Press, Cambridge.

Isla, F. I. (1989). Holocene sea-level fluctuation in the southern hemisphere. *Quaternary Science Reviews*, 8, 359–368.

Issar, A. S. and Zohar, M. (2004). *Climate Change – Environment and Civilization in the Middle East*, Springer, Berlin.

Ivanoff, J. (April 2005). Sea gypsies of Myanmar. *National Geographic Magazine*, pp. 36–55.

Ivens, W. G. (1927). *Melanesians of the South-East Solomon Islands*, Benjamin Blom, New York.

Jenkins, J. (1975). *Diary of a Welsh Swagman, 1869–1894*, Macmillan, London.

Jennings, A. E. and Weiner, N. J. (1996). Environmental change in eastern Greenland during the last 1300 years: evidence from foraminifera and lithofacies in Nansen Fjord, 68°N. *The Holocene*, 6, 179–191.

Jenny, B., Valero-Garces, B. L., Urrutia, R., Kelts, K., Veit, H., Appleby, P. G. and Geyh, M. (2002). Moisture changes and fluctuations of the Westerlies in Mediterranean Central Chile during the last 2000 years: The Laguna Aculeo record (33°50'S). *Quaternary International*, 87, 3–18.

Jin, L. and Su, B. (2000). Natives or immigrants: Modern human origin in East Asia. *Nature Reviews Genetics*, 1, 126–133.

Johannes, R. E. (1982). Traditional conservation methods and protected marine areas in Oceania. *Ambio*, 2, 258–261.

Johnson, T. C., Barry, S., Chan, Y. and Wilkinson, P. (2001). Decadal record of climate variability spanning the past 700 yr in the Southern Tropics of East Africa. *Geology*, 29, 83–86.

Jokiel, P. L. and Brown, E. K. (2004). Global warming, regional trends and inshore environmental conditions influence coral bleaching in Hawaii. *Global Change Biology*, 10, 1627–1688.

Jones, A. T. (1997). Late Holocene shoreline development in the Hawaiian Islands. *Journal of Coastal Research*, 14, 3–9.

Jones, P. D., Briffa, K. R., Barnett, T. P. and Tett, S. F. B. (1998). High-resolution palaeoclimatic records for the last millennium: Interpretation, integration and comparison with general circulation model control-run temperatures. *The Holocene*, 8, 455–471.

Jones, R. (1977). Man as an element of a continental fauna: The case of the sundering of the Bassian Bridge, in Allen, J., Golson, J. and Jones, R. (eds), *Sunda and Sahul*, Academic Press, London, pp. 317–386.

Jones, T. L., Brown, G. M., Raab, M., McVickar, J. L., Spaulding, W. G., Kennett, D. J., York, A. and Walker, P. L. (1999). Environmental imperatives reconsidered: Demographic crises in western North America during the Medieval Climatic Anomaly. *Current Anthropology*, 40, 137–170.

Jordan, J. W. and Maschner, H. D. G. (2000). Coastal paleogeography and human occupation of the western Alaska Peninsula. *Geoarchaeology*, 15, 385–414.

Jordan, W. C. (1996). *The Great Famine: Northern Europe in the Early Fourteenth Century*, Princeton University Press, Princeton.

Juillet-Leclerc, A., Thiria, S., Naveau, P., Delcroix, T., Le Bec, N., Blamart, D., and Corrège, T. (2006). SPCZ migration and ENSO events during the 20th century as revealed by climate proxies from a Fiji coral. *Geophysical Research Letters*, 33, L17710, doi:10.1029/2006GL025950.

Kahn, J. (2003). Maohi social organization at the micro-scale: household archaeology in the 'Opunohu Valley, Mo'orea, Society Islands (French Polynesia, in Sand, C. (ed). *Pacific Archaeology: Assessments and Prospects (Proceedings of the International Conference for the 50th Anniversary of the First Lapita Excavation, Kone-Nouméa 2002)*, Services des Musées et du Patrimoine, Nouméa, New Caledonia, pp. 353–367.

Kaluwin, C. and Smith, A. (1997). Coastal vulnerability and integrated coastal zone management in the Pacific Island region. *Journal of Coastal Research*, 24, 95–106 Special Issue.

Kaplan, H. and Hill, K. (1992). The evolutionary ecology of food acquisition, in Smith, E. A. and Winterhalder, B. (eds), *Evolutionary Ecology and Human Behavior*, Aldine de Bruyter, New York, pp. 167–201.

Karlén, W., Fastook, J. L., Holmgren, K., Malmstrom, M., Matthews, J. A., Odada, E., Risberg, J., Rosqvist, G., Per, S., Westberg, A. and Ove, L. (1999). Glacier fluctuations on Mount Kenya since 6000 cal. years B.P: implications for Holocene climatic change in Africa. *Ambio*, 28, 409–418.

Kasper, J. N. and Allard, M. (2001). Late-Holocene climatic changes as detected by the growth and decay of ice wedges on the southern shore of Hudson Strait, northern Québec, Canada. *The Holocene*, 11, 563–577.

Katayama, K., Nunn, P. D., Kumar, R., Matararaba, S. and Oda, H. (2003). Reconstruction of a Lapita lady skeleton unearthed from the Moturiki Island, Fiji. *Anthropological Science*, 111, 404–405 (preliminary report).

Kaufmann, R. K., Kauppi, H. and Stock, J. H. (2006). Emissions, concentrations, and temperature: A time series analysis. *Climatic Change*, 77, 249–278.

Kawamura, Y. (1998). Daiyonki ni okeru Nihon retto e no honyurui no ido (Immigration of mammals into the Japanese islands during the Quaternary). *Daiyonki Kenkyu (The Quaternary Research)*, 37, 251–257 (in Japanese with English abstract).

Kawana, T., Miyagi, T., Fujimoto, K. and Kikuchi, T. (1995). Late Holocene sea-level changes and mangrove development in Kosrae Island, the Carolines, Micronesia, in Kikuchi, T. (ed), *Rapid Sea Level Rise and Mangrove Habitat*, Institute for Basin Ecosystem Studies, Gifu University, Japan, pp. 1–7.

Kealhofer, L. and Piperno, D. R. (1994). Early agriculture in Southeast Asia: phytolith evidence from the Bang Pakong Valley, Thailand. *Antiquity*, 68, 564–572.

Keally, C. T., Taniguchi, Y. and Kuzmin, Y. (2003). Understanding the beginnings of pottery technology in Japan and neighboring East Asia. *The Review of Archaeology*, 24, 3–14.

Kearney, M. S. (1996). Sea-level change during the last thousand years in Chesapeake Bay. *Journal of Coastal Research*, 12, 977–983.

Keeling, C. D. and Whorf, T. P. (2000). The 1,800-year oceanic tidal cycle: A possible cause of rapid climate change. *Proceedings of the National Academy of Sciences*, 97, 3814–3819.

Keigwin, L. D. (1996). The Little Ice Age and Medieval Warm Period in the Sargasso Sea. *Science*, 274, 1504–1508.

Keigwin, L. D. and Boyle, E. A. (2000). Detecting Holocene changes in thermohaline circulation. *Proceedings of the National Academy of Sciences*, 97, 1343–1346.

Khazanov, A. M. and Wink, A. (eds). (2001). *Nomads in the Sedentary World*. Curzon, Richmond.

Kikuchi, W. (1976). Prehistoric Hawaiian fishponds. *Science*, 193, 295–299.

Kiladis, G. N. and Diaz, H. F. (1986). An analysis of the 1877–1878 ENSO episode and comparison with 1982–83. *Monthly Weather Review*, 114, 1035–1047.

King, L. (1990). *Evolution of Chumash Society*, Garland, New York.

King, J. C. and Harangozo, S. A. (1998). Climate change in the western Antarctic Peninsula since 1945: Observations and possible causes. *Annals of Glaciology*, 27, 571–575.

Kingdon, J. (1993). *Self-Made Man: Human Evolution from Eden to Extinction?* Wiley, New York.

Kingdon, J. (2003). *Lowly Origin: Where, When, and Why Our Ancestors First Stood Up*, Princeton University Press, Princeton.

Kirch, P. V. (1984a). *The Evolution of the Polynesian Chiefdoms*, Cambridge University Press, Cambridge.

Kirch, P. V. (1984b). The Polynesian Outliers: Continuity, change, and replacement. *The Journal of Pacific History*, 4, 224–238.

Kirch, P. V. (1986). Rethinking east Polynesian prehistory. *Journal of the Polynesian Society*, 95, 9–40.

Kirch, P. V. (1990). The evolution of sociopolitical complexity in prehistoric Hawaii: An assessment of the archaeological evidence. *Journal of World Prehistory*, 4, 311–345.

Kirch, P. V. (1997a). *The Lapita Peoples: Ancestors of the Oceanic World*, Blackwell, Oxford.

Kirch, P. V. (1997b). Changing landscapes and sociopolitical evolution in Mangaia, central Polynesia, in Kirch, P. V. and Hunt, T. L. (eds), *Historical Ecology in the Pacific Islands*, Yale University Press, New Haven, pp. 147–165.

Kirch, P. V. (2000a). *On the Road of the Winds: An Archaeological History of the Pacific Islands Before European Contact*, University of California Press, Berkeley.

Kirch, P. V. (2000b). Pigs, humans, and trophic competition on small oceanic islands, in Anderson, A. and Murray, T. (eds), *Australian Archaeologist: Collected Papers in Honour of Jim Allen*, Coombs Academic Publishing, The Australian National University, Canberra, pp. 427–440.

Kirch, P. V., Hartshorn, A. S., Chadwick, O. A., Vitousek, P. M., Sherrod, D. R., Coil, J., Holm, L. and Sharp, W. D. (2004). Environment, agriculture, and settlement patterns in a marginal Polynesian landscape. *Proceedings of the National Academy of Sciences*, 101, 9936–9941.

Kirch, P. V. and Rosendahl, P. H. (1976). Early Anutan settlement and the position of Anuta in the prehistory of the Southwest Pacific. *Bulletin of the Royal Society of New Zealand*, 11, 225–244.

Kirch, P. V. and Yen, D. E. (1982). *Tikopia – The Prehistory and Ecology of a Polynesian Outlier*, B.P. Bishop Museum, Honolulu, Bulletin 238.

Kitagawa, H. and Matsumoto, E. (1995). Climatic implications of $\delta^{13}C$ variations in a Japanese cedar *Cryptomeria japonica* during the last two millenia. *Geophysical Research Letters*, 22, 2155–2158.

Knecht, R. (1995). *The late prehistory of the Alutiiq people: Culture change on the Kodiak Archipelago from 1200–1750 A.D.* Unpublished Ph.D. dissertation. Bryn Mawr College, PA, USA.

Knowles, N., Dettinger, M. D. and Cayan, D. R. (2006). Trends in snowfall versus rainfall in the western United States. *Journal of Climate*, 19, 4545–4560.

Knutson, T. R., Tuleya, R. E. and Kurihara, Y. (1998). Simulated increase of hurricane intensities in a CO_2-warmed climate. *Science*, 279, 1018–1020.

Koch, J. and Kilian, R. (2005). 'Little Ice Age' glacier fluctuations, Gran Campo Nevado, southernmost Chile. *The Holocene*, 15, 20–28.

Kolata, A. L., Binford, M. W., Brenner, M., Janusek, J. W. and Ortloff, C. (2000). Environmental thresholds and the empirical reality of state collapse: A response to Erickson (1999). *Antiquity*, 74, 424–426.

Kolata, A. L. and Ortloff, C. R. (1996). Agroecological perspectives on the decline of the Tiwanaku State, in Kolata, A. L. (ed), *Tiwanaku and its Hinterland: Archaeology and Palaeoecology of an Andean Civilization*, Smithsonian Institution, Washington, pp. 181–199.

Kononenko, N. A. (ed), (1996). *Late Paleolithic–Early Neolithic of Eastern Asia and Northern America (Materials of the International Symposium)*. Archaeology and Ethnography Press, Institute of History, Vladivostok.

Korotky, A. M., Razjigaeva, N. G., Grebennikova, T. A., Ganzey, L. A., Mokhova, L. M., Bazarova, V. B., Sulerzhitsky, L. D. and Lutaenko, K. A. (2000). Middle- and late-Holocene environments and vegetation history of Kunashir Island, Kurile Islands, northwestern Pacific. *The Holocene*, 10, 311–331.

Kouwenberg, L., Wagner, R., Kürschner, W. and Visscher, H. (2005). Atmospheric CO_2 fluctuations during the last millennium reconstructed by stomatal frequency analysis of *Tsuga heterophylla* needles. *Geology*, 33, 33–36.

Kraft, J. C., Kayan, I. and Aschenbrenner, S. E. (1985). Geological studies of coastal change applied to archaeological settings, in Rapp, G. Jr. and Gifford, J. A. (eds), *Archaeological Geology*, Yale University Press, New Haven, pp. 57–84.

Kreutz, K. J., Mayewski, P. A., Meeker, L. D., Twickler, M. D., Whitlow, S. I. and Pittalwala, I. I. (1997). Bipolar changes in atmospheric circulation during the Little Ice Age. *Science*, 277, 1294–1296.

Krinner, G., Magand, O., Simmonds, I., Genthon, C., and Dufresne, J.-L. (2006). Simulated Antarctic precipitation and surface mass balance at the end of the twentieth and twenty-first centuries. *Climate Dynamics*, doi:10.1007/s00382-006-0177-x.

Kuhlken, R. (1999). Warfare and intensive agriculture in Fiji, in Gosden, C. and Hather, J. (eds), *The Prehistory of Food: Appetites for Change*, Routledge, London, pp. 270–287.

Kuhlken, R. and Crosby, A. (1999). Agricultural terracing at Nakauvadra, Viti Levu: A late prehistoric irrigated agrosystem in Fiji. *Asian Perspectives*, 38, 62–89.

Kumar, R., Nunn, P. D., Field, J. E. and de Biran, A. (2006). Human responses to climate change around AD 1300: A case study of the Sigatoka Valley, Viti Levu Island, Fiji. *Quaternary International*, 151, 133–143.

Kumar, R., Nunn, P. D., Katayama, K., Oda, H., Matararaba, S. and Osborne, T. (2004). The earliest-known humans in Fiji and their pottery: The first dates from the 2002 excavations at Naitabale (Naturuku), Moturiki Island. *South Pacific Journal of Natural Science*, 22, 15–21.

Kuzmin, Y. V., Levchuk, L. K., Burr, G. S. and Jull, A. J. T. (2004). AMS ^{14}C dating of the marine Holocene key section in Peter the Great Gulf, Sea of Japan. *Nuclear Instruments and Methods in Physics Research*, B 223–224, 451–454.

Ladefoged, T. and Pearson, R. (2000). Fortified castles on Okinawa Island during the Gusuki Period, A.D. 1200–1600. *Antiquity*, 74, 404–412.

Ladefoged, T. N. (1995). The evolutionary ecology of Rotuman political integration. *Journal of Anthropological Archaeology*, 14, 341–358.

Laird, K. (1996). A 2300-year sub-decadal record of drought severity, duration and frequency in the northern Great Plains. *AMQUA (American Quaternary Association) 1996: Program and Abstracts of the 14th Biennial Meeting*. Flagstaff Arizona, 14, 172.

Laird, K. R., Fritz, S. C., Grimm, E. C. and Mueller, P. G. (1996a). Century-scale paleoclimate reconstruction from Moon Lake, a closed-basin lake in the northern Great Plains. *Limnology and Oceanography*, 41, 890–902.

Laird, K. R., Fritiz, S. C., Maasch, K. A. and Cumming, B. F. (1996b). Greater drought intensity and frequency before A.D. 1200 in the northern Great Plains, USA. *Nature*, 384, 552–554.

Lamb, H., Darbyshire, I. and Verschuren, D. (2003). Vegetation response to rainfall variation and human impact in central Kenya during the past 1100 years. *The Holocene*, 13, 285–292.

Lamb, H. H. (1965). The early medieval warm epoch and its sequel. *Palaeogeography, Palaeoclimatology, Palaeoecology*, 1, 13–37.

Lamb, H. H. (1977). *Climate, Past, Present and Future. Volume 2. Climate History and the Future*, Methuen, London.

Lamb, H. H. (1982). *Climate, History and the Modern World*, Methuen, London.

Lara, A. and Villalba, R. (1993). A 3260-year temperature record from *Fitzroya cupressoides* tree rings in southern South America. *Science*, 260, 1104–1106.

Larocque, S. J. and Smith, D. J. (2003). Little Ice Age glacial activity in the Mt. Waddington area, British Columbia Coast Mountains, Canada. *Canadian Journal of Earth Sciences*, 40, 1413–1436.

Larson, D. O., Neff, H., Graybill, D. A., Michaelson, J. and Ambos, E. (1996). Risk, climatic variability, and the study of southwestern prehistory: An evolutionary perspective. *American Antiquity*, 61, 217–241.

Laughlin, W. S. and Harper, A. B. (1988). Peopling of the continents: Australia and America, in Mascie-Taylor, C. G. N. and Lasker, G. W. (eds), *Biological Aspects of Human Migration*, Cambridge University Press, Cambridge, pp. 14–40.

Le Roy Ladurie, E. L. (1971). *Times of Feast, Times of Famine: A History of Climate since the Year 1000*, George, Allen and Unwin, London.

Lea, D. W. (2004). The 100,000-yr cycle in tropical SST, greenhouse forcing, and climate sensitivity. *Journal of Climate*, 17, 2170–2180.

Leach, B. F. (1981). The prehistory of the Southern Wairarapa. *Journal of the Royal Society of New Zealand*, 11, 11–33.

Leach, B. F., and Ward, G. (1981). *Archaeology on Kapingamarangi Atoll: A Polynesian Outlier in the Eastern Caroline Islands*. Manuscript, Pacific Collection, The University of the South Pacific Library.

Leach, H. M. and Leach, B. F. (1979). Environmental change in Palliser Bay, in Leach, B. F. and Leach, H. M. (eds), *Prehistoric Man in Palliser Bay*, National Museum of New Zealand, Wellington, pp. 229–240 (Bulletin 21).

Leavitt, S. W. (1994). Major wet interval in White Mountains medieval warm period evidenced in δ^{13}C of bristlecone pine tree rings. *Climatic Change*, 26, 299–307.

Lehodey, P., Alheit, J., Barange, M., Baumgartner, T., Beaugrand, G., Drinkwater, K., Fromentin, J.-M., Hare, S. R., Otterson, G., Perry, R. I., Roy, C., Van der Lingen, D. D. and Werner, F. (2006). Climate variability, fish and fisheries. *Journal of Climate*, 19, 5009–5031.

Lemcke, G. and Sturm, M. (1997). $\delta^{18}O$ and trace element measurements as proxy for the reconstruction of climate changes at Lake Van (Turkey): Preliminary results, in Dalfes, H. N., Kukla, G. and Weiss, H. (eds), *Third Millennium BC Climate Change and Old World Collapse*, Springer, Berlin, pp. 653–678.

Lepofsky, D. (1988). The environmental context of Lapita settlement locations, in Kirch, P. V. and Hunt, T. L. (eds), *Archaeology of the Lapita Cultural Complex: A Critical Review*, Burke Museum, Seattle, pp. 33–47.

Lepofsky, D., Kirch, P. V. and Lertzman, K. P. (1996). Stratigraphic and paleobotanical evidence for prehistoric human-induced environmental disturbance on Mo'orea, French Polynesia. *Pacific Science*, 50, 253–273.

Levitus, S., Antonov, J. I., Boyer, T. P. and Stephens, C. (2000). Warming of the world ocean. *Science*, 287, 2225–2229.

Lewis, D. (1994). *We, The Navigators: The ancient art of landfinding in the Pacific* 2nd Edition. University of Hawaii Press, Honolulu.

Lewis, D. H. and Smith, D. J. (2004). Little Ice Age glacial activity in Strahcona Provincial Park, Vancouver Island, British Columbia, Canada. *Canadian Journal of Earth Sciences*, 41, 285–297.

Li, H.-C., Bischoff, J. L., Ku, T.-L., Lund, S. P. and Stott, L. D. (2000). Climate variability in east-central California during the past 1000 years reflected by high-resolution geochemical and isotopic records from Owens Lake sediments. *Quaternary Research*, 54, 189–197.

Liew, P.-M., Huang, S.-Y. and Kuo, C.-M. (2006). Pollen stratigraphy, vegetation and environment of the last glacial and Holocene – a record from Toushe Basin, central Taiwan. *Quaternary International*, 147, 16–33.

Lilley, I. (1988). Prehistoric exchange across the Vitiaz Strait, Papua New Guinea. *Current Anthropology*, 29, 513–516.

Linsley, B. K., Dunbar, R. B., Wellington, G. M. and Mucciarone, D. A. (1994). A coral-based reconstruction of intertropical convergence zone variability over central America since 1707. *Journal of Geophysical Research*, 99, 9977–9994.

Linsley, B. K., Wellington, G. M. and Schrag, D. P. (2000). Decadal sea surface temperature variability in the sub-tropical South Pacific from 1726 to 1997 A.D. *Science*, 290, 1145–1148.

Liu, K., Reese, C. A. and Thompson, L. G. (2005). Ice-core pollen record of climatic changes in the central Andes during the last 400 yr. *Quaternary Research*, 64, 272–278.

Liu, Z.-X., Berne, S., Saito, Y., Lericolais, G. and Marsset, T. (2000). Quaternary seismic straigraphy and paleoenvironments on the continental shelf of the East China Sea. *Journal of Asian Earth Sciences*, 18, 441–452.

Loeb, E. M. (1926). *History and Traditions of Niue*, B.P. Bishop Museum, Honolulu, Bulletin 32.

Lowe, D. J., Newnham, R. M., McFadgen, B. G. and Higham, T. F. G. (2000). Tephras and New Zealand archaeology. *Journal of Archaeological Science*, 27, 859–870.

Lucking, L. J. (1984). *An archaeological investigation of prehistoric Palauan terraces*. Unpublished Ph.D. thesis. University of Minnesota, Minneapolis.

Luckman, B. H. (1994). Evidence for climatic conditions between ca. 900–1300 A.D. in the southern Canadian Rockies. *Climatic Change*, 26, 171–182.

Luckman, B. H., Briffa, K. R., Jones, P. D. and Schweingruber, F. H. (1997). Tree-ring based reconstruction of summer temperatures at the Columbia Icefield, Alberta, Canada, A.D. 1073–1983. *The Holocene*, 7, 375–389.

Luckman, B. H., Holdsworth, G. and Osborn, G. D. (1993). Neoglacial glacial fluctuations in the Canadian Rockies. *Quaternary Research*, 39, 144–153.

Luckman, B. H. and Wilson, R. J. S. (2005). Summer temperatures in the Canadian Rockies during the last millennium: a revised record. *Climate Dynamics*, 24, 131–144.

Luders, D. (1996). Legend and history: did the Vanuatu-Tonga kava trade cease in AD 1447? *Journal of the Polynesian Society*, **105**, 287–310.

Lyons, M. (1986). *The Totem and the Tricolour: A Short History of New Caledonia since 1774*, New South Wales University Press, Kensington.

MacDonald, G. M., and Case, R. A. (2005). Variations in the Pacific Decadal Oscillation over the past millennium. *Geophysical Research Letters*, 32, L08703, doi:10.1029/2005GL022478.

Mackay, R. and White, J. P. (1987). Musselling in on the NSW coast. *Archaeology in Oceania*, 22, 107–111.

Magilligan, F. J. and Goldstein, F. J. (2001). El Niño floods and culture change: A late Holocene flood history for the Rio Moquegua, southern Peru. *Geology*, 29, 431–434.

Malamud-Roam, F. P., Ingram, B. L., Hughes, M. and Florsheim, J. L. (2006). Holocene paleoclimate records from a large California estuarine system and its watershed region: Linking watershed climate and bay conditions. *Quaternary Science Reviews*, 25, 1570–1598.

Mann, D., Chase, J., Edwards, J., Beck, W., Reanier, R. and Mass, M. (2003). Prehistoric destruction of the primeval soils and vegetation of Easter Island, in Trancredi, J. T. and Loret, J. (eds), *Easter Island, Scientific Exploration into the World's Environmental Problems in Microcosm*, Kluwer Academic/Plenum Publishers, New York, pp. 133–153.

Mann, D. H., Heiser, P. A. and Finney, B. P. (2002). Holocene history of the Great Kobuk Sand Dunes, northwestern Alaska. *Quaternary Science Reviews*, 21, 709–731.

Mann, M. E., Bradley, R. S. and Hughes, M. K. (1999). Northern hemisphere temperatures during the past millennium: Inferences, uncertainties and limitations. *Geophysical Research Letters*, 26, 759–762.

Mannion, A. M. (1999). Domestication and the origins of agriculture. *Progress in Physical Geography*, 23, 37–56.

Mantua, N. J., Hare, S. R., Zhang, Y., Wallace, J. M. and Francis, R. C. (1997). A Pacific interdecadal climate oscillation with important impacts on salmon production. *Bulletin of the American Meteorological Society*, 78, 1069–1079.

Marker, M. E. (1997). Evidence for a Holocene low sea level at Knysna. *South African Geographical Journal*, 79, 106–107.

Marshall, Y., Crosby, A., Matararaba, S. and Wood, S. (2000). *Sigatoka: the shifting sands of Fijian prehistory*, Oxbow Books, Oxford (University of Southampton, Department of Archaeology, Monograph 1).

Maschner, H. D. G. and Reedy-Maschner, K. L. (1998). Raid, retreat, defend (repeat): The archaeology and ethnohistory of warfare on the North Pacific Rim. *Journal of Anthropological Archaeology*, 17, 19–51.

Mass, J. P. (1997). *The Origins of Japan's Medieval World: Courtiers, Clerics, Warriors, and Peasants in the Fourteenth Century*, Stanford University Press, Stanford.

Masse, W. B., Liston, J., Carucci, J. and Athens, J. S. (2006). Evaluating the effects of climate change on environment, resource depletion, and culture in the Palau Islands between A.D. 1200 and 1600. *Quaternary International*, 151, 106–132.

Masse, W. B., Snyder, D. and Gumerman, G. J. (1984). Prehistoric and historical settlement in the Palau Islands, Micronesia. *New Zealand Journal of Archaeology*, 6, 107–127.

Mathis, M. A. (2000). The middle to late woodland shift on the central coast of North Carolina, in Gunn, J. D. (ed), *The Years without Summer: Tracing A.D. 536 and its Aftermath*, Archaeopress, Oxford, pp. 111–118 (British Archaeological Reports 872).

Matsuda, F., Fujiwara, O., Sakai, T., Araya, T., Tamura, T. and Kamataki, T. (2001). Chiba-ken Kujukurihama heiya no Kanshinto no hattatsu katei (Progradation of the Holocene beach-shoreface system in the Kujukuri strand plain, Pacific coast of the Boso Peninsula, central Japan). *Daiyonki Kenkyu (The Quaternary Research)*, 40, 223–233 (in Japanese with English abstract).

Mayewski, P. A., Meeker, L. D., Morrison, M. C., Twickler, M. S., Whitlow, S. I., Ferland, K. K., Meese, D. A., Legrand, M. R. and Steffenson, J. P. (1993). Greenland ice core "signal" characteristics: An expanded view of climate change. *Journal of Geophysical Research*, 98(D7), 12839–12847.

McCall, G. (1980). *Rapanui: Tradition and Survival on Easter Island*, University of Hawaii Press, Honolulu.

McCall, G. (1993). Little Ice Age: Some speculations for Rapanui. *Rapa Nui Journal*, 7, 65–70.

McCall, G. (1994). *Rapanui: Tradition and Survival on Easter Island*, Allen and Unwin, St. Leonards.

McCoy, P. C. (1979). Easter Island, in Jennings, J. (ed), *The Prehistory of Polynesia*, Harvard University Press, Cambridge, pp. 135–166.

McDermott, F., Mattey, D. P. and Hawkesworth, C. (2001). Centennial-scale Holocene climate variability revealed by a high-resolution speleothem $\delta^{18}O$ record from SW Ireland. *Science*, 294, 1328–1331.

McFadgen, B. G. (1994). Archaeology and Holocene sand dune stratigraphy on Chatham Island. *Journal of the Royal Society of New Zealand*, 24, 17–44.

McGlone, M. S. (1983). Polynesian deforestation of New Zealand: A preliminary synthesis. *Archaeology in Oceania*, 18, 11–25.

McGlone, M. S., Anderson, A. J. and Holdaway, R. N. (1994). An ecological approach to the Polynesian settlement of New Zealand, in Sutton, D. (ed), *The Origins of the First New Zealanders*, Auckland University Press, Auckland, pp. 136–163.

McGoodwin, J. R. (1992). Human responses to weather-induced catastrophes in a west Mexican fishery, in Glantz, M. H. (ed), *Climate Variability, Climate Change, and Fisheries*, Cambridge University Press, Cambridge, pp. 168–184.

McIntyre, S. and McKitrick, R. (2003). Corrections to the Mann et al. (1998) proxy data base and Northern Hemisphere average temperature series. *Energy and Environment*, 14, 751–771.

McMichael, A. J., Haines, A. and Slooff, R. (eds). (1996). *Climate Change and Human Health*. World Health Organization, Geneva.

McNeill, J. R. (1994). Of rats and men: A synoptic environmental history of the island Pacific. *Journal of World History*, 5, 299–349.

McNeill, J. R. (1999). Islands in the rim: Ecology and history in and around the Pacific, 1521–1996, in Flynn, D. O., Frost, L. and Latham, A. J. H. (eds), *Pacific Centuries: Pacific and Pacific Rim History since the Sixteenth Century*, Routledge, London, pp. 70–84.

McNiven, I. (1999). Fissioning and regionalisation: The social dimensions of changes in Aboriginal use of the Great Sandy Region, southeast Queensland, in: J. Hall, & I. McNiven (eds). *Australian Coastal Archaeology*. Canberra: Australian National University, pp. 157–168 (Research Papers in Archaeology and Natural History 31).

Meacham, W. (1996). Defining the hundred yue. *Indo-Pacific Prehistory Association, Bulletin*, 15, 93–99.

Meko, D. M., Therrell, M. D., Baisan, C. H. and Hughes, M. K. (2001). Sacramento River flow reconstructed to AD 869 from tree rings. *Journal of the American Water Resources Association*, 37, 1029–1039.

Meleisea, M. (1987). *Lagaga: A Short History of Western Samoa*, Institute of Pacific Studies, The University of the South Pacific, Suva.

Metcalfe, S. E. (1987). Historical data and climatic change in México: A review. *The Geographical Journal*, 153, 211–222.

Meyerson, E. A., Mayewski, P. A., Sneed, S. B., Kurbatov, A. V., Kreutz, K. J., Zielinski, G. A., Taylor, K. C., and Brook, E. J. (2003). *Bipolar synchroneity and latitudinal timing of Holocene climate change*. Unpublished conference poster, Sterling, Virginia, accessed in December 2006 at http://igloo.gsfc.nasa.gov/wais/pastmeetings/abstracts03/Meyerson.html.

Midun, Z. and Lee, S-C. (1995). Implications of a greenhouse-induced sea-level rise: a national assessment for Malaysia. *Journal of Coastal Research, Special Issue*, 14, 96–115.

Mimura, N. and Pelesikoti, N. (1997). Vulnerability of Tonga to future sea-level rise. *Journal of Coastal Research*, 24, 117–132 Special Issue.

Mingram, J., Allen, J. R. M., Brüchmann, C., Liu, J., Luo, X., Negendank, J. F. W., Nowaczyk, N. and Schettler, G. (2004). Maar- and crater lakes of the Long Gang Volcanic Field (N.E. China) – overview, laminated sediments, and vegetation history of the last 900 years. *Quaternary International*, 123-125, 135–147.

Mitchell, P. (2005). *African Connections: An Archaeological Perspective on Africa and the Wider World*, AltaMira Press, Walnut Creek.

Mitrovica, J. X. and Peltier, W. R. (1991). On postglacial geoid subsidence over the equatorial oceans. *Journal of Geophysical Research*, 96, 20053–20071.

Moberg, A., Sonechkin, D. M., Holmgren, K., Datsenko, N. M. and Karlén, W. (2005). Highly variable Northern Hemisphere temperatures reconstructed from low- and high-resolution proxy data. *Nature*, 433, 613–617.

Mochanov, Y. A. (1980). Early migrations to America in the light of study of the Dyuktai Paleolithic Culture in northeast Asia, in Browman, D. L. (ed), *Early Native Americans: Prehistoric Demography, Economy, and Technology*, Mouton, The Hague, pp. 174–177.

Moerenhout, J. A. (1837). *Voyages aux Îles du Grand Océan*. Bertrand, Paris (2 volumes).

Monzier, M., Robin, C. and Eissen, J. P. (1994). Kuwae (c. 1425): The forgotten caldera. *Journal of Volcanology and Geothermal Research*, 59, 207–218.

Mooney, S. D. and Maltby, E. L. (2006). Two proxy records revealing the late Holocene fire history at a site on the central coast of New South Wales, Australia. *Austral Ecology*, 31, 682–695.

Moore, C. A. (January–February 2002). Awash in a rising sea – how global warming is overwhelming the islands of the tropical Pacific. *International Wildlife*, pp. 1–5.

Moore, G. W. K., Holdsworth, G. and Alverson, K. (2002). Climate change in the North Pacific region over the past three centuries. *Nature*, 420, 401–403.

Moore, J. J., Hughen, K. A., Miller, G. H. and Overpeck, J. T. (2001). Little Ice Age recorded in summer temperature reconstruction from varved sediments of Donard Lake, Baffin Island, Canada. *Journal of Paleolimnology*, 25, 503–517.

Morgan, V. I. (1985). An oxygen isotope – climate record from the Law Dome, Antarctica. *Climatic Change*, 7, 415–426.

Moriwaki, H., Chikamori, M., Okuno, M. and Nakamura, T. (2006). Holocene changes in sea level and coastal environments on Rarotonga, Cook Islands, South Pacific Ocean. *The Holocene*, 16, 839–848.

Morrill, C., Overpeck, J. T. and Cole, J. E. (2003). A synthesis of abrupt changes in the Asian summer monsoon since the last deglaciation. *The Holocene*, 13, 465–476.

Morris, C. and von Hagen, A. (1993). *The Inka Empire and its Andean Origins*, Abbeville Press, New York.

Moseley, M. (1997). Climate, culture, and punctuated change: New data, new challenges. *The Review of Archaeology*, 18, 19–27.

Mosley-Thompson, E. (1992). Paleoenvironmental conditions in Antarctica since A.D. 1500: ice core evidence, in Bradley, R. S. and Jones, P. D. (eds), *Climate since A.D. 1500*, Routledge, New York, pp. 572–591.

Mote, P. W., Canning, D. J., Fluharty, D. L., Francis, R. C., Franklin, J. F., Hamlet, A. F., Hershman, M., Holmberg, M., Ideker, K. N., Keeton, W. S., Lettenmaier, D. P., Leung, L. R., Mantua, N. J., Miles, E. L., Noble, B., Parandvash, H., Peterson, D. W., Snover, A. K. and Willard, S. R. (1999). *Impacts of Climate Variability and Change, Pacific Northwest*, National Atmospheric and Oceanic Administration, Office of Global Programs, and JISAO/SMA Climate Impacts Group, Seattle.

Moy, C. M., Seltzer, G. O., Rodbell, D. T. and Anderson, D. M. (2002). Variability of El Niño/Southern Oscillation activity at millennial timescales during the Holocene epoch. *Nature*, 420, 162–165.

Moyle, R. M. (ed), (1984). *The Samoan Journals of John Williams 1830 and 1832*. Australian National University Press, Canberra.

Munk, W. (2002). Twentieth century sea level: An enigma. *Proceedings of the National Academy of Sciences*, 99, 6550–6555.

Nakada, M. (1986). Holocene sea levels in oceanic islands: Implications for the rheological structure of the Earth's mantle. *Tectonophysics*, 21, 263–276.

Nance, R. D., Worsley, T. R. and Moody, J. B. (1988). The supercontinent cycle. *Scientific American*, 259, 72–79.

National Research Council (2002). *Abrupt Climate Change: Inevitable Surprises*, National Academy Press, Washington.

Naurzbaev, M. M. and Vaganov, E. A. (2000). Variation of early summer and annual temperature in east Taymir and Putoran (Siberia) over the last two millennia inferred from tree rings. *Journal of Geophysical Research*, 105, 7317–7326.

Naurzbaev, M. M., Vaganov, E. A., Sidorava, O. V. and Schweingruber, F. H. (2002). Summer temperatures in eastern Taimyr inferred from a 2427-year late-Holocene tree-ring chronology and earlier floating series. *The Holocene*, 12, 727–736.

Nederbragt, A. J. and Thurow, J. W. (2001). A 6000 yr varve record of Holocene climate in Saanich Inlet, British Columbia, from digital sediment colour analysis of ODP Leg 169S cores. *Marine Geology*, 174, 95–110.

Neff, H., Pearsall, D. M., Jones, J. G., Arroyo de Pieters, B. and Freidel, D. E. (2006). Climate change and population history in the Pacific lowlands of Mesoamerica. *Quaternary Research*, 65, 390–400.

Nesje, A., Dahl, S. O., Matthews, J. A. and Berrisford, M. S. (2001). A ~4500-yr record of river floods obtained from a sediment core in Lake Atnsjoen, eastern Norway. *Journal of Paleolimnology*, 25, 329–342.

Neumann, A. C., and MacIntyre, I. (1985). Reef response to sea-level rise: keep-up, catch-up or give-up. In: *Proceedings of the 5th International Coral Reef Congress*. Tahiti, French Polynesia, 3, 105–110.

Neumann, J. and Sigrist, R. M. (1978). Harvest dates in ancient Mesopotamia as possible indicators of climatic variations. *Climatic Change*, 1, 239–252.

Nicholson, S. E. and Yin, X. (2001). Rainfall conditions in equatorial East Africa during the nineteenth century as inferred from the record of Lake Victoria. *Climatic Change*, 48, 387–398.

Niggemann, S., Mangini, A., Richter, D. K. and Wurth, G. (2003). A paleoclimate record of the last 17,600 years in stalagmites from the B7 cave, Sauerland, Germany. *Quaternary Science Reviews*, 22, 555–567.

Nordt, L., Hayashida, F., Hallmark, T. and Crawford, C. (2004). Late prehistoric soil fertility, irrigation management, and agricultural production in northwest coastal Peru. *Geoarchaeology*, 19, 21–46.

Núñez, L., Grosjean, M. and Cartajena, I. (2002). Human occupations and climate change in the Puna de Atacama, Chile. *Science*, 298, 821–824.

Nunn, P. D. (1988). Recent environmental changes along south-west Pacific coasts and the prehistory of Oceania: Developments of the work of the late John Gibbons. *Journal of Pacific Studies*, 14, 42–58.

Nunn, P. D. (1990). Recent environmental changes on Pacific islands. *The Geographical Journal*, 156, 125–140.

Nunn, P. D. (1994). *Oceanic Islands*, Blackwell, Oxford.

Nunn, P. D. (1995). Holocene sea-level changes in the South and West Pacific. *Journal of Coastal Research*, 17, 311–319 Special Issue.

Nunn, P. D. (1998). *Pacific Island Landscapes*, Institute of Pacific Studies, The University of the South Pacific, Suva.

Nunn, P. D. (1999). *Environmental Change in the Pacific Basin: Chronologies, Causes, Consequences*, Wiley, London.

Nunn, P. D. (2000a). Environmental catastrophe in the Pacific Islands about A.D. 1300. *Geoarchaeology*, 15, 715–740.

Nunn, P. D. (2000b). Illuminating sea-level fall around A.D. 1220–1510 (730–440 cal yr BP) in the Pacific Islands: Implications for environmental change and cultural transformation. *New Zealand Geographer*, 56, 4–12.

Nunn, P. D. (2001). Sea-level change in the Pacific, in Noye, J. and Grzechnik, M. (eds), *Sea-Level Changes and their Effects*, World Scientific Publishing, Singapore, pp. 1–23.

Nunn, P. D. (2003a). Nature–society interactions in the Pacific Islands. *Geografiska Annaler*, 85B, 219–229.

Nunn, P. D. (2003b). Revising ideas about environmental determinism: Human–environment relations in the Pacific Islands. *Asia-Pacific Viewpoint*, 44, 63–72.

Nunn, P. D. (2004). Understanding and adapting to sea-level change, in Harris, F. (ed), *Global Environmental Issues*, Wiley, Chichester, pp. 45–64.

Nunn, P. D. (2005). Reconstructing tropical paleoshorelines using archaeological data: examples from the Fiji Archipelago, southwest Pacific. *Journal of Coastal Research, Special Issue*, 42, 15–25.

Nunn, P. D. (2007a). Holocene sea-level change and human response in Pacific Islands. *Transactions of the Royal Society of Edinburgh: Earth and Environmental Sciences*, in press.

Nunn, P. D. (2007b). Space and place in an ocean of islands: Thoughts on the attitudes of the Lapita people towards islands and their colonization. *South Pacific Studies*, 27, 24–35.

Nunn, P.D. (2007c). The AD 1300 Event in the Pacific Basin: Overview and Teleconnections. *The Geographical Review*. forthcoming.

Nunn, P. D. and Britton, J. M. R. (2001). Human–environment relationships in the Pacific Islands around A.D. 1300. *Environment and History*, 7, 3–22.

Nunn, P. D., Hunter-Anderson, R., Carson, M. T., Thomas, F., Ulm, S., and Rowland, M. (2007). Times of plenty, times of less: chronologies of last-millennium societal disruption in the Pacific Basin. *Human Ecology: An Interdisciplinary Journal*, in press, doi: 10.1007/S10745-006-9090-5.

Nunn, P. D., Keally, C. T., King, C., Wijaya, J. and Cruz, R. (2006). Human responses to coastal change in the Asia-Pacific region, in Harvey, N. (ed), *Global Change and Integrated Coastal Management: The Asia-Pacific Region*, Springer, Berlin, pp. 117–161.

Nunn, P. D. and Kumar, R. (2006). Coastal history in the Asia-Pacific region, in Harvey, N. (ed), *Global Change and Integrated Coastal Management: The Asia-Pacific Region*, Springer, Berlin, pp. 93–116.

Nunn, P. D., Kumar, R., Matararaba, S., Ishimura, T., Seeto, J., Rayawa, S., Kuriyawa, S., Nasila, A., Oloni, B., Rati Ram, A., Saunivalu, P., Singh, P. and Tegu, E. (2004). Early Lapita settlement site at Bourewa, southwest Viti Levu Island, Fiji. *Archaeology in Oceania*, 39, 139–143.

Nunn, P. D., Matararaba, S., Ishimura, T., Kumar, R. and Nakoro, E. (2005). Reconstructing the Lapita-era geography of northern Fiji: A newly-discovered Lapita site on Yadua Island and its implications. *New Zealand Journal of Archaeology*, 26, 41–55.

Nunn, P. D., Matararaba, S. and Ramos, J. (2000). Investigations of anthropogenic sediments in Qaranilaca (cave), Vanuabalavu Island, Fiji. *Archaeology in New Zealand*, 43, 125–156.

Nunn, P. D. and Mimura, N. (2006). Promoting sustainability on vulnerable island coasts: A case study of the smaller Pacific Islands, in McFadden, L. (ed), *Managing Coastal Vulnerability: An Integrated Approach*, Elsevier, Amsterdam, pp. 193–220.

Nunn, P. D. and Peltier, W. R. (2001). Far-field test of the ICE-4G (VM2) model of global isostatic response to deglaciation: Empirical and theoretical Holocene sea-level reconstructions for the Fiji Islands, Southwest Pacific. *Quaternary Research*, 55, 203–214.

Nydick, K. R., Bidwell, A. B., Thomas, E. and Varekamp, J. C. (1995). A sea-level rise curve from Guilford, Connecticut, U.S.A.. *Marine Geology*, 124, 137–159.

O'Brien, S. R., Mayewski, P. A., Meeker, L. D., Meese, D. A., Twickler, M. S. and Whitlow, S. I. (1995). Complexity of Holocene climate as reconstructed from a Greenland ice core. *Science*, 270, 1962–1964.

O'Hara, S. L. (1993). Historical evidence of fluctuations in the level of Lake Pátzcuaro, Michoacán, México over the last 600 years. *The Geographical Journal*, 159, 51–62.

Obeyesekere, G. (2001). Narratives of the self: Chevalier Peter Dillon's Fijian cannibal adventures, in Horn, J. and Creed, B. (eds), *Body Trade: Captivity, Cannibalism and Colonialism in the Pacific*, Routledge, New York, pp. 69–125.

Ogilvie, A. E. J. and Jónsson, T. (2001). 'Little Ice Age' research: A perspective from Iceland. *Climatic Change*, 48, 9–52.

Ogilvie, A. E. J. and McGovern, T. H. (2000). Sagas and science: Climate and human impacts in the North Atlantic, in Fitzhugh, W. W. and Ward, E. I. (eds), *Vikings: The North Atlantic Saga*, Smithsonian Institution Press, Washington, pp. 385–393.

Okajima, T. (1999). Cannibalism of Rarotonga, Cook Islands in A.D. 13 c. *Abstract for 33rd Congress of Ethnology of Japan*, Tokyo, Japan, p. 59.

Oouchi, K., Yoshimura, J., Yoshimura, H., Mizuta, R., Kusunoki, S. and Noda, A. (2006). Tropical cyclone climatology in a global-warming climate as simulated in a 20 km-mesh global atmospheric model: Frequency and wind intensity analyses. *Journal of the Meteorological Society of Japan*, 84, 259–276.

Oppenheimer, C. (2003). Climatic, environmental and human consequences of the largest known historic eruption: Tambora volcano (Indonesia) 1815. *Progress in Physical Geography*, 27, 230–259.

Oppenheimer, S. (1999). *Eden in the East: The Drowned Continent of Southeast Asia*, Phoenix, London.

Oppenheimer, S. (2003). *Out of Eden: The Peopling of the World*, Constable, London.

Orliac, C. (2003). Ligneus et palmiers de l'île de Pâques du XIème au XVIIème siècle de notre ère, in Orliac, C. (ed), *Archéologie en Océanie Insulaire, Peuplement, Sociétés et Paysages*, Éditions Artcom, Paris, pp. 184–199.

Orliac, M. (1997). Human occupation and environmental modifications in the Papeno'o valley, Tahiti, in Kirch, P. V. and Hunt, T. L. (eds), *Historical Ecology in the Pacific Islands*, Yale University Press, New Haven, pp. 200–229.

Orlove, B. (2005). Human adaptation to climate change: A review of three historical cases and some general perspectives. *Environmental Science and Policy*, 8, 589–600.

Ortlieb, L. (2000). The documented historical record of El Niño events in Peru: An update of the Quinn record (sixteenth through nineteenth centuries), in Diaz, H. F. and Markgraf, V. (eds), *El Niño and the Southern Oscillation: Multiscale Variability and Global and Regional Impacts*, Cambridge University Press, Cambridge, pp. 207–295.

Ortlieb, L., Escribano, R., Follegati, R., Zuñiga, O., Kong, I., Rodriguez, L., Valdes, J., Guzman, N. and Iratchet, P. (2000). Recording of ocean-climate changes during the last 2,000 years in a hypoxic marine environment off northern Chile (23°S). *Revista Chilena de Historia Natural*, 73, 221–242.

Ortlieb, L., Fournier, M. and Macharé, J. (1995). Beach ridges and major late Holocene El Niño events in northern Peru. *Journal of Coastal Research*, 17, 109–117 Special Issue.

Ozanne-Rivierre, F. (1994). Iaai loanwords and phonemic changes in Fagauvea, in Dutton, T. and Tryon, D. T. (eds), *Language Contact and Change in the Austronesian World*, Mouton de Gruyter, Berlin, pp. 523–549.

Page, M. J. and Trustrum, N. A. (1997). A late Holocene lake sediment record of the erosion response to land use change in a steepland catchment, New Zealand. *Zeitschrift für Geomorphologie*, 41, 369–392.

Pang, K. D. (1993). Climatic impact of the mid-fifteenth century Kuwae caldera formation as reconstructed from historical and proxy data. *Eos, Transactions of the American Geophysical Union*, 74, 106.

Parkes, A. (1997). Environmental change and the impact of Polynesian colonization: Sedimentary records from central Polynesia, in Kirch, P. V. and Hunt, T. L. (eds), *Historical Ecology in the Pacific Islands*, Yale University Press, New Haven, pp. 166–199.

Parkes, A., and Flenley, J. (1990). *Hull University Mo'orea Expedition, 1985.* School of Geography and Earth Resources, Miscellaneous Series 37, University of Hull, UK.

Parmentier, R. J. (1987). *The Sacred Remains: Myth, History, and Polity in Belau*, The University of Chicago Press, Chicago.

Patzold, J. (1986). *Temperature and CO_2 Changes in Tropical Surface Waters of the Philippines during the Past 120 Years: Record in the Stable Isotopes of Hermatypic Corals.* Report (Berichte) 12, Geologie-Paläontologie, University of Kiel, Germany.

Paulay, G. (1996). Dynamic clams: Changes in the bivalve fauna of Pacific islands as a result of sea level fluctuations. *American Malacological Bulletin*, 12, 45–57.

Paulsen, A. C. (1976). Environment and empire: Climatic factors in prehistoric Andean culture change. *World Archaeology*, 8, 121–132.

Paulsen, D. E., Li, H.-C. and Ku, T.-L. (2003). Climate variability in central China over the last 1270 years revealed by high-resolution stalagmite records. *Quaternary Science Reviews*, 22, 691–701.

Pavlides, C. and Gosden, C. (1994). 35,000-year-old sites in the rainforests of west New Britain, Papua New Guinea. *Antiquity*, 68, 604–610.

Pearl, F. B. (2004). The chronology of mountain settlements on Tutuila, American Samoa. *Journal of the Polynesian Society*, 114, 331–348.

Pearl, F. B. (2006). Late Holocene landscape evolution and land-use expansion in Tutuila, American Samoa. *Asian Perspectives*, 45, 48–68.

Pelling, M. and Uitto, J. I. (2001). Small island developing states: Natural disaster vulnerability and global change. *Environmental Hazards*, 3, 49–62.

Peltier, W. R. (1998). Postglacial variations in the level of the sea: Implications for climate dynamics and solid earth geophysics. *Reviews of Geophysics*, 36, 603–689.

Perry, C. A. and Hsu, K. J. (2000). Geophysical, archaeological, and historical evidence support a solar-output model for climate change. *Proceedings of the National Academy of Sciences*, 97, 12433–12438.

Peteet, D. (1995). Global Younger Dryas? *Quaternary International*, 28, 93–104.

Petersen, K. L. (1994). A warm and wet Little Climatic Optimum and a cold and dry Little Ice Age in the southern Rocky Mountains, U.S.A. *Climatic Change*, 26, 243–269.

Pfister, C., Luterbacher, J., Schwarz-Zanetti, G. and Wegman, M. (1998). Winter air temperature variations in western Europe during the early and high Middle Ages (A.D. 750–1300). *The Holocene*, 8, 535–552.

Pielou, E. C. (1991). *After the Ice Age: The Return of Life to Glaciated North America*, The University of Chicago Press, Chicago.

Pierrehumbert, R. T. (2000). Climate change and the tropical Pacific: The sleeping dragon wakes. *Proceedings of the National Academy of Sciences*, 97, 1355–1358.

Pirazzoli, P. A., and Montaggioni, L. F. (1985). Lithospheric deformation in French Polynesia (Pacific Ocean) as deduced from Quaternary shorelines. In: *Proceedings of the Fifth International Coral Reef Congress*, Tahiti. 3, 195–200.

Pirazzoli, P. A. and Montaggioni, L. F. (1987). Les îles Gambier et l'atoll de Temoe (Polynésie franç aise): Anciennes lignes de rivage et comportement géodynamique. *Géodynamique*, 2, 13–25.

Pirazzoli, P. A. and Montaggioni, L. F. (1988). Holocene sea-level changes in French Polynesia. *Palaeogeography, Palaeoclimatology, Palaeoecology*, 68, 153–175.

Pirazzoli, P. A., Montaggioni, L. F., Salvat, B. and Faure, G. (1988). Late Holocene sea level indicators from twelve atolls in the central and eastern Tuamotus (Pacific Ocean). *Coral Reefs*, 7, 57–68.

Pisias, N. G. (1978). Paleoceanography of the Santa Barbara Basin during the last 8000 years. *Quaternary Research*, 10, 366–384.

Polissar, P. J., Abbott, M. B., Wolfe, A. P., Bezada, M., Rull, V., and Bradley, R. S. (2006). Solar modulation of Little Ice Age climate in the tropical Andes. *Proceedings of the National Academy of Sciences*, doi:10.1073/pnas.0603118103.

Pope, G. G. (1992). Replacement versus regionally continuous models: the paleobehavioral and fossil evidence from East Asia, in Akazawa, T., Aoki, K. and Kimura, T. (eds), *The Evolution and Dispersal of Modern Humans in Asia*, Hokusen-sha, Japan, pp. 3–14.

Power, S., Casey, T., Folland, C., Colman, A. and Mehta, V. (1999). Inter-decadal modulation of the impact of ENSO on Australia. *Climate Dynamics*, 15, 319–324.

Pugh, D. T. (2004). *Changing Sea Levels: Effects of Tides, Weather and Climate*, Cambridge University Press, Cambridge.

Qian, W. and Zhu, Y. (2002). Little Ice Age climate near Beijing, China, inferred from historical and stalagmite records. *Quaternary Research*, 57, 109–119.

Qin, X., Tan, M., Liu, T., Wang, X., Li, T. and Lu, J. (1999). Spectral analysis of a 1000-year stalagmite lamina-thickness record from Shihua Cave, Beijing, China, and its climatic significance. *The Holocene*, 9, 689–694.

Quinn, T. M., Crowley, T. J. and Taylor, F. W. (1996). New stable isotope results from a 173-year coral record from Espiritu Santo, Vanuatu. *Geophysical Research Letters*, 23, 3413–3416.

Quinn, T. J., Crowley, T. M., Taylor, F. W., Henin, C., Joannot, P. and Join, Y. (1998). A multicentury stable isotope record from a New Caledonia coral: Interannual and decadal SST variability in the southwest Pacific since 1657. *Paleoceanography*, 13, 412–426.

Raab, L. M. and Larson, D. O. (1997). Medieval climate anomaly and punctuated cultural evolution in coastal southern California. *American Antiquity*, 62, 319–336.

Rainbird, P. (2004). *The Archaeology of Micronesia*, Cambridge University Press, Cambridge.

Rampino, M. R. and Ambrose, S. H. (2000). Volcanic winter in the Garden of Eden: The Toba supereruption and the Late Pleistocene human population crash, in McCoy, F. W. and Heiken, G. (eds), *Volcanic Hazards and Disasters in Human Antiquity*, Geological Society of America, Boulder, pp. 71–82 (Special Paper 345).

Rampino, M. R., Sanders, J. E., Newman, W. S. and Konigsson, L. K. (1987). *Climate History, Periodicity, and Predictability*, Van Nostrand Reinhold, New York.

Ravuvu, A. (1987). *The Fijian Ethos*, Institute of Pacific Studies, The University of the South Pacific, Suva.

Razjigaeva, N. G., Grebennikova, T. A., Ganzey, L. A., Mokhova, L. M. and Bazarova, V. B. (2004). The role of global and local factors in determining the middle to late Holocene environmental history of the South Kurile and Komandar Islands, northwestern Pacific. *Palaeogeography, Palaeoclimatology, Palaeoecology*, 209, 313–333.

Razjigaeva, N. G., Korotky, A. M., Grebennikova, T. A., Ganzey, L. A., Mokhova, L. M., Bazarova, V. B., Sulerzhitsky, L. D. and Lutaenko, K. A. (2002). Holocene climatic changes and environmental history of Iturup Island, Kurile Islands, northwestern Pacific. *The Holocene*, 12, 469–480.

Rechtman, R. B. (1992). *The evolution of sociopolitical complexity in the Fiji Islands.* Unpublished Ph.D. thesis. University of California, Los Angeles, CA.

Reid, A. C. (1977). The fruit of the Rewa: Oral traditions and the growth of the pre-Christian Lakeba state. *Journal of Pacific History,* 12, 1–24.

Ren, G. (1998). Pollen evidence for increased summer rainfall in the Medieval Warm Period at Maili, northeast China. *Geophysical Research Letters,* 25, 1931–1934.

Renssen, H., Isarin, R. F. B., Vandenberghe, J., Lautenschlager, M. and Schlese, U. (2000). Permafrost as a critical factor in palaeoclimate modelling: the Younger Dryas case in Europe. *Earth and Planetary Science Letters,* 176, 1–5.

Repinski, P., Holmgren, K., Lauritzen, S. E. and Lee-Thorp, J. A. (1999). A late Holocene climate record from a stalagmite, Cold Air Cave, Northern Province, South Africa. *Palaeogeography, Palaeoclimatology, Palaeoecology,* 150, 269–277.

Reyes, A. V. and Clague, J. J. (2004). Stratigraphic evidence for multiple Holocene advances of Lillooet Glacier, southern Coast Mountains, British Columbia. *Canadian Journal of Earth Sciences,* 41, 903–918.

Richerson, P. J., Boyd, R. and Bettinger, R. L. (2001). Was agriculture impossible during the Pleistocene but mandatory during the Holocene? A climate change hypothesis. *American Antiquity,* 66, 387–411.

Rigozo, N. R., Echer, E., Vieira, L. E. A. and Nordemann, D. J. R. (2001). Reconstruction of Wolf sunspot numbers on the basis of spectral characteristics and estimates of associated radio flux and solar wind parameters for the last millennium. *Solar Physics,* 203, 179–191.

Rigsby, C. A., Baker, P. A. and Aldenderfer, M. S. (2003). Fluvial history of the Rio Ilave valley, Peru, and its relationship to climate and human history. *Palaeogeography, Palaeoclimatology, Palaeoecology,* 194, 165–185.

Riley, T. J. (1987). Archaeological survey and testing, Majuro Atoll, Marshall Islands, in Dye, T. (ed), *Marshall Islands Archaeology,* Bishop Museum, Honolulu, pp. 169–270 (Pacific Anthropological Records 58).

Ritter, L. and Ritter, P. (1982). *The European Discovery of Kosrae Island,* Historical Preservation Office, Saipan, Micronesian Archaeological Survey Reports 13.

Robichaux, H. R. (2000). The Maya hiatus and the A.D. 536 atmospheric event, in Gunn, J. D. (ed), *The Years without Summer: Tracing AD 536 and its Aftermath,* Archaeopress, Oxford, pp. 45–53 (British Archaeological Reports 872).

Rodbell, D. T., Seltzer, G. O., Anderson, D. M., Abbott, M. B., Enfield, D. B. and Newman, J. H. (1999). A ~15,000-year record of El Niño-driven alluviation in southwestern Ecuador. *Science,* 283, 516–520.

Rolett, B. V. (1998). *Hanamiai: Prehistoric Colonization and Cultural Change in the Marquesas Islands (East Polynesia),* Department of Anthropology and The Peabody Museum, Yale University, New Haven, (Yale University Publications in Anthropology 81).

Rolett, B. V. (1989). *Hanamiai: Changing subsistence and ecology in the prehistory of Tahuata (Marquesas Islands, French Polynesia).* Ph.D. dissertation. Yale University. University Microfilms, Ann Arbor.

Rolett, B. V. (2002). Voyaging and interaction in ancient east Polynesia. *Asian Perspectives,* 41, 182–194.

Rolett, B. V., Conte, E., Pearthree, E. and Sinton, J. M. (1997). Marquesan voyaging: Archaeometric evidence for inter-island contact, in Weisler, M. I. (ed), *Prehistoric Long-Distance Interaction in Oceania: An Interdisciplinary Approach,* New Zealand Archaeological Association, Auckland, pp. 134–148 (Monograph 21).

Rooney, J. J. B. and Fletcher, C. H. (2005). Shoreline change and Pacific climatic oscillations in Kihei, Maui, Hawaii. *Journal of Coastal Research,* 21, 535–547.

Rothman, M. (ed), (2001). *Uruk Mesopotamia and its Neighbors: Cross-Cultural Interactions in the Era of State Formation.* Society for American Research, Santa Fe.

Routledge, D. (1985). *Matanitu: The Struggle for Power in Early Fiji,* Institute of Pacific Studies. The University of the South Pacific, Suva.

Rowland, M. J. (1976). *Cellana denticulata* in middens on the Coromandel coast, NZ – possibilities for a temporal horizon. *Journal of the Royal Society of New Zealand,* 6, 1–15.

Rowland, M. J. (1985). Further radiocarbon dates from Mazie Bay, North Keppel Island. *Australian Archaeology,* 21, 113–118.

Rowland, M. J. (1999). Holocene environmental variability: have its impacts been underestimated in Australian pre-history? *The Artefact*, 22, 11–48.

Roy, P. S. and Connell, J. (1991). Climatic change and the future of atoll states. *Journal of Coastal Research*, 7, 1057–1075.

Sabels, B. E. (1966). Climatic variations in the tropical Pacific as evidenced by trace element analysis of soils, in Blumenstock, D. I. (ed), *Pleistocene and Post-Pleistocene Climatic Variations in the Pacific Area*, Bishop Museum Press, Honolulu, pp. 131–151.

Sagan, C. (1996). *The Demon-Haunted World: Science as a Candle in the Dark*, Ballantine, New York.

Sahlins, M. D. (1972). *Stone Age Economics*, Aldine, Chicago.

Saito, Y., Yang, Z. and Hori, K. (2001). The Huanghe (Yellow River) and Changjiang (Yangtze River) deltas: A review on their characteristics, evolution and sediment discharge during the Holocene. *Geomorphology*, 41, 219–231.

Sakaguchi, Y. (1983). Warm and cold stages in the past 7600 years in Japan and their global correlation. *University of Tokyo, Bulletin of the Department of Geography*, 15, 1–31.

Salinger, M. J. (1976). New Zealand temperatures since A.D. 1300. *Nature*, 260, 310–311.

Salinger, M. J., Basher, R. E., Fitzharris, B. B., Hay, J. E., Jones, P. D., Macveigh, J. P. and Schmidely-Leleu, I. (1995). Climate trends in the south-west Pacific. *International Journal of Climatology*, 15, 285–302.

Salinger, M. J., Renwick, J. A. and Mullan, A. B. (2001). Interdecadal Pacific Oscillation and South Pacific climate. *International Journal of Climatology*, 21, 1705–1722.

Sampei, Y., Matsumoto, E., Dettman, D. L., Tokuoka, T. and Abe, O. (2005). Paleosalinity in a brackish lake during the Holocene based on stable oxygen and carbon isotopes of shell carbonate in Nakaumi Lagoon, southwest Japan. *Palaeogeography, Palaeoclimatology, Palaeoecology*, 224, 352–366.

Sanchez, W. A. and Kutzbach, J. E. (1974). Climate of the American tropics and subtropics in the 1960s and possible comparisons with climatic variations of the last millennium. *Quaternary Research*, 4, 128–135.

Sand, C. (1995). *'Le Temps d'Avant': La Préhistoire de la Nouvelle-Calédonie*, L'Harmattan, Paris.

Sandweiss, D. H., Maasch, K. A., Burger, R. L., Richardson, J. B., Rollins, H. B. and Clement, A. (2001). Variation in Holocene El Niño frequencies: climate records and cultural consequences in ancient Peru. *Geology*, 29, 603–606.

Sato, H. (2001). Nihon Retto no Zenki-Chuki Kyusekki Jidai o kangaeru – Fujimura-shi hikanyo shiryo kara no mitoshi (Considering the Japanese Early and Middle Palaeolithic from materials not associated with Fujimura). *Dai-15-kai Tohoku Nihon no Kyusekki Bunka o Kataru Kai – Yokoshu (Resumes from the 15th Meeting for the Discussion of the Palaeolithic Culture in Northeastern Japan)*, held in Akita City, December 22–23, 127–142 (in Japanese).

Satterlee, D. R., Moseley, M. E., Keefer, D. M. and Tapia, J. E. A. (2000). The Miraflores El Niño disaster: Convergent catastrophes and prehistoric agrarian change in southern Peru. *Andean Past*, 6, 95–116.

Savoskul, O. S. (1999). Holocene glacier advances in the headwaters of Sredniaya Avacha, Kamchatka, Russia. *Quaternary Research*, 52, 14–26.

Schilt, A. R. (1984). *Subsistence and Conflict in Kona, Hawaii*. B.P. Bishop Museum, Department of Anthropology, Honolulu, Report 84-1.

Schimmelmann, A., Lange, C. B. and Meggers, B. J. (2003). Palaeoclimatic and archaeological evidence for a ~200-yr recurrence of floods and droughts linking California, Mesoamerica and South America over the past 2000 years. *The Holocene*, 13, 763–778.

Schmidt, M. (1996). The commencement of pa construction in New Zealand prehistory. *Journal of the Polynesian Society*, 105, 441–460.

Schwimmer, R. A., Pizzuto, J. E. and Carey, W. L. (1997). Little Ice Age (?) regressive expansion of salt marshes of southeastern Delaware. *Geological Society of America Abstracts with Programs*, 29, 257.

Scuderi, L. A. (1993). A 2000-year tree ring record of annual temperatures in the Sierra Nevada mountains. *Science*, 259, 1433–1436.

Sellars, J. R. (1998). The Natufian of Jordan, in Henry, D. O. (ed), *The Prehistoric Archaeology of Jordan*, Archaeopress, Oxford, pp. 83–101 (BAR International Series).

Seltzer, G. and Hastorf, C. (1990). Climatic change and its effects on Prehispanic agriculture in the central Peruvian Andes. *Journal of Field Archaeology*, 17, 397–414.

Serre-Bachet, F. (1994). Middle Ages temperature reconstructions in Europe, a focus on northeastern Italy. *Climatic Change*, 26, 213–224.

Shepherd, M. J. (1990). The evolution of a moderate energy coast in Holocene time, Pacific Harbour, Viti Levu, Fiji. *New Zealand Journal of Geology and Geophysics*, 33, 547–556.

Sheppard, P. J., Walter, R. and Nagaoka, T. (2000). The archaeology of head-hunting in Roviana Lagoon, New Georgia. *Journal of the Polynesian Society*, 109, 9–39.

Sher, A. V. (1997). Late-Quaternary extinction of large mammals in northern Eurasia: a new look at the Siberian contribution, in Huntley, B., Cramer, W., Morgan, A. V., Prentice, H. C. and Allen, J. R. M. (eds), *Past and Rapid Future Environmental Changes: The Spatial and Evolutionary Responses of Terrestrial Biota*, Springer-Verlag, Berlin, pp. 319–339.

Shimada, I. (2000). The late prehistoric coastal states, in Minelli, L. L. (ed), *The Inca World: The Development of Pre-Columbian Peru*, A.D. 1000–1534, University of Oklahoma Press, Norman, pp. 49–110.

Shimada, I., Schaaf, C. B., Thompson, L. G. and Mosley-Thompson, E. (1991). Cultural impacts of severe droughts in the prehistoric Andes – application of a 1,500-year ice core precipitation record. *World Archaeology*, 22, 247–270.

Shindell, D. T., Schmidt, G. A., Mann, M. E., Rind, D. and Waple, A. (2001). Solar forcing of regional climate change during the Maunder Minimum. *Science*, 294, 2149–2152.

Shindell, D. T., Schmidt, G. A., Miller, R. L. and Mann, M. E. (2003). Volcanic and solar forcing of climate change during the preindustrial era. *Journal of Climate*, 16, 4094–4107.

Shnirelman, V. A. (1992). Crises and economic dynamics in traditional societies. *Journal of Anthropological Archaeology*, 11, 25–46.

Shun, K. and Athens, J. S. (1990). Archaeological investigations on Kwajalein Atoll, Marshall Islands, Micronesia. *Micronesica Supplement*, 2, 231–240.

Siegert, M. J. (2001). *Ice Sheets and Late Quaternary Environmental Change*, Wiley, Chichester.

Sinoto, Y. (1983). Archaeological investigations of the Vaito'otia and Fa'ahia sites on Huahine Island, French Polynesia. *National Geographic Society, Research Reports*, 15, 583–599.

Siringan, F. P., Maeda, Y., Rodolfo, K. S. and Omura, A. (2000). Short-term and long-term changes of sea level in the Philippine Islands, in Mimura, N. and Yokoki, H. (eds), *Global Change and Asia Pacific Coasts*, Asia-Pacific Network for Global Change Research, Kobe, pp. 143–149.

Sivan, D., Lambeck, K., Toueg, R., Raban, A., Porath, Y. and Shirman, B. (2004). Ancient coastal wells of Caesarea Maritima, Israel, an indicator for relative sea level changes during the last 2000 years. *Earth and Planetary Science Letters*, 222, 315–330.

Smith, R. A. (1980). Golden Gate tidal measurements, 1854–1978. *Journal of the Waterway, Port, Coastal and Ocean Division, Proceedings of the American Society of Civil Engineers*, 106, 407–410.

Soden, B. J., Wetherald, R. T., Stenchikov, G. L. and Robock, A. (2002). Global cooling after the eruption of Mount Pinatubo: A test of climate feedback by water vapor. *Science*, 296, 727–730.

Soon, W., Baliunas, S., Idso, C., Idso, S. and Legates, D. R. (2003). Reconstructing climatic and environmental changes of the past 1000 years: a reappraisal. *Energy and Environment*, 14, 233–296.

Sopher, D. E. (1977). *The Sea Nomads*, National Museum of Singapore, Singapore.

Sorrenson, M. P. K. (1979). *Maori Origins and Migrations*, Auckland University Press, Auckland.

Spencer, T., Stoddart, D. R. and Woodroffe, C. D. (1987). Island uplift and lithospheric flexure: Observations and cautions from the South Pacific. *Zeitschrift für Geomorphologie, Supplementband*, 63, 87–102.

Spriggs, M. (1986). Landscape, land use, and political transformation in southern Melanesia, in Kirch, P. V. (ed), *Island Societies: Archaeological Approaches to Evolution and Transformation*, Cambridge University Press, Cambridge, pp. 6–19.

Spriggs, M. (1988). The Hawaiian transformation of Ancestral Polynesian society: Conceptualizing chiefly states, in Gledhill, J., Bender, B. and Larsen, M. T. (eds), *State and Society: The Emergence and Development of Social Hierarchy and Political Centralization*, Unwin Hyman, London, pp. 57–73.

Spriggs, M. (1997). *The Island Melanesians*, Blackwell, Oxford.

Spriggs, M. (2000a). Out of Asia? The spread of Pleistocene and Neolithic maritime cultures in Island SE Asia and the Western Pacific, in O'Connor, S. and Veth, P. (eds), *East of Wallace's Line: Studies of Past and Present Maritime Cultures of the Indo-Pacific Region*, Balkema, Rotterdam, pp. 51–75.

Spriggs, M. (2000b). Can hunter-gatherers live in tropical rain forests? The Pleistocene island Melanesian evidence, in Schweitzer, P., Biesele, M. and Hitchcock, R. K. (eds), *Hunters and Gatherers in the Modern World: Conflict, Resistance and Self-Determination*, Berghahn, New York, pp. 287–304.

Spriggs, M. and Anderson, A. (1993). Late colonization of east Polynesia. *Antiquity*, 67, 200–217.

Stahle, D. W., D'Arrigo, R. D., Krusic, P. J., Cleaveland, M. K., Cook, E. R., Allan, R. J., Cole, J. E., Dunbar, R. B., Therrell, M. D., Gay, D. A., Moore, M. D., Stokes, M. A., Burns, B. T., Villanueva-Diaz, J. and Thompson, L. G. (1998). Experimental dendroclimatic reconstruction of the Southern Oscillation. *Bulletin of the American Meteorological Society*, 79, 2137–2153.

Stainforth, D. A., Aina, T., Christensen, C., Collins, M., Faull, N., Frame, D. J., Kettleborough, J. A., Knight, S., Martin, A., Murphy, J. M., Piani, C., Sexron, D., Smith, L. A., Spicer, R. A., Thorpe, A. J. and Allen, M. R. (2005). Uncertainty in predictions of the climate response to rising levels of greenhouse gases. *Nature*, 433, 403–406.

Steadman, D. W., Antón, S. C. and Kirch, P. V. (2000). Ana Manuku: A prehistoric ritualistic site on Mangaia, Cook Islands. *Antiquity*, 74, 873–883.

Steig, E. J., Brook, E. J., White, J. W. C., Sucher, C. M., Bender, M. L., Lehman, S. J., Morse, D. L., Waddington, E. D. and Clow, G. D. (1998). Synchronous climate changes in Antarctica and the North Atlantic. *Science*, 282, 92–95.

Steinitz-Kannan, M., Riedinger, M. A., Last, W., Brenner, M. and Miller, M. C. (1997). Un registro de 6000 años de manifestos del fenomeno de El Niño en sedimentos de lagunas de las islas Galapagos, in Cadier, E. and Galarrage, R. (eds), *Seminario Internacional "Consequencias Climaticas e Hidrologicas del Evento El Niño a Escala Regional y Local"*, Memorias Tecnicas, Edicion Prelimina, ORSTOM/INAMHI, Mexico City, pp. 79–88.

Stenni, B., Proposito, M., Gragnani, R., Flora, O., Jouzel, J., Falourd, S. and Frezzotti, M. (2002). Eight centuries of volcanic signal and climate change at Talos Dome (East Antarctica). *Journal of Geophysical Research*, 107 doi:10.1029/2000JD000317.

Stewart, P. J. and Strathern, A. (1999). Feasting on my enemy: images of violence and change in the New Guinea Highlands. *Ethnohistory*, 46, 645–669.

Stine, S. (1990). Late Holocene fluctuations of Mono Lake, eastern California. *Palaeogeography, Palaeoclimatology, Palaeoecology*, 78, 333–381.

Stine, S. (1994). Extreme and persistent drought in California and Patagonia during Mediaeval time. *Nature*, 369, 546–549.

Stine, S. (1998). Medieval climate anomaly in the Americas, in Issar, A. S. and Brown, N. (eds), *Water, Environment and Society in Times of Climatic Change*, Kluwer, Amsterdam, pp. 43–67.

Stott, L. (2002). SST variability in the Western Pacific Warm Pool during the past 2000 years. *Eos, Transactions of the American Geophysical Union*, 83(Suppl.), F915. (abstract).

Stott, L., Poulsen, C., Lund, S. and Thunell, R. (2002). Super ENSO and global climate oscillations at millennial time scales. *Science*, 297, 222–226.

Streck, C. S. (1990). Prehistoric settlement in eastern Micronesia: archaeology on Bikini Atoll, Republic of the Marshall Islands. *Micronesica, Supplement*, 2, 247–260.

Stuiver, M., Grootes, P. M. and Braziunas, T. F. (1995). The GISP2 δ^{18}O records of the past 16,500 years and the role of the sun, ocean, and volcanoes. *Quaternary Research*, 44, 341–354.

Stuiver, M., Reimer, P. J., Bard, E., Beck, J. W., Burr, G. S., Hughen, K. A., Kromer, B., McCormac, G., van der Plicht, J. and Spurket, M. (1998). Intcal98 radiocarbon age calibration, 24,000-0 cal BP. *Radiocarbon*, 40, 1041–1083.

Suggs, R. C. (1961). *The Archaeology of Nuku Hiva, Marquesas Islands, French Polynesia*, American Museum of Natural History, New York, Anthropology Papers, 49, Part 1.

Sullivan, M. E. (1987). The recent prehistoric exploitation of edible mussel in Aboriginal shell middens in southern New South Wales. *Archaeology in Oceania*, 22, 97–106.

Sutton, D. G. (1980). A culture history of the Chatham Islands. *Journal of the Polynesian Society*, 89, 67–93.

Swetnam, T. W. (1993). Fire history and climate change in giant sequoia groves. *Science*, 262, 885–889.

Szabó, K. (2001). Molluscan evidence for late Holocene climate change on Motutapu Island, Hauraki Gulf. *Journal of the Polynesian Society*, 110, 79–87.

Tacon, P. and Chippindale, C. (1994). Australia's ancient warriors. Changing depictions of fighting in the rock art of Arnhem Land, N.T. *Cambridge Archaeological Journal*, 4, 211–248.

Tagami, Y. (1996). Some remarks on the climate of the Medieval Warm Period of Japan, in Mikami, T., Matsumoto, E., Ohta, S. and Sweda, T. (eds), *Paleoclimate and Environmental Variability in Austral-Asian Transect during the Past 2000 Years*, International Geosphere-Biosphere Programme, Nagoya, pp. 115–119.

Tainter, J. A. (1988). *The Collapse of Complex Societies*, Cambridge University Press, Cambridge.

Takamiya, H. (1996). Initial colonization and subsistence adaptation processes in the late prehistory of the island of Okinawa, in Glover, I. C. and Bellwood, P. (eds), Indo-Pacific Prehistory: The Chiang Mai Papers. Australian National University, Canberra, Vol. 2, pp. 143–180.

Takamiya, H. (2004). Population dynamics in the prehistory of Okinawa, in Fitzpatrick, S. M. (ed), *Voyages of Discovery: The Archaeology of Islands*, Praeger, Westport, pp. 111–128.

Takayama, J. and Intoh, M. (1978). *Archaeological Investigation at Chukienu Shell Midden on Tol, Truk*, Tezukayama University, Nara, Reports of Pacific Archaeological Survey 5.

Tanner, W. F. (1991). The "Gulf of Mexico" late Holocene sea level curve and river delta history. *Transactions: Gulf Coast Association of Geological Societies*, 49, 583–589.

Tanner, W. F. (1992). 3000 years of sea-level change. *Bulletin of the American Meteorological Society*, 73, 297–304.

Tanner, W. F. (1993). An 8000-year record of sea-level change from grain-size parameters: Data from beach ridges in Denmark. *The Holocene*, 3, 220–231.

Taylor, D., Robertshaw, P. and Marchant, R. A. (2000). Environmental change and political-economic upheaval in precolonial western Uganda. *The Holocene*, 10, 527–536.

Terrell, J. E. (2002). Tropical agroforestry, coastal lagoons, and Holocene prehistory in Greater Near Oceania, in Yoshida, S. and Matthews, P. J. (eds), *Vegeculture in Eastern Asia and Oceania*, Centre for Area Studies, Osaka, pp. 195–216.

Terry, J. P., Garimella, S. and Kostaschuk, R. A. (2002). Rates of floodplain accretion in a tropical island river system impacted by cyclones and large floods. *Geomorphology*, 42, 171–183.

Ters, M. (1987). Variations in Holocene sea level on the French Atlantic coast and their climatic significance, in Rampino, M. R., Sanders, J. E., Newman, W. S. and Konigsson, L. K. (eds), *Climate: History, Periodicity, and Predictability*, Van Nostrand Reinhold, New York, pp. 204–236.

Thiel, B. (1987). Early settlement of the Philippines, Eastern Indonesia, and Australia–New Guinea: A new hypothesis. *Current Anthropology*, 28, 236–241.

Thomas, F. R., Nunn, P. D., Osborne, T., Kumar, R., Areki, F., Matararaba, S., Steadman, D. and Hope, G. (2004). Recent archaeological findings at Qaranilaca Cave, Vanuabalavu Island, Fiji. *Archaeology in Oceania*, 39, 42–49.

Thomas, N. (1986). *Planets around the Sun: Dynamics and Contradictions of the Fijian Matanitu*. University of Sydney, Sydney, Oceania Monograph 31.

Thompson, L. (1945). *The native culture of the Marianas Islands*, B.P. Bishop Museum, Honolulu, Bulletin 185.

Thompson, L. G. and Mosley-Thompson, E. (1987). Evidence of abrupt climatic change during the last 1,500 years recorded in ice cores from the tropical Quelccaya ice cap, Peru, in Berger, W. H. and Labeyrie, L. D. (eds), *Abrupt Climatic Change*, Reidel, Dordrecht, pp. 99–110.

Thompson, L. G., Mosley-Thompson, E., Bolzan, J. F. and Koci, B. R. (1985). A 1500 year record of tropical precipitation in ice cores from the Quelccaya Ice Cap, Peru. *Science*, 229, 971–973.

Thompson, L. G., Mosley-Thompson, E., Dansgaard, W. and Grootes, P. M. (1986). The Little Ice Age as recorded in the stratigraphy of the tropical Quelccaya Ice Cap. *Science*, 234, 361–364.

Thompson, R. S., Whitlock, C., Bartlein, P. J., Harrison, S. P. and Spaulding, W. G. (1993). Climatic changes in the western United States since 18,000 yr B.P., in Wright, H. E., Kutzbach, J. E., Webb, T., Ruddiman, W. F., Street-Perrott, F. A. and Bartleim, P. J. (eds), *Global Climates since the Last Glacial Maximum*, University of Minnesota, Minneapolis, pp. 468–513.

Thompson, L. G., Mosley-Thompson, E., Davis, M. E., Henderson, K. A., Brecher, H. H., Zagorodnov, V. S., Mashiotta, T. A., Lin, P.-N., Mikhalenko, V. N., Hardy, D. R. and Beer, J. (2002). Kilimanjaro ice core records: Evidence of Holocene climate change in tropical Africa. *Science*, 298, 589–593.

Thompson, L. G., Mosley-Thompson, E., Davis, M. E., Lin, P.-N., Henderson, K. and Mashiotta, T. A. (2003). Tropical glacier and ice core evidence of climate change on annual to millennial time scales. *Climatic Change*, 59, 137–155.

Thompson, L. G., Mosley-Thompson, E., Brecher, H., Davis, M., León, B., Lin, P.-N., Mashiotta, T. and Mountain, K. (2006). Abrupt tropical climate change: Past and present. *Proceedings of the National Academy of Sciences*, 103, 10536–10543.

Till, C. and Guiot, J. (1990). Reconstruction of precipitation in Morocco since 1100 A.D. based on *Cedrus atlantica* tree-ring widths. *Quaternary Research*, 33, 337–351.

Tippett, A. (1954). The nature and social function of Fijian war. *Transactions of the Fiji Society*, 5, 137–155.

Trenberth, K. E. (1976). Spatial and temporal variations of the Southern Oscillation. *Quarterly Journal of the Royal Meteorological Society*, 102, 639–653.

True, D. L. (1990). Site locations and water supply: A perspective from northern San Diego County, California. *New World Archaeology*, 4, 37–60.

Tuchman, B. (1978). *A Distant Mirror: The Calamitous 14th Century*, Knopf, New York.

Tudhope, A. W., Chilcott, C. P., McCulloch, M. T., Cook, E. R., Chappell, J., Ellam, R. M., Lea, D. W., Lough, J. M. and Shimmield, G. B. (2001). Variability in the El Nino-Southern Oscillation through a glacial–interglacial cycle. *Science*, 291, 1511–1517.

Turner, G. (1861). *Nineteen Years in Polynesia: Missionary Life, Travels and Researches in the Islands of the Pacific*, John Snow, London.

Turner, C. G. (1985). The dental search for native American origins, in Kirk, R. and Szathmáry, E. J. E. (eds), *Out of Asia: Peopling the Americas and the Pacific*, Journal of Pacific History, Canberra, pp. 31–78.

Tyson, P. D., Karlen, W., Holmgren, K. and Heiss, G. A. (2000). The Little Ice Age and medieval warming in South Africa. *South African Journal of Science*, 96, 121–126.

Ulm, S. (2004). *Investigations towards a late Holocene archaeology of Aboriginal Lifeways on the Southern Curtis Coast, Australia*. Unpublished Ph.D. thesis. School of Social Science, University of Queensland, Brisbane.

Urban, F. E., Cole, J. E. and Overpeck, J. T. (2000). Influence of mean climate change on climate variability from a 155-year tropical Pacific coral record. *Nature*, 407, 989–993.

Vaganov, E. A., Briffa, K. R., Naurzbaev, M. M., Schweingruber, F. H., Shiyatov, S. G. and Shishov, V. V. (2000). Long-term climatic changes in the arctic region of the northern hemisphere. *Doklady Earth Sciences*, 375, 1314–1317.

Valdes, J., Ortlieb, L. and Sifeddine, A. (2003). Variaciones del sistema de surgencia de Punta Angamos (23°S) y la Zona de Mínimo Oxígeno durante el pasado reciente: Una aproximación desde el registro sedimentario de la Bahía Medjillones del Sur. *Revista Chilena de Historia Natural*, 76, 347–362.

van de Plassche, O., van der Borg, K. and de Jong, A. F. M. (1998). Sea level – climate correlation during the past 1400 yr. *Geology*, 26, 319–322.

van de Plassche, O., van der Schrier, G., Weber, S.L., Gehrels, W.R. and Wright, A.J. (2003). Sea-level variability in the northwest Atlantic during the past 1500 years: a delayed response to solar forcing? *Geophysical Research Letters*, 30, doi:10.1029/2003GL017558.

van Geel, B. and Renssen, H. (1998). Abrupt climate change around 2,650 BP in north-west Europe: Evidence for climatic teleconnections and a tentative explanation, in Issar, A. S. and Brown, N. (eds), *Water, Environment and Society in Times of Climatic Change*, Kluwer, Dordrecht, pp. 21–41.

Varekamp, J. C., Thomas, E. and Thompson, W. G. (1999). Sea-level climate correlation during the past 1400 yr: Comment and reply. *Geology*, 27, 189–190.

Vayda, A. P. (1976). *War in Ecological Perspective*, Plenum, New York.

Verschuren, D., Laird, K. R. and Cumming, B. F. (2000). Rainfall and drought in equatorial east Africa during the past 1,100 years. *Nature*, 403, 410–414.

Villalba, R. (1990). Climatic fluctuations in northern Patagonia during the last 1000 years as inferred from tree-ring records. *Quaternary Research*, 34, 346–360.

Villalba, R., Leiva, J. C., Rubulis, S., Suarez, J. and Lenzano, L. (1990). Climate, tree-ring, and glacial fluctuations in the Rio Frias valley, Rio Negro, Argentina. *Arctic and Alpine Research*, 22, 215–232.

Von Storch, H., Zorita, E., Jones, J. M., Dimitriev, Y., González-Rouco, F. and Tett, S. F. B. (2004). Reconstructing past climate from noisy data. *Science*, 306, 679–682.

Vuille, M. and Bradley, R. S. (2000). Mean annual temperature trends and their vertical structure in the tropical Andes. *Geophysical Research Letters*, 27, 3885–3888.

Waddell, E. (1975). How the Enga cope with frost: Responses to climatic perturbations in the central highlands of New Guinea. *Human Ecology*, 3, 249–273.

Walker, P. L. (1986). Porotic hyperostosis in a marine-dependent California Indian population. *American Journal of Physical Anthropology*, 69, 345–354.

Walker, P. L. (1989). Cranial injuries as evidence of violence in prehistoric southern California. *American Journal of Physical Anthropology*, 80, 313–323.

Walker, P. L., & Lambert, P. M. (1989). Skeletal evidence for stress during a period of cultural change in prehistoric California. In: L. Capasso (Ed). *Advances in Paleopathology (Proceedings of the VII European Meeting of the Paleopathology Association)*. Marino Solfanelli, Chieti, Italy, pp. 207–212.

Walter, R. (1990). *The southern Cook Islands in eastern Polynesian prehistory*. Ph.D. dissertation. University of Auckland, Auckland.

Walter, R. (1996). Settlement pattern archaeology in the southern Cook Islands: A review. *Journal of the Polynesian Society*, 105, 63–99.

Wang, B., Chen, S., Zhang, K. and Shen, J. (1995). Potential impacts of sea-level rise on the Shanghai area. *Journal of Coastal Research*, 14, 151–166 Special Issue.

Wang, Y., Cheng, H., Edwards, R. L., He, Y., Kong, X., An, Z., Wu, J., Kelly, M. J., Dykoski, C. A. and Li, X. (2005). The Holocene Asian monsoon: Links to solar changes and North Atlantic climate. *Science*, 308, 854–857.

Ward, P. D. (2001). *Rivers in Time: The Search for Clues to Earth's Mass Extinctions*, Columbia University Press, New York.

Ward, R. G. (1989). Earth's empty quarter? The Pacific islands in the Pacific century. *The Geographical Journal*, 155, 235–246.

Ward, R. G. and Brookfield, M. (1992). The dispersal of the coconut: did it float or was it carried to Panama? *Journal of Biogeography*, 19, 467–480.

Warrick, R. A. and Oerlemans, J. (1990). Sea level rise, in Houghton, J. T., Jenkins, G. J. and Ephraums, J. J. (eds), *Climate Change: The IPCC Scientific Assessment*, Cambridge University Press, Cambridge, pp. 257–281.

Waterhouse, J. (1866). *The King and People of Fiji*, Wesleyan Conference Office, London.

Waters, M. R., Byrd, B. F. and Reddy, S. N. (1999). Geoarchaeological investigations of San Mateo and Las Flores Creeks, California: Implications for coastal settlement models. *Geoarchaeology*, 14, 289–306.

Watson, E. and Luckman, B. H. (2001). Dendroclimatic reconstruction of precipitation for sites in the southern Canadian Rockies. *The Holocene*, 11, 203–213.

Weber, S. L., Crowley, T. J. and van der Schrier, G. (2004). Solar irradiance forcing of centennial climate variability during the Holocene. *Climate Dynamics*, 22, 539–553.

Webster, D. (2002). *The Fall of the Ancient Maya*, Thames and Hudson, London.

Weisler, M. (1993). *Long-distance interaction in prehistoric Polynesia: Three case studies*. Unpublished Ph.D. dissertation. University of California at Berkeley, CA.

Weisler, M. I. (1995). Henderson Island prehistory: Colonization and extinction on a remote Polynesian Island. *Biological Journal of the Linnean Society*, 56, 377–404.

Weisler, M. I. (1996a). Taking the mystery out of the Polynesian 'mystery islands': A case study from Mangareva and the Pitcairn group, in Davidson, J. M., Irwin, G., Leach, B. F., Pawley, A. and Brown, D. (eds), *Oceanic Culture History: Essays in Honour of Roger Green*, New Zealand Journal of Archaeology Special Publication, Auckland, pp. 615–629.

Weisler, M. I. (1996b). An archaeological survey of Mangareva: Implications for regional settlement models and interaction studies. *Man and Culture in Oceania*, 12, 61–85.

Weisler, M. (1999). The antiquity of aroid pit agriculture and significance of buried A horizons on Pacific atolls. *Geoarchaeology*, 14, 621–654.

Weisler, M. (2002). Centrality and the collapse of long-distance voyaging in east Polynesia, in Glascock, M. D. (ed), *Geochemical Evidence for Long-Distance Exchange*, Bergin and Garvey, Westport, pp. 257–273.

Weisler, M. and Green, R. C. (2001). Holistic approaches to interaction studies: A Polynesian example. *Research in Anthropology and Linguistics*, 5, 417–457.

Weisler, M. and Kirch, P. V. (1985). The structure of settlement space in a Polynesian chiefdom: Kawela, Molokai, Hawaiian Islands. *New Zealand Journal of Archaeology*, 7, 129–158.

Weiss, H. (2000). Beyond the Younger Dryas: collapse as adaptation to abrupt climate change in ancient west Asia and the eastern Mediterranean, in Bawden, G. and Reycraft, R. M. (eds), *Environmental Disaster and the Archaeology of Human Response*, Maxwell Museum of Anthropology, Albuquerque, pp. 75–95 (Anthropological Papers 7).

Weiss, H. and Bradley, R. (2001). What drives societal collapse? *Science*, 291, 609–610.

Weiss, H., Courty, M.-A., Wetterstrom, W., Guichard, F., Senior, L., Meadow, R. and Curnow, A. (1993). The genesis and collapse of third millennium North Mesopotamian civilization. *Science*, 261, 995–1004.

Wells, L. E. (1992). Holocene landscape change on the Santa Delta, Peru: Impact on archaeological site distributions. *The Holocene*, 2–3, 193–204.

Wells, L. E. (1996). The Santa Beach Ridge Complex: Sea-level and progradational history of an open gravel coast in central Peru. *Journal of Coastal Research*, 12, 1–17.

Wells, L. E. and Noller, J. S. (1999). Holocene coevolution of the physical landscape and human settlement in northern coastal Peru. *Geoarchaeology*, 14, 755–789.

Weng, Q. (1994). The relationship between the environmental change of the Zhujiang River Delta in Holocene and its cultural origins and propagation. *Chinese Geographical Science*, 4, 303–309.

Weng, Q. (2000). Human–environment interactions in agricultural land use in a South China's wetland region [sic]: A study on the Zhujiang Delta in the Holocene. *GeoJournal*, 51, 191–202.

White, T. D., Suwa, G. and Asfaw, B. (1994). *Australopithecus ramidus*, a new species of early hominid from Aramis, Ethiopia. *Nature*, 371, 306–312.

Whyte, A. L. H., Marshall, S. J. and Chambers, G. K. (2005). Human evolution in Polynesia. *Human Biology*, 77, 157–177.

Wiles, G. C., Barclay, D. J. and Calkin, P. E. (1999a). Tree-ring-dated 'Little Ice Age' histories of maritime glaciers from western Prince William Sound, Alaska. *The Holocene*, 9, 163–173.

Wiles, G. C., Post, A., Muller, E. H. and Molnia, B. F. (1999b). Dendrochronology and late Holocene history of Bering Piedmont Glacier, Alaska. *Quaternary Research*, 52, 185–195.

Wilkes, C. (1845). *Narrative of the United States Exploring Expedition during the years 1838, 1839, 1840, 1841, 1842. Volume 3. Tongataboo, Feejee Group, Honolulu*, Lea and Blanchard, Philadelphia.

Wilkinson, C. R. (ed), (2000). *Status of Coral Reefs of the World 2000*. Australian Institute of Marine Science, Global Coral Reef Monitoring Network, Townsville.

Willard, D. A., Cronin, T. M. and Verardo, S. (2003). Late-Holocene climate and ecosystem history from Chesapeake Bay sediment cores, USA. *The Holocene*, 13, 201–214.

Williams, M. A. J., Dunkerley, D. L., De Deckker, P., Kershaw, A. P. and Stokes, T. (1993). *Quaternary Environments*, Edward Arnold, London.

Williams, P. W., King, D. N. T., Zhao, J.-X. and Collerson, K. D. (2004). Speleothem master chronologies: Combined Holocene ^{18}O and ^{13}C records from the North Island of New Zealand and their palaeoenvironmental interpretation. *The Holocene*, 14, 194–208.

Williams, P. W., King, D. N. T., Zhao, J.-X. and Collerson, K. D. (2005). Late Pleistocene to Holocene composite speleothem ^{18}O and ^{13}C chronologies from South Island, New Zealand – did a global Younger Dryas really exist? *Earth and Planetary Science Letters*, 230, 301–317.

Williams, P. W., Marshall, A., Ford, D. C. and Jenkinson, A. V. (1999). Palaeoclimatic interpretation of stable isotope data from Holocene speleothems of the Waitomo district, North Island, New Zealand. *The Holocene*, 9, 649–657.

Wilmshurst, J. M. and Higham, T. F. G. (2004). Using rat-gnawed seeds to independently date the arrival of Pacific rats and humans in New Zealand. *The Holocene*, 14, 801–806.

Wilson, A. T., Hendy, C. H. and Reynolds, C. P. (1979). Short-term climate change and New Zealand temperatures during the last millennium. *Nature*, 279, 315–317.

Winkler, S. (2004). Lichenometric dating of the 'Little Ice Age' maximum in Mt Cook National Park, Southern Alps, New Zealand. *The Holocene*, 14, 911–920.

Winter, A., Oba, T., Ishioroshi, H., Watanabe, T. and Christy, J. (2000). Tropical sea surface temperatures: two-to-three degrees cooler than present during the Little Ice Age. *Geophysical Research Letters*, 27, 3365–3368.

Woodroffe, C. D. (2002). *Coasts: Form, Process and Evolution*, Cambridge University Press, Cambridge.

Woodroffe, C. D. and Grindrod, J. (1991). Mangrove biogeography: The role of Quaternary environ-mental and sea-level fluctuations. *Journal of Biogeography*, 18, 479–492.

Woodroffe, C. D., Murray-Wallace, C. V., Bryant, E. A., Brooke, B. P. and Heijnis, H. (1995). Late Quaternary sea-level highstands in the Tasman Sea: Evidence from Lord Howe Island. *Marine Geology*, 125, 61–72.

Wright, H. E., Kutzbach, J. E., Webb, T., Ruddiman, W. F., Street-Perrott, F. A. and Bartleim, P. J. (eds). (1993). *Global Climates since the Last Glacial Maximum*. University of Minnesota, Minneapolis.

Wu, W. and Liu, T. (2004). Possible role of the "Holocene Event 3" on the collapse of Neolithic cultures around the Central Plain of China. *Quaternary International*, 117, 153–166.

Wyatt, S. (2004). Ancient transpacific voyaging to the New World via Pleistocene South Pacific Islands. *Geoarchaeology*, 19, 511–529.

Wyrtki, K. (1990). Sea level rise: The facts and the future. *Pacific Science*, 44, 1–16.

Yadav, R. R. and Singh, J. (2002). Tree-ring-based spring temperature patterns over the past four centuries in western Himalaya. *Quaternary Research*, 57, 299–305.

Yalcin, K., Wake, C. P., Kreutz, K. J. and Whitlow, S. I. (2006). A 1000-yr record of forest fire activity from Eclipse Icefield, Yukon, Canada. *The Holocene*, 16, 200–209.

Yamaguchi, M., Sugai, T., Fujiwara, O., Ohmori, H., Kamataki, T. and Sugiyama, Y. (2003). Nobi Heiya boringu koa kaiseki ni motozuku Kanshinto no taiseki katei (Depositional process of the Holocene Nobi Plain, Central Japan, reconstructed from drilling core analysis). *Daiyonki Kenkyu (The Quaternary Research)*, 42, 335–346 (in Japanese with English abstract).

Yang, B., Braeuning, A., Johnson, K. R. and Yafeng, S. (2002). General characteristics of temperature variation in China during the last two millennia. *Geophysical Research Letters*, 29doi:10.1029/2001GL014485.

Yasuda, Y. (1976). Early historic forest clearance around the ancient castle site of Tagajo, Miyagi Prefecture, Japan. *Asian Perspectives*, 19, 42–58.

Yasuda, Y. and Shinde, V. (eds). (2004). *Monsoon and Civilization: Proceedings of the Second International Workshop of the Asian Lake Drilling Programme (ALDP)*. Pune, India. Lustre Press, Roli Books, New Delhi.

Yi, S. and Saito, Y. (2003). Palynological evidence for late Holocene environmental change on the Gimhae Fluvial Plain, southern Korean Peninsula: Reconstructing the rise and fall of the Golden Crown Gaya State. *Geoarchaeology*, 18, 831–850.

Yim, W. W.-S. (1995). Implications of sea-level rise for Victoria Harbour, Hong Kong. *Journal of Coastal Research*, 14, 167–189 Special Issue.

Yu, K.-F., Zhao, J.-X., Collerson, K. D., Shi, Q., Chen, T.-G., Wang, P.-X. and Liu, T.-S. (2004). Storm cycles in the last millennium recorded in Yongshu Reef, southern South China Sea. *Palaeogeography, Palaeoclimatology, Palaeoecology*, 210, 89–100.

Yu, Z. and Ito, E. (1999). Possible solar forcing of century-scale drought frequency in the northern Great Plains. *Geology*, 27, 263–266.

Yulianto, E., Rahardjo, A. T., Noeradi, D., Siregar, D. A. and Hirakawa, K. (2005). A Holocene pollen record of vegetation and coastal environmental changes in the coastal swamp forest at Batulicin, South Kalimantan, Indonesia. *Journal of Asian Earth Sciences*, 25, 1–8.

Zaro, G. and Alvarez, A. U. (2005). Late Chiribaya agriculture and risk management along the arid Andean coast of southern Perú, A.D. 1200–1400. *Geoarchaeology*, 20, 717–737.

Zhang, D. (1994). Evidence for the existence of the Medieval Warm Period in China. *Climatic Change*, 26, 289–297.

Zhang, Q., Zhu, C., Liu, C. L. and Jiang, T. (2005). Environmental change and its impacts on human settlement in the Yangtze Delta, P.R. China. *Catena*, 60, 267–277.

Zhang, Y., Wallace, J. M. and Battisti, D. S. (1997). ENSO-like interdecadal variability: 1900–93. *Journal of Climate*, 10, 1004–1020.

Zhao, X. (1998). Origin of rice paddy cultivation at the Hemudu site. *Agricultural Archaeology*, 1998, 131–137 (in Chinese).

Zhou, H. and Zheng, X. M. (2000). Role of environmental changes on prehistoric human civilization, taking the collapse of Liangzhu culture in the north part of the Changjiang delta plain as an example. *Journal of East Normal University*, 4, 71–77 (in Chinese).